THE DECLINE OF AMERICAN STEEL

THE DECLINE OF AMERICAN STEEL

How Management, Labor, and Government Went Wrong

PAUL A. TIFFANY

New York Oxford
OXFORD UNIVERSITY PRESS
1988

Oxford University Press

Oxford New York Toronto
Delhi Bombay Calcutta Madras Karachi
Petaling Jaya Singapore Hong Kong Tokyo
Nairobi Dar es Salaam Cape Town
Melbourne Auckland

and associated companies in
Beirut Berlin Ibadan Nicosia

Published by Oxford University Press, Inc.,
200 Madison Avenue, New York, New York 10016

Library of Congress Cataloging-in-Publication Data
Tiffany, Paul A.
The decline of American steel.
Revision of thesis (Ph.D.)—University of California at Berkeley.
Bibliography: p.
Includes index.
1. Steel industry and trade—United States.
2. Steel industry and trade—Government policy—United States.
3. Trade-unions—Iron and steel workers—United States.
I. Title.
HD9515.T54 1988 338.4′7669142′0973 87-5782
ISBN 0-19-504382-0

9 8 7 6 5 4 3 2 1

Printed in the United States of America
on acid-free paper

Although they were a distraction from the timely meeting of deadlines, this book is nevertheless dedicated to Janet, Rafael, Roland, and Brandon.

Preface

In July of 1959 an event occurred that marked the beginning of the end of American dominance in the global steel industry—a position that this nation had enjoyed over most of the twentieth century. The United Steel Workers of America went out on the longest industrial strike in the country's history. As a direct consequence of that strike, annual steel imports into the U.S. in 1959 exceeded exports for the first time in the century. This ratio never again reversed itself, eventually plunging the steel industry into a shattering collapse from its proud heritage.

Today the evidence of decline in the American carbon steel industry is everywhere apparent. Plants have been permanently closed, workers terminated by the thousands, and capacity and output significantly scaled back. At the same time, foreign steel imported into the United States has achieved widespread acceptance, with even further expansion impeded only by delicately arranged trade embargoes with supplier nations. While some progress that would arrest these trends has at last begun to manifest itself, the conditions underlying the domestic industry's global competitiveness remain suspect. Moreover, it also appears that the stabilization that recent times have witnessed, primarily having to do with company financial matters, has not resulted in changes so striking that the American industry can again challenge for global dominance.

The basic question this book addresses is deceivingly simple, yet I believe it is also contentiously complex: Why did the steel industry enter into decline? Some might argue that, along with the U.S. economy in general, the steel industry was faced with a simple matter of market economics: changing consumer preferences, long-term shifts in demand, the introduction of new technologies, changing comparative advantages among global competitors, and the like. Others might point to political theories of corporate power that result in a failure by monopolistic suppliers to innovate while their competitors do so. While conventional economic and political analysis can bring powerful tools to bear on the problem, they are insufficient for the deeper task of this book, that is, an explanation of the changes over time in the American political economy that accounts for the decline of the steel industry.

I will argue that this outcome is the direct result of policies implemented in the postwar period bounded by the years 1945–60. I will also argue that these policies—acts of both commission and omission—were not the result of isolated initiatives on the part of any single player, be it the managers of the firms, organized labor, or government officials. Rather, they grew out of the collective interaction of all the participants. However, there is no dearth of blame to be meted out in the steel case; the supply is abundant and can be liberally applied to all involved.

My analysis, however, does differ in several striking ways from more traditional conclusions that have been reached by prior investigators who have written on the problems of the steel industry.[1] According to the consensus of most studies of the subject, the primary reason for the postwar decline of steel can be found in managerial inefficiency and arrogance. Greedy and narrow-minded corporate managers, attempting to maximize short-term shareholder profits in a near-monopolistic setting, led the industry down a path to destruction because they refused to acknowledge the longer-term implications of this policy. Or so we are told. Yet I believe that while steel managers were no doubt a contributing cause of the industry's decline, they were not the only cause. Labor demands initiated by union leadership also had a major influence on the policies undertaken by the steel companies during the critical 1945–60 period. At the same time, governmental officials, attempting to respond to new global challenges never experienced by their predecessors, and unsure of the implications of their actions in this untested arena, retreated into patterns of the past, leaving the steel industry to seek its own solutions. Meanwhile, governments in other steel-producing nations were rushing to help their producers in this vital industrial sector.

The culmination of this mixture of public and private policies was the strike in 1959, a symbolically important year for steel that saw all three of the major actors deeply involved in the event. Their collective failure to resolve positively the problems at hand, indeed to even understand the deeper implications of their situation, has had a devastating effect on the subsequent fortunes of both the industry and the nation.

The premier institutional economist of our century, Joseph A. Schumpeter, said that to understand economic phenomena one must know history, statistics, and theory. "Of these fundamental fields," he said, "economic history—which issues into and includes present-day facts—is by far the most important."[2] This book attempts to apply that dictum to an understanding of the decline of the American steel industry. While the book does not explicitly attempt to extrapolate its conclusions about steel to the present problems in the remainder of the American manufacturing sector (the fundamental factors involved are quite different in each industry in most cases), I would nevertheless assert that one can learn much about why America is a declining economic power through a close analysis of the history of steel.

Philadelphia
August 1987

P.T.

Acknowledgments

Unlike some scholarly pursuits, historical research is one in which many debts are assumed. This present study is no different, and I am pleased to acknowledge the following individuals and institutions for their commitments of time, money, and interest in my behalf.

My book began as a dissertation at the University of California at Berkeley, and the three members of my dissertation committee were instrumental in its development: Joe Pratt opened up the domain of business history to me, especially the rich Chandler/Galambos traditions as they prevailed at Johns Hopkins; Richard Abrams showed me in excruciating detail what it meant to write honest history; and Edwin Epstein provided constant encouragement and the affection with which he is so richly endowed. I am heavily indebted to all three of these scholars and gentlemen.

Other associates at Berkeley were also giving of their time when called upon. David Vogel has been a constant source of assistance (including his repeated proddings to finish the project). Dow Votaw, Robert Harris, David Teece, and John Zysman read (or listened patiently) to parts of the original manuscript and offered helpful comments; Christine Rosen was always willing to hear my excuses. Finally, three doctoral student colleagues—David Palmer, Jeanne Logsdon, and Stephanie Lenway—made life both bearable and at times even worthwhile.

Parts of the general material presented in this book were discussed at one time or another at various meetings. These included seminars and scholarly conventions at the University of Michigan (thanks to Jim Reece), the Harvard Business School (thanks to Tom McCraw, Ron Fox, and Richard H. K.

Vietor), MIT (thanks to Mel Horwitch), Ohio State University and the Ohio Historical Society (thanks to Mansel Blackford), the University of Pennsylvania at a seminar on the history and sociology of science (thanks to Thomas P. Hughes), and at the business schools of Stanford and Boston universities. Papers presented at recent annual meetings elicited helpful comments: at the Economic and Business Historical Society from Edwin Perkins, Theodore Kovaleff, William Carlisle, and James Soltow; at the Economic History Association from Claudia Goldin; and at the Social Science History Association from Kim McQuaid—all contributed to the improvement of my study. I am happy to acknowledge their assistance.

I would be woefully remiss were I not to acknowledge the substantial commitment of time and energy provided me by Louis Galambos. He carefully read every page of my original manuscript, offering comprehensive commentary and suggestions for improvement (all at a time when he was laboring under his usual overwhelming work load). Although I have not necessarily incorporated all of his suggestions into this study, I am nevertheless sure that it is a much better work because of his deft hand.

Many of my current colleagues have been continuously giving in their encouragement of this project. They include R. Edward Freeman; Edward H. Bowman; Thomas Dunfee; Lawrence Hrebiniak; Bruce Kogut; Franklin R. Root; William Evan; Balaji Chakravarthy; Ian Maitland; Robert Hessen; Terri Langan; Richard Schubert; Alfred D. Chandler, Jr.; Albro Martin; Marta Porter; and David A. MacDonald. They offered assistance, material and otherwise, over the course of my research, and I thank them again.

Numerous libraries have been the source of my research, and not once was I disappointed in the kindness shown to me by their staff personnel; indeed, most went out of their way to provide help. These include staffs at the U.S. Library of Congress and National Archives in Washington, D.C.; the presidential libraries of Franklin D. Roosevelt, Truman, Eisenhower, Kennedy, Johnson, and Ford; the several libraries of the University of California at Berkeley; the, unfortunately, now out-of-existence Charles Schwab Memorial Library of the Bethlehem Steel Corporation in Bethlehem, Pennsylvania; the Seeley G. Mudd Library of Princeton University; the Baker Library at the Harvard Business School; the library of the State Historical Society of Wisconsin in Madison; the Eleutherian Mills-Hagley Foundation Historical Library in Wilmington, Delaware; and the Lippincott Library of the University of Pennsylvania.

There is never enough funding for historical research. I am thus grateful to acknowledge the following sources for their financial assistance in one phase or another of this book: the Rovensky Fellowship Program of the Lincoln Foundation; the University of Pennsylvania Research Foundation; the Reginald H. Jones Center for Management Policy, Strategy, and Organization (and its director, Edward H. Bowman); the Harry S. Truman Presidential Library Research Fund; the Management Department of the Wharton School (and its then-chair, Peter Lorange); and the Business and Public Policy Program of the School of Business of the University of California at Berkeley.

Dr. Ann Bohara of the Wharton School provided expert editing assistance as well as encouragement at just a time when it was needed; Carol Morrison, also at Wharton, spent many hours typing manuscripts for me when she had far more interesting things to do; and Herb Addison of Oxford University Press—although he probably gave up on me several times—was never far from my consciousness for well over a year. I thank them all (including the several anonymous reviewers provided by the publisher, who gave many helpful comments for improvements in the manuscript).

Contents

Prolog 3

1 Steel and the State: A Legacy of Conflict 5

2 The Postwar Debate on the Expansion of Steel Capacity 21

3 Truman and the Steelmakers, 1945–48 42

4 Domestic Steel and the New International Economic Order 64

5 The Triumph of Conflict: Truman and Steel, 1949–52 83

6 Eisenhower and Reconstruction of the International Steel Industry 103

7 A New Strategy for Competition: Eisenhower and the Steelmakers, 1953–56 128

8 The Beginning of the End: Eisenhower and Steel, 1956–60 153

9 An Industry in Decline: The Growth of Steel Imports into America 167

10 History, Ideology, and Steel 185

Notes 191

Bibliography 257

List of Acronyms 275

Index 277

THE DECLINE OF AMERICAN STEEL

Prolog

The American steel industry has degenerated to an apparent state of permanent contraction. Steel mills, once surging with orders to be filled, are now forever shuttered in many parts of the nation. Numerous companies have filed for bankruptcy, while others escaped by only the thinnest of margins. Workers, once "idled," were then permanently eliminated, and steel imports displaced an ever larger share of the domestic market.

Though the crisis came to national attention only in the past decade or so, it actually has its roots in events that began in the first years of this century and came to a head in the fateful decade and a half following World War II. To understand the recent crisis, it is necessary to look at the past history of twisted relations between steel industry management, the government, and the steel workers. In perhaps more than any other major domestic industry, government policies that affect the current management and performance of steel producers have been largely dependent on earlier patterns of relationships with various public officials, agencies, and policies. Though these complex relationships also affected the performance of the steel industry in the nineteenth century, they became more decidedly pronounced in the period since 1900. As a result, any contemporary analysis of the steel industry that purports to explain the recent crisis runs the risk of failure if it does not invoke the rich and compelling history that has led to the events of today.

The major part of this book will focus specifically on the post-World War II period of 1945–60. In this brief span the domestic steel industry slipped from a position of undisputed international dominance to one of confusion and weakness. At the same time, the leading foreign producers, thought to be hopelessly burdened by the consequences of a devastating war in their home territories, were moving in the opposite direction. By 1960, for example, the imminent comparative advantage of the Japanese producers was apparent to all but the most obtuse observers. My book attempts to explain this dramatic shift in economic fortunes by examining the nature and course of the steel industry's business–government–labor relationship over these critical and often turbulent years. It is within the confines of this period and these relations, I shall assert, that the roots of the industry's present economic decline can be found.

The first chapter of this book, intended as an introduction, presents the historical context from 1900 to the end of World War II and describes the evolving conflict between the steel industry, government, and labor. The second chapter examines the crucial debate in the immediate postwar years between the government, wanting a vast expansion of steel capacity to meet anticipated pent-up demand, and the industry, fearing that overexpansion would be disastrous if demand failed to materialize. Chapters 3, 4, and 5 look at the years of the Truman presidency that culminated in his failed attempt to seize the steel mills during the industry-wide strike of 1952. Chapters 6 through 9 examine the Eisenhower years, which also culminated in a national strike in 1959. In that year, steel imports for the first time in the twentieth century exceeded exports, as they have in every year since then. Chapter 10 draws conclusions about why events in the steel industry happened as they did and attempts to identify steps that might assist other industries facing similar crises.

This book does not portray villians on one side and heroes on the other, nor does it find simple reasons for what happened to the steel industry. It does present a complex web of people and events that I believe can deepen our understanding of a major economic phenomenon of our time. With the United States in the midst of what is arguably one of the most critical transitions in its economic history, it is imperative that we learn as much as we can from the lessons of the past, such as they are. It is in this spirit that the present book was written.

1

Steel and the State: A Legacy of Conflict

The modern American steel industry was inaugurated in April 1901 through the formation of the United States Steel Corporation (USS) by the financier J. P. Morgan. The firm was the product of mergers and, indeed, capped one of the most frenzied periods of industrial amalgamation in history.[1] While a number of factors shaped these events, the principal reason behind U.S. Steel's creation was an unusual set of economic conditions that underlay competition between firms in the industry. It was the steelmakers' perception of these conditions that would animate how they responded to both external and internal forces over the ensuing years. Correspondingly, it was a strikingly different interpretation of these conditions that would foster much of the public suspicion toward the industry over the same period. It is thus necessary to review the conditions of competition in steel to understand better the resulting conflicts in business–government–labor relations in this turbulent industry.

By the turn of the century, the manufacture of tonnage carbon steel had evolved into a capital intensive business.[2] The construction of a large, competitively sized steel mill required an investment of many millions of dollars. Although the cost of labor and raw materials necessary to actual production was also high, it was not as important as capital in the final pricing of the product. As such, steelmaking is a capital intensive, high fixed-cost industry in which high break-even levels of output have to be achieved before profits are realized. Market conditions, meanwhile, are equally challenging. Steel is essentially a producer's good that is bought by a wide variety of other manufacturers who incorporate it into products that are eventually sold to ultimate consumers. Yet demand is not disproportionately influenced by any one group of users; rather, steel is so pervasive to the products of an industrialized economy that its demand rises and falls with larger economic conditions that exist at any given time.

This combination of being a producer's good that is subject to cyclic demand has important implications for steel pricing. In general the product is price inelastic—that is, a given reduction in steel's price will not result in any comparable increase in demand. In addition, transportation costs (thus mill location) can substantially influence pricing. Regardless of the high capital costs required in steel production, the product sells commercially for a relatively few cents per pound. The direct result of this is that transportation costs involved in moving bulk loads of raw materials—from their source to the mill and then transporting the finished product from the mill to customers—are usually a large component of the final price charged to the buyer. Given the large fixed investment necessary for steel production—an immobile factory with an unusually long economic life—any geographical shifts in markets or resource sites can drastically upset company pricing strategies no matter how well they may originally have been conceived.[3]

When these economic factors are considered in the context of a freely competitive marketplace, the results may prove volatile. Indeed, this is precisely what happened after 1893 in American markets, leading to the celebrated merger that created U.S. Steel. As we noted, high fixed-cost industries such as steel have high break-even points of production; that is, they must maintain mill utilization rates at nearly full capacity to recoup investment capital as well as pay for variable labor, raw material, and related costs. In periods of strong demand, mill profits can be handsome once break-even levels are surpassed. But in cyclic downturns, the situation can become precarious owing to the same high fixed-cost structure of production. Without any artificial means to stabilize prices or output among competing producers, mills can be expected to slash prices during a downturn in order to increase demand to break-even levels. The pressures attendant on high fixed-cost production naturally bring this about since independently acting steel producers believe they can maintain acceptable utilization rates at the expense of competing mills that are presumed to refrain from such price-cutting tactics. Yet when this behavior occurs, competing firms naturally slash their prices in response because they are driven by similar motivations. During recessionary periods, the competitive frenzy of price-cutting will drive the least efficient firms out of business; as the recessionary period lengthens, even some well-managed firms will resort to new sources of capital or cease operations. This was the situation that beset the steel industry in the United States following the Panic of 1893 and the long depression that it generated.[4]

For steelmakers, the implications of these economic imperatives, termed "cutthroat competition," are substantial. Whenever a market downturn develops, even the hardiest free-market advocates are forced to seriously reconsider their beliefs. Economic theorists, on the other hand, have generally concluded that such conditions are not only normal, but healthy: "Creative destruction," Schumpter's term for capitalism's process of constant innovation, although unpleasant for its victims, can be expected to lead to greater benefits for the economy as a whole once consumers obtain the fruits of competition.[5] But in the vastly restructured steel sector that emerged after 1901, the new leaders of the industry—essentially managers rather than entrepreneurs—reasoned that substantially similar social welfare outcomes might be derived from a less debilitating process. Instead of creative destruction, it was now held that cooperation and conciliation among producers could yield equal results: Both lowered cost and less risk to shareholders, managers, labor, and the general public alike.

The propagation of this theme was to become the motive force behind the industry's dominant personality over the first quarter of the new century: Judge Elbert H. Gary, chairman of USS from its founding until his death in 1927. The creation of U.S. Steel had, of course, been engineered by the investment banker J. P. Morgan in order to protect financial positions he had taken earlier in the industry. But it was to Gary that he turned over managerial control of the firm, and it was to this man's philosophy of competition that the industry* would turn for strategic guidance over the next fifty years.[6]

Throughout his long career at U.S. Steel, Gary strove unceasingly to obtain industrywide stabilization, especially in matters of price policy. He first used the massive size of his firm to bring about compliance by his domestic competitors (USS, the nation's first billion-dollar enterprise, was formed with over two-thirds of the industry's crude steel capacity within its domain).[7] When this failed to achieve satisfactory results, he turned to a basing-point pricing scheme (Pittsburgh Plus) that had been utilized in one form or another by steel producers since the 1890s.[8] When tightened antitrust enforcement threatened this device, he formed a new trade association (the American Iron and Steel Institute [AISI]) to carry out his cooperative goals.[9] The vigor he applied to the achievement of fixed steel prices was also brought to bear on other issues involving industrywide stabilization, including capacity expansion, labor relations, and the international steel trade.[10] In all of these areas, Gary fervently insisted that only through coordinated action could the industry survive cutthroat competition and thus spare the nation from economic ruin. "There is," he stated,

> only so much business to go around . . . [and] you cannot increase it by trying to get something for yourself that naturally belongs to some one else. But competition is the life of trade. It is not only desirable; it is necessary. And we should not try in any respect to suppress fair and reasonable and active competition. But destructive competition

*In this context, "industry" refers generally to those large, integrated steel firms that dominated production. The eight largest of these firms never held less than 50% of total industry ingot capacity from 1901 to 1960 and always held over 70% after 1925. See G. G. Schroeder, *The Growth of Major Steel Companies, 1900–1950* (Baltimore: Johns Hopkins University Press, 1953), p. 197.

> is not reasonable, not desirable, and never beneficial in the long run to anyone.[11]

It was to the eradication of "destructive competition" in steel that Gary devoted his life. In its place, he counseled, the "visible hand" of management would see to the industry's (and by extension the nation's) economic prosperity.

Considered purely in the context of the industry's high fixed costs and price inelasticity, there is perhaps nothing unusual about Judge Gary's competitive beliefs (many other businesspersons confronted by similar economic circumstances held similar views). But most public officials had a wholly different opinion regarding competition; one direct consequence was that the industry and government would experience a near-constant state of tension from 1901 onward because of their conflicting attitudes. In the minds of many, the steel industry had been monopolized by virtue of the creation of U.S. Steel and the proper role of public policy should be to dissolve that monopoly or at least insure that it did not enlarge any further. By 1911, with the steel industry already the subject of intense public scrutiny and mounting public antipathy, President Taft's Justice Department filed an antitrust suit seeking the dismemberment of U.S. Steel.[12] Although the company eventually survived this particular judicial threat—in a highly controversial four-to-three Supreme Court decision handed down in 1920—it was never able to achieve the public acceptance and legitimization that Gary had so assiduously cultivated throughout his long career.[13] The balance of relations between steel and the government seemed no closer to resolution at the time of his passing a few years later than it had in 1901 when the firm was created.

I

The stabilization of price in the domestic steel market was the primary objective behind the formation of U.S. Steel. Another factor, however, must not be overlooked. This was the opportunity to expand export sales and thus provide the new firm with a safety valve to offset cyclic slumps in domestic steel demand.[14] The threat of foreign competition had been the primary impetus for industry political activity throughout the nineteenth century. Iron and steelmakers were constantly fearful that more advanced European producers might enter domestic markets in volume, and they played a pivotal role in the imposition of high protective tariffs from their initial establishment in 1789 through each succeeding consideration of such legislation.[15] By the turn of the century, however, large American steel producers were as efficient as any in the world, and they no longer needed excessive import protection. Accordingly, U.S. Steel began to retreat from the industry's traditional protectionist position as it sought to capture offshore markets for the domestic stabilization benefits they might provide.[16] The firm's annual report in 1906 stated:

> The aim has been to build up a permanent and continuous export trade with a view to providing markets which at all times may be relied upon to absorb a fair proportion of the total production, rather than to sell material in foreign countries only at times when the domestic market is unable to take the entire output of the mills.[17]

To implement this new policy, an export subsidiary was formed in 1903. Foreign sales, under $100 million in 1903 climbed to $305 million by 1913 (representing 16.5% of the firm's total revenue).[18] But as U.S. Steel's foreign sales began to expand, they reached a point where fluctuations in offshore markets could materially affect company performance, thus exerting dangerous pressures on the order and stability that Gary had sought to establish in his home market. Moreover, the growing influence of American steel in foreign markets was stirring both defensive and retaliatory moves by offshore competitors who had previously dominated the international steel trade.[19]

As the firm was drawn further into the dynamics of international trade, it began to confront a new set of competitive conditions—quite apart from purely domestic problems—that could influence company performance. Gary and his successors found that foreign competitors often worked in close cooperation with their home governments. Because steel was frequently the leading industrial sector in these foreign nations and because of historical patterns of industrial development that were substantially different from those in America, offshore steelmakers received a number of public benefits that were not available to domestic American producers. This, of course, created dilemmas for domestic managers. They might seek similar cooperation at home in order to counter the foreign advantage or they might seek direct contact with foreign competitors in order to pursue on their own the international stabilization policies and agreements that they felt were essential to success.

In the formative years of the modern American steel industry, beset with serious business–government tensions, the largest firms (led, of course, by U.S. Steel) chose the route of private accommodation with their offshore rivals. Thus was established a pattern that proved difficult to break for many years to come. The problems of international competition under such circumstances would play a pivotal role in the American steel industry's development throughout the twentieth century.

Gary's initial effort in global cooperation was the formation of a worldwide trade association designed to achieve goals similar to those pursued in the domestic arena by the AISI. The attempt, however, foundered amid governmental fears that the proposed organization would lead to anticompetitive behavior by steel firms.[20] Nevertheless, U.S. Steel did expand its foreign business during the years of the Wilson administration, as did many other American manufacturers—but this was due more to market opportunities created by World War I than anything else.[21] With the election in 1920 of a new president from a party deemed far more favorable to steel's particular needs than the previous Democratic administration, the industry felt it would now be able to consolidate its enhanced global presence and perhaps even achieve the international stabilization that Gary had long sought. As early as 1916 the AISI had emphasized:

> When peace is again secured we will, for a while at least, be the leading export nation of the world; whether we retain this supremacy will depend largely upon the efficiency of our merchandising and industrial organization and the cooperation of our Government with business.[22]

This supremacy, however, was not to be retained.

The reason for this outcome rested not with governmental policy (at least not directly), but rather with the actions of another constituency that had long exercised considerable influence over industry affairs, that is, the investment banking community. J. P. Morgan & Company, the dominant firm in the field, had been the instrumental force behind the creation of U.S. Steel in 1901, and its partners continued to administer a strong interest over the financial affairs of the steelmaker through their presence on both its board of directors as well as on its powerful finance committee.[23]

Operational managers of USS (as well as other producers) sought to expand their foreign business by exploiting the void in European production caused by the destruction of capacity during the war. This objective came into conflict, however, with American investment bankers who also foresaw great opportunities in Europe through profits that could be secured from underwriting loans to reconstruct European industry. Following implementation of the Dawes Plan in 1924 (with which the Morgan firm was closely involved), the United States became the principal source of such reconstruction funding. The European steelmakers—the leading industrialists in nearly every nation on the Continent—received their fair share of these funds, and the result was a rebuilt and rationalized industry that was poised for reentry into global markets earlier abandoned to U.S. suppliers. "Should the American steel industry let the foreign steel producers take away its world markets?" asked one banker bluntly in 1927. "It should," he answered.[24]

With this development, the foreign expansion plans of domestic U.S. steelmakers were shelved, including some proposals for the construction of new mills in Europe. The industry, it was determined by its bankers, should not be allowed to compete with the investments in offshore steel production undertaken by the bankers themselves; if these foreign loans were to be repaid, then as few impediments as possible should be permitted to get in the way—including U.S. competition. The domestic steel industry's plans for global hegemony were thus dashed by the very force that brought it to life:

the nation's powerful private investment banking community. The steelmakers, for all their supposed power and influence, in fact, were the subordinate actors in this drama.[25]

II

The end of the Prosperity Decade found the domestic steel industry in a state of disorder. The hopes of the larger firms for a liberalized world trade regime in which American manufacturing interests could predominate were shattered by both the investment bankers as well as the resurrection of high protectionist tariff barriers that had been promoted by smaller firms. Internally, cooperative modes of behavior were subjected to new pressures by legal restrictions (the Pittsburgh Plus basing-point price scheme was declared illegal in 1924), mounting overcapacity, and a consequent erosion of pricing discipline. The Great Crash of 1929 and its resultant economic chaos only added to industry instability. Clearly, it was time for a fundamental reformulation of industry strategy if the larger producers were to survive and prosper.

It appeared to some that perhaps the government should now play a more important role in restructuring the industry. Steel was, after all, the largest employer in the nation and generally considered to be its leading manufacturing sector. But such public assistance would not come to pass. The new Democratic successor to the White House, Franklin D. Roosevelt, came into office with a perceived bias against the larger firms in the industry, especially its primary entrant, U.S. Steel.[26] His administration's ambivalence concerning the amount of power, both economic and political, that was vested in these larger firms resulted in a relationship between steel and the government that would be rife with suspicion and hostility well into the 1940s. Three issues would preoccupy the agenda of distrust during this period: prices, foreign competition, and labor relations. They would therefore be the focus of the industry's strategic planning throughout the decade.

With foreign markets now seemingly beyond the industry's reach, the issue of domestic price control reappeared as the chief hope for steelmakers to achieve stability. The problems following the 1929 crash severely increased competitive pressures throughout the industry, but these fell most heavily on U.S. Steel and Bethlehem Steel, the two largest firms. This was due to the reliance these firms had traditionally placed on the production of heavy steel, which was typically used in capital goods, such as machinery and locomotives; during the depression, this turned out to be the industry's least-popular product line. Thus with their markets eroding and competitive pressures increasing, the larger firms sought means to maintain stable prices until a rebound occurred.

Oddly enough, an answer to this crisis would be provided by the new administration. Though obviously not by President Roosevelt's design, its

National Industrial Recovery Act (NIRA) effectively breathed new life into Gary's waning program of steel industry cooperation and conciliation. Under cover of the economic uncertainty of the period, the AISI was able to exploit the situation and usurp the role of industry leadership that Roosevelt clearly intended to be filled by all segments of this sector, small and large producer alike. The institute, however, remained firmly in the hands of only the larger industry interests, and they consequently were able (through the National Recovery Administration [NRA]) to implement policies specifically favorable to their ends. The result was price stability that benefited U.S. Steel and Bethlehem Steel to the detriment of the smaller firms.[27] By early 1934 Eugene G. Grace, president of Bethlehem Steel, would be able to state:

> In the progress [under the NRA] made to date, . . . I think it may be said that a sounder basis has been developed for industry . . . than it has enjoyed at any time during the post-war period.[28]

With domestic price stability temporarily in hand, the large steelmakers could now devote greater attention to their other top-priority items on the strategic agenda: the twin threats of foreign competition and labor unrest. The modernization of the European steel industry continued to challenge domestic markets by the mid-1930s. Although the 1930 tariff act had maintained the steep rates of the past, there were still sizable iron and steel imports in certain American markets. Another foreign development also caused the industry worry. In 1926—amid the massive reconstruction of European steel mills that American funds had fueled—France, Belgium, and Germany formed a new steel cartel in order to stabilize their chaotic market conditions. After 1929, however, the cartel, known as the Entente internationale de l'acier (International Steel Cartel [ISC]), broke down under the burden of the crash and its subsequent uncertainties.[29] But by mid-1933 the ISC was revitalized through German leadership as an export cartel, and by 1935—with the participation of Great Britain, Poland, Czechoslovakia, South Africa, and several other steel-producing nations—the group became the dominant force in world steel markets.[30]

American steel exporters had maintained some contact with the ISC through the Steel Export Association (SEA), a company organized by U.S. Steel in 1928 under the Webb-Pomerene Act.[31] This legislation, however, put limitations on local participation in international cartel agreements. Consequently, domestic producers viewed the growing ISC with some trepidation. This fear was exacerbated when the administration began to turn less friendly to big business in the mid-1930s following the initial phase of cooperation that characterized Roosevelt's presidency. Two issues in particular seemed to threaten the steelmakers at this time: tariff reform and labor reform. Liberalized free trade might have been appropriate in the first two decades of the century, reasoned the larger steelmakers, but now that the international competitive situation had changed so much in their disfavor, protection once again seemed the best policy to promote. Under the leadership of Secretary of State Cordell Hull, however, the administration introduced the Reciprocal Trade Agreement Act (RTAA) in 1934, which was easily passed by large Democratic majorities in both houses of Congress. This legislation promised to revolutionize America's historical reliance on protective tariffs as the nation's basic foreign economic policy.[32] The act allowed the president to unilaterally revise existing tariff rates up or down by 50% for countries that would agree to reciprocate on trade concessions; it also permitted insertion of most-favored-nation clauses in the agreements.[33] Although this new legislation would actually do little to foster international economic recovery during the 1930s, nevertheless it struck many observers—steel leaders included—as a symbolically radical step that was sure to result in a flood of cheaper imports into home markets.

Soon after passage of the RTAA, the industry was dealt further blows. In May 1935 the courts ruled the NRA unconstitutional; this was followed only one month later by passage of the National Labor Relations Act (NLRA).[34] For the first time in history, the weight of the federal government would now be placed solidly behind collective bargaining for labor. To the steelmakers—who had one of the most contentious records of labor relations in America—this new law appeared foreboding.

Indeed, labor relations had long been a serious problem in the industry. A series of confrontations in the late nineteenth century, capped by the bloody Homestead strike of 1892, underscored the steelmakers' rigid determination to resist unionization at any cost.[35] Failed organizing drives in the early years of U.S. Steel's dominance, followed by well-publicized federal investigations into industry working conditions, only added to this image of corporate arrogance.[36] The World War I years saw a temporary improvement in labor conditions in the mills, but this was brought about only by governmental insistence; after the war, a long and bitter strike in 1919—a battle lost by the workers—resulted in a resumption of prior hostilities. During the early 1920s the long hours worked by some steel mill laborers (seven days per week, twelve-hour shifts) again aroused public antipathy toward the employers, leading in 1922 to private intervention by the Harding administration to ameliorate working conditions. By 1923 this resulted in a reluctant shift to the eight-hour day in the steel mills.[37]

Yet while labor finally did win this concession, part of the price was effective abandonment of its goal of an independent union for steelworkers. Instead, employers now turned to organizing their workers into company unions.[38] These structures remained in place through 1935 when passage of the NLRA revitalized the movement for independent labor.

Industry fears of labor instability stemming from the new law proved correct when organizers quickly began a union drive among steelworkers. Under the leadership of John L. Lewis of the United Mine Workers of America (UMWA), a Steel Workers Organizing Committee (SWOC) was formed in June 1936 to organize the mills.[39] At first the employers maintained a tightly united stance against this new threat to their control; in an open letter to both workers and the public, the AISI declared:

> The Steel Industry will oppose any attempt to compel its employees to join a union or to pay tribute for the right to work.[40]

But a later paragraph in this industry manifesto gave a more telling reason for the opposition:

> The Steel Industry is recovering from six years of depression and huge losses, and the employees are now beginning to receive the benefits of increased operations. Any interruption of the forward movement will seriously injure the employees and their families and all business dependent upon the Industry, and will endanger the welfare of the country. The announced [union] drive, with its accompanying agitation for industrial strife, threatens such interruption.[41]

The industry, after steady market depression since 1929, was by mid-1936 finally beginning to see an upturn in business. This was especially true for USS and Bethlehem, the two steel firms most vulnerable to the economic decline by virtue of their reliance on heavy-steel markets. The reason for optimism lay in events in Europe: Spearheaded by the new National Socialist (Nazi) regime in Germany, there was a massive resurgence of defense orders in many European nations. Since steel (especially heavy steel) was perhaps the single most important industrial commodity to war-making, it was only natural that steelmakers would benefit from this activity. With European mills beginning to fill their order books in the face of booming demand, idle American mills were offered production contracts. The major domestic suppliers of heavy-steel shapes conducive to war-goods, U.S. Steel and Bethlehem Steel, would receive the bulk of these offshore orders.[42] Some steel leaders concluded that SWOC might fatally dash this perhaps-fleeting opportunity for economic recovery if it took the workers out on strike.

Labor thus became the strategic fulcrum on which the complex decisions facing the larger firms in the industry were balanced. Simply put, leaders of U.S. Steel decided that labor unrest would have to be contained if their firm expected to work its way out of the depression. Two basic options were available: recognize the union or fight it into submission. Most companies in the industry favored the latter approach, based primarily on historical patterns of conflict in this arena. U.S. Steel alone, however, chose the first alternative. On March 1, 1937, the company's chairman, Myron C. Taylor, shocked both the industry and the nation by announcing that he would sit down with Lewis to negotiate a contract.[43]

To understand why U.S. Steel chose this surprising stance toward labor, we must first return to events in the international arena, for the underlying logic of Taylor's move went beyond domestic concerns. The smaller firms that refused to recognize the new labor organization were in general inland producers, protected both by natural transportation barriers from the threat of imports as well as by generally more efficient mills than those of USS and Bethlehem. Of course, U.S. Steel was highly vulnerable to these offshore competitors and at the same time stood to gain the most from the new defense contracts flowing from Europe. This factor allowed the revitalized ISC to play a key role in the firm's emerging political–economic strategy. The ISC was a powerful force in itself, but it realized that without U.S. participation global stability could never be assured. Consequently, European cartel executives pressed their American counterparts to cooperate with them beyond the limits of the Webb-Pomerene Act.

The strategy of the ISC representatives was to obtain cooperation from the entire American industry by offering concessions to the leaders, USS and Bethlehem. Reasoning that these two firms could exercise powerful influence over their domestic colleagues, the cartel offered them a number of enticements. A major condition to the grant of these lucrative concessions, however, was that USS and Bethlehem would convince the smaller domestic firms to join the agreement too. In past years several of these smaller firms, led by National Steel and Armco Steel, had innovated new processes that gave them advantages over foreign competitors in certain "light" steel product lines. These advantages had been put to use in some "neutral" export markets, and as a result the Americans were undercutting cartel bids and winning orders. The ISC wanted U.S. Steel and Bethlehem to force a stop to this practice by the smaller firms because it was unsettling the global stabilization policies that the cartel was designed to achieve.[44]

This would prove a tough order for the American leaders to fulfill. Nevertheless, Myron Taylor of U.S. Steel began active negotiations with the cartel in mid-1936; dark hints regarding retaliation against U.S. markets were the entente not joined apparently proved compelling.[45] As Taylor considered his options, two factors seemed to constrict his freedom of choice: the antitrust laws and the tense labor situation. Webb-Pomerene stipulations did not allow domestic firms to participate in international marketing arrangements if the domestic market would thereby be restricted from foreign competition. Regarding labor, John L. Lewis was obviously moving toward confrontation with the industry over his demands for recognition. Thus Taylor's position was not enviable: If American participation in the ISC were not forthcoming, the corporation's markets might be ravaged by cheap imports and its access to lucrative foreign orders diminished; with such a result would go any chance of pulling the firm out of the depression into which it had been stalled for so long. And if the labor question were not resolved amicably, the firm might lose all control over employee costs and further alienate an already hostile government and public.

Taylor decided to move boldly on both fronts. Gambling that he could control his lesser domestic rivals, he struck the historic March 1, 1937, agreement with Lewis to guarantee labor peace in U.S. Steel's mills. Just before that date, on February 26, he secretly committed U.S. Steel to the terms discussed previously with European cartel leaders regarding participation in the ISC (he also persuaded Bethlehem to join him in this).[46] The language of collaboration with the cartel was drawn in such a way that U.S. antitrust laws would ostensibly not be violated, but its clear intent (as later investigations would prove) was to stabilize domestic markets by fixing import volumes and prices—a direct violation of both the spirit and letter of the Webb-Pomerene Act.[47] Taylor's apparent reasoning was not only to agree to labor peace with Lewis and pay the marginal labor costs this might entail, but also to compensate by raising domestic steel prices a comparable—if not greater—amount. He felt this strategy would pay off because he now knew that cheaper imports would not threaten his markets, and he assumed that he could persuade his domestic colleagues to go along with the higher prices.

Taylor's bold stratagems would seem to have resolved all of his firm's problems at once, yet, as had occurred so often in the past, these moves did not yield results exactly as anticipated. Two problems emerged to confound matters: Most of the other members of the steel industry would not go along with U.S. Steel's recognition of SWOC (even cohort Bethlehem refused) and New Deal officials soon became alarmed at the corporation's apparent ability to administer steel prices in the domestic market. In consequence of the first problem, Taylor could not deliver the industry to the ISC on the terms promised, which in turn posed renewed threats of dumped foreign steel in USS markets. As to the second problem, it contributed to the formation of the congressional Temporary National Economic Committee (TNEC), which was designed to ferret out alleged abuses of power by concentrated industries in the American economy and to rectify such abuse as necessary.

U.S. Steel's carefully laid plans thus began to unravel. Not only did the smaller integrated firms in the industry refuse to recognize SWOC, but they also would not agree to the export quotas that the larger firms had negotiated for them through the ISC. Under threats from the cartel to enforce discipline, U.S. Steel and Bethlehem Steel were forced to give up part of their own quotas to compensate for the smaller firms' exports.[48] To make matters worse, the seeming end to the depression forecast by the economic recovery of 1936 soon evaporated when a new plunge occurred in late 1937. Some in the administration, stunned by this turn of events and groping for a public response, began to claim that rigid high prices had brought on the reversal. The large steel producers—perhaps the easiest and most vulnerable target for public policymakers to isolate—bore the brunt of the attack from the government's rising antimonopolist faction.[49] Economic science, it must be recalled, was even less advanced a discipline than it is today; as a consequence, public policymakers had few reliable tools with which to gauge or repair systemwide economic behavior. Roosevelt soon accepted the administered-price hypothesis and by April 1938 sent a message to Congress demanding an investigation into "a concealed cartel system" and the "disappearance of price competition" in the American economy. Congress responded by establishing the

TNEC to make a "thorough study of the concentration of economic power" in America.[50]

Against this background of increasing public suspicion, U.S. Steel became unnerved in its strategic price policies. Throughout 1938 price cuts were common in the industry. By that year's end U.S. Steel went into the red for the first time since 1934, and Myron Taylor's bold gamble for stability, seemingly within his grasp only months before, now lay in considerable disarray. He retired from his position with the firm in April 1938. It would remain for others to regroup the industry's forces against the tides of depression.

III

The prevailing gloom of 1938 eventually culminated in the steel industry's revival. The source, unfortunately, was war. When German military preparation turned to invasion in September 1939, it remained only a matter of time before America would be drawn into the conflict. The financial reprieve of the industry soon followed. But while financial statements would brighten, industry relations with the government would not. Where the White House once demanded that the firms lower prices to stimulate demand, public policymakers now turned to the advocacy of supply stimulation. Yet, just as they had on the prior issue, the steelmakers would again disagree with the administration's position.

With the approach of war, the specific dimension of business–government conflict in steel turned to the industry's capacity levels. Government planners believed that producers should rapidly expand their mills to turn out the steel necessary for the presumed defense effort. The larger firms, however—fearful of the long-term competitive consequences of a capacity buildup—were reluctant to cooperate on a purely voluntary basis (a situation that arose in other key industrial sectors that were also urged to expand for war production).[51] Although this somewhat-technical capacity issue never reached the level of public confrontation that accompanied clashes over labor and pricing in the 1930s, nevertheless it was a point of strategic significance to both private decisionmakers and public policymakers.

The basic problem of capacity expansion had its roots in industry fears of international competition that developed during the 1920s and 1930s. As we saw, these fears were not overcome until 1936 when U.S. Steel and Bethlehem Steel engineered American participation in the ISC. But the coming of the war obviously exerted destructive forces on cartel stability; by March 1940 the ISC reluctantly was forced to disband.[52] This was not an event greeted with enthusiasm by the domestic producers. With termination of the international entente, there would now be no vehicle to stabilize future world steel trade. Prewar capacity increases that could not be filled by postwar demand would thus burden critical mill utilization rates and perhaps even plunge the industry back into the financial distress out of which it had only recently emerged. Moreover, without the ISC to police and coordinate world trade, vulnerable American markets might again fall prey to cheaper foreign imports. All matters considered, concluded the major steelmakers, capacity expansion was to be avoided as much as possible.

But public officials were now demanding action in the face of the escalating European war.[53] By December 1940 Henry Morgenthau, Jr., secretary of the treasury, wrote to Roosevelt urging that "an immediate expansion program for the steel industry [be] called for." The president eagerly concurred with this advice.[54] The industry, however, for reasons enumerated earlier, was less enthusiastic. Irving S. Olds, now chairman of U.S. Steel, stated to an AISI audience in 1941:

> There can be no doubt that the steel industry has a heavy responsibility in insuring an adequate supply of steel to meet all proper demands of the future. These needs may well involve some enlargement of existing steel capacity. Long-range planning of a statesmanlike nature, however, is needed in the formulation of any extensive expansion program. In my opinion plans for enlargements of facilities at this time will not be sound, if based chiefly upon an estimated civilian steel demand growing out of an abnormal period of extraordinary defense effort and resulting extraordinary expenditures.[55]

Within the bounds of this brief statement can be found the fundamental source of industry reluctance to join with President Roosevelt in his call for more capacity. The war effort, as Olds and many others believed, was an abnormal period, and one could not properly extrapolate from such an extraordinary demand to future needs. If the steel industry—whose investment was characterized by expensive and long-lived assets—were to undergo a massive expansion to meet the defense effort, industry leaders asked, what then would happen to that capacity later—say within "two years, by which time, it is reasonable to expect, the critical emergency will have been met or passed" (to quote Walter Tower, president of the AISI).[56]

As far as the steelmakers could estimate, expansion would lead only to dangerous ends: overcapacity, higher costs, and lower profits. Moreover, the United States was not the only nation participating in this capacity buildup; Europe, Japan, and other nations were also rapidly adding to existing capacities in iron and steel production.[57] Assuming, as industry leaders did, that much of this new global capacity would survive the war, then these executives could conjure up only one fearful scenario: Intensive postwar global competition that would result in a massive broadside attack against American domestic markets by cheaper foreign imports. To vividly animate these fears, steel leaders had only to recall the past generation of their industry's history. In this context the present appeared to be but a reincarnation of the post-World War I experience. Then the industry had suffered severe economic setbacks from the aftermath of that period—indeed, it was only now recovering through the heated-up demands of war—and steelmakers obviously had no desire to relive such memories.

In consequence, industry interests fought vigorously the government's demand to expand. But delaying tactics would not hold up for very long as the nation began to chew up iron and steel much faster than it could be produced. By May 1941 Roosevelt would write aides:

> In spite of all the steel experts, I am concerned about additional future steel production needs. . . . I think therefore, that immediate steps should be taken to bring me in a plan covering ten million tons additional production East of the Rockies and five million tons additional production on the Pacific Coast—the production to be available as early in 1942 as possible.[58]

Eventually, the industry was forced to concede that its expectations regarding immediate demand were understated and it fell into line with the president's wishes. But it would not necessarily be patriotism alone that compelled such behavior nor would it be any change in postwar competitive expectations. Rather, it was brought about only through massive federal underwriting of the necessary expansion program; by September 1941 the administration had pledged $1.1 billion in public funds to help add some 15.2 million tons of new steel capacity to existing levels.[59] The government would provide for construction funds and would own the new mills it financed, but they would be leased to, and operated by, firms already in the industry. Moreover, the management contracts called for the mills to be sold to the private operators at war's end at a considerable reduction from original costs.[60] The industry leader in the development of this program was U.S. Steel, which took steps to insure the maintenance of its superior competitive position in the industry that would emerge from the war as a result of the forced expansion.[61]

The steelmakers' position, of course, was no different from that of many other industries asked to expand for wartime needs. There was widespread belief that the European conflict would be brief and relatively noninjurious to capital assets.[62] As such, these industrial interests were rightly concerned with the shape of postwar competition that expansion would bring. But perhaps more than any other lesson, the nature of the differences between the major steelmakers and the Roosevelt administration on this vital capacity issue demonstrates how far apart the two sides had drifted since 1933. Even though the defense of the nation was arguably at stake, the steel industry was initially reluctant to cooperate with the president's commands. But the government, too, failed to consider the long-term economic implications of its demands on producers, who after the war would again be forced to seek economic prosperity on their own. There was no cooperative planning, no industrial policymaking to attempt to chart a course of action that would be mutually beneficial to all sides. Instead—reflecting by now some forty years of suspicion and conflict—the steelmakers and the government viewed each other as wary adversaries, each pursuing its own very different agendas of action.

This was hardly a propitious foundation for the construction of industrial programs to meet the national interest. It was, however, quite indicative of the long history of conflict that animated steel–government relations in America. It would be against this distressing background that postwar developments in the steel industry would emerge.

2

The Postwar Debate on the Expansion of Steel Capacity

The crucial issue that initially divided steel strategists and public policymakers in the early postwar years was the expansion of industry capacity. By 1943 governmental planners had begun to turn attention to reconversion problems that would face the nation at war's end. A basic objective, they concluded, was encouragement of a full-employment economy to prevent repetition of the disastrous economic experience of the 1930s. Quite naturally, steel output was high on these officials' list of critical sectors: If full-employment goals were to be realized, they theorized, then it was imperative that sufficient supplies of steel be available to support overall economic growth. Any failures to achieve this goal could create bottleneck effects that might well jeopardize postwar expansion and welfare.[1]

Industry strategists, however, remained skeptical of the outlook for long-term steel demand. Although they might have agreed that some pent up consumer purchasing could be expected immediately after the war, they were highly doubtful about the prospects of sustaining this trend over time. Steel production, they reminded, required massive investment in long-lived plants that had few, if any, alternative uses; short-term abnormal-demand patterns did not justify such investment. This theme was in evidence as early as the American Iron and Steel Institute's (AISI's) annual meeting in the spring of 1943:

> As you know, completion of the [war-time] expansion program will leave your industry with about 95,000,000 tons of ingot capacity. . . . I do not need . . . to dwell on past performance of steelmaking in non-

> war years. You will recall readily enough the rare times when output for a full year exceeded 75 per cent of capacity. . . . Will the most diligent search turn up any likely growth of old users, or spontaneous creation of new users, on a scale to absorb any such tonnage?[2]

President Walter S. Tower of AISI continued by supplying perhaps the best perspective on such questions when he responded to the gathered industry audience:

> You probably will have to decide which of two courses you prefer to follow. The first may seem simple and easy—to operate such part of steelmaking capacity as domestic needs will support, and let the rest lie idle. . . . The other course is to seek a higher rate of activity through a drive for world markets. That course is complicated and difficult, because it probably would involve some sort of wide control over steelmaking in those countries which are now our enemies.
>
> . . . Members of an industry like yours well may ask: Shall the United States stick to a policy of unconcern, to let all happen that may? How far shall we go to rebuild or try to remake the rest of the world? . . .
>
> I question whether you want to see this country assume the burden of directing an international P.W.A. [Public Works Administration], whatever it might mean in demand for steel. And I doubt whether you would deem it wise for our foreign policy, responding to a surge of economic altruism, to embark on full rehabilitation of those who have fought to destroy us.[3]

Thus, in response to public planners, industry spokespersons counseled that steel capacity necessary for domestic requirements was already sufficient, whereas expansion for international demand was highly problematic. Part of their argument was grounded in historical factors. Recalling the industry expansion following World War I (when steelmakers were publicly encouraged to increase output to meet optimistic projections of demand both at home and abroad), Tower noted:

> [In] all this talk of great shortages and pentup demands, there is much that recalls the experience of 27 years ago [i.e., 1920]. It was then that the idea of "accumulated shortages" gained popular acceptance. Then, too, we heard about a steel-starved world. Some prophets of that day saw visions of at least "ten years of unbroken and unparalleled prosperity" for the steel industry. The very next year your furnaces operated at 35 per cent of capacity.[4]

In essence, Tower's counsel seemed to be that the industry should do nothing more than sit tight for the time being. The war, it seemed to him, was nothing so much as a relatively brief interruption to the throes of depression that had already cost the industry so much. With that war soon to end—and the thought of some 95 million tons of domestic steel capacity already saddled to the balance sheet—the industry was hardly receptive to any notions of further expansion. Domestic markets could not possibly be expected to absorb such tonnages nor did it seem "the time or place to try to see through the fog which obscures the future export markets for steel."[5] Enshrouded in that mist, of course, were visages of European steelmakers, the International Steel Cartel (ISC), and the potential output from newly industrialized countries that had been rapidly building their own capacity since the 1920s.[6] By 1945 the Allied powers alone had over 138 million tons of usable aggregate steelmaking capacity; by contrast, worldwide demand had been only 133.5 million tons in 1937, the last year before the wartime buildup began in earnest.[7] Capacity additions thus seemed foolhardy to industry interests at that point in time.

But to some outside the industry, steel output was considered too important to national well-being to be left entirely in the hands of private producers. According to these views (held primarily by governmental and academic economists), there would be a drastic and severe shortfall of steel output that could well lead to renewed depression unless more mill construction were undertaken immediately. The problem of steel capacity was soon elevated to a matter of national importance because of this growing public concern. The resulting politicization of the issue would dwarf the dimensions of the debate that surrounded similar expansion considerations in 1939–41.

I

The steel-expansion controversy had its roots in public fears that the nation would not have a sufficient supply of steel to meet postwar demand. While the rationale behind the steel producers' reluctance to expand had some clear ties to their prewar position on the matter, the controversy that developed after 1945 was embedded in a more complex tangle of competing economic philosophies. The stakes, in consequence, were viewed as higher by both sides. Basically, the revolutionary macroeconomic principles formulated by John Maynard Keynes (along with similar theories promoted by the American economist Alvin Hansen) were being widely disseminated within governmental circles during the late 1930s and early 1940s as the only realistic means to break out of the Great Depression once the war was brought to an end.[8] What these principles would necessitate, according to their proponents, was a degree of formalized and systematic governmental involvement in the economy far greater than had ever occurred before in peacetime America. This would naturally extend at least to economic planning efforts over those essential, or so-called commanding heights, sectors that were crucial to overall macroeconomic performance. Agriculture, critical natural resources, credit, and energy (among others) were considered important sectors, yet few could excite as much contemporary interest as iron and steel. This latter area was widely believed to be the backbone of any industrialized economy. Whether it was technically true or not, there was a great deal of merit to the traditional notion that "as steel goes, so goes the nation."[9] It was thus inevitable that as the government's interest in economic management grew, so would the fears of a private steel industry long antagonistic to any incursions by outsiders into what it considered to be purely proprietary matters.

An important source of encouragement and activism in the trend toward a publicly managed American economy was the Fiscal Division of the Bureau of the Budget, organized in 1939 to examine "questions of fiscal policy and [to give] staff assistance in the formulation of the President's financial program."[10] In conjunction with such other groups as the National Resources Planning Board (a public agency) and the National Planning Associates and the National Farmers Union (both private agencies), the Fiscal Division came to play a key role in drafting legislation that was intended to institutionalize Keynesian policies within the economy. The most significant result, of course, was the Full Employment Act of 1945, introduced into Congress in January of that year and eventually emerging (though in highly emasculated form) as the Employment Act of 1946.[11]

Among the more prominent exponents of full-employment planning policies in the Fiscal Division was Louis H. Bean, an economist serving as chief fiscal analyst for the unit. Bean was a Russian-born immigrant who, at the age of ten, had come to the United States with his parents; following graduation from the Harvard Business School in 1922, he took his first governmental position as a research economist with the Department of Agriculture. He had been in governmental service ever since.[12] In the course of Bean's work, he recollected, "one of the side interests I had developed had to do with the amount of steel that this country would have to have to support a full employment economy after the war. And there was conflict among economists and over the country as to what level of steel production we should have."[13] Bean's analysis in the early 1940s led him to the conclusion that there would be a serious shortfall of steel unless capacity were expanded immediately,[14] a finding supported by other governmental planners.[15] But as Bean himself noted, not all parties were in agreement with his outlook; some within the government were quite wary of federal meddling in economic matters.[16] This un-Keynesian view was also shared by the steel industry itself. Typical was the comment of Tom Girdler, chairman of the Republic Steel Corporation, who stated that "our long range stability may be clouded if industry expands too rapidly in an effort to meet an abnormal accumulated demand for goods."[17]

It was over this alleged abnormal nature of postwar steel demand that the capacity issue would be argued. In principle, the controversy revolved around two widely divergent interpretations of how the depression era had affected steel demand. On the one hand, public-sector economists like Bean viewed the 1930s as merely an interlude in the rising secular trend of steel consumption in the United States from 1900 through 1929; this group assumed that after the war—and given the proper stimulation available through Keynesian policies—the high-demand trend line of the 1920s could be resumed. On the opposite side, industry economists and their supporters believed that 1929 represented a peak year of domestic steel demand that would never be seen again. Thus the low level of steel consumption that characterized the 1930s represented not an interlude, but rather a fundamentally new demand schedule that would persist into the foreseeable future. To them, the World War II years were the real interlude: Demand reached abnormally high levels in both domestic and international markets owing to defense procurement needs. All matters considered, the industry maintained, one could reasonably expect the situation to revert back to the stagnant patterns of the 1930s once the postwar reconversion process was complete, an event that would take perhaps two years at most. The only unknown variable would be export demand, an area where the larger firms in the industry probably desired little governmental oversight given their earlier involvement with the ISC.[18]

The central point of contention was the nature of declining per capita consumption evident during the depression years: Was this only a temporary aberration from the high-growth experience of the 1920s (as government planners maintained), or was it the onset of a new stage of long-term secular decline (as industry economists believed)? If the companies were correct in their pessimistic growth assumptions but went ahead with expansion, the financial consequences for the future could be disastrous. If the government forecasters were right—and no steel expansion was forthcoming—then the entire performance of the American economy might be jeopardized given the "bottleneck" and "leverage" effects that steel output was believed to produce in macroeconomic behavior.[19]

II

The proximate cause that brought the capacity issue to national attention was congressional concern over the condition of small business. The proper role that this sector should play in the nation's economy had been a topic of debate throughout the New Deal era.[20] Yet, rather than being the central focus of that debate, small business problems usually were treated as a reflection of an opposite condition: big business manipulation of the economy. This controversy carried over into Truman's administration, where it unfortunately was to receive no clearer definition than before. The reason for this was that public officials could not reach a strong consensus about the social welfare effects of concentrated enterprise.[21] Nevertheless, Truman administration advocates for a more prosperous small business sector demanded more responsive governmental assistance toward this constituency. There was general agreement in most quarters that recent years had not been kind to small business; "one of the casualties of the War," Truman noted, "was Small Business."[22] The need for rapid mobilization and coordinated production and distribution during the hostilities translated naturally into reliance on those firms best able to meet such demands—namely, large corporations.[23] But with the war at an end, agitation developed to reassert the traditional prerogatives of the small businessperson, long the symbol of economic pluralism so cherished (if not actually assisted) by the nation's policymakers.[24] The president, of course, was hardly unsympathetic to these goals: For some years his name, too, could be found on the dwindling roster of Main Street entrepreneurs trying to squeeze out a living between the tightening grip of the large corporations over the American economy. At the end of World War I he had operated a haberdashery in Kansas City, which failed after only two years of life.[25] "I never believed in the concentration of economic power in the hands of a few," Truman wrote in early 1946, "[and] I am exceedingly anxious to use every means at our command to reinstate the place of Small Business where it should be."[26] Thus when the Republican-controlled Senate approved a resolution in January 1946 to undertake an extended review of the small business sector—to be headed by Senator Kenneth S. Wherry (R-Nebraska)—the president gave his support.[27]

The leaders of the steel industry—though initially not paying much attention to this legislative activity—would soon come to regret the Wherry Committee's formation. As in previous debates, it became impossible to discuss small business without also addressing the role of the large firm. As Wherry's two-year-long investigation began to unfold (eventually to consume nearly five thousand pages of testimony gathered in nationwide hearings), the steel and petroleum industries emerged as the primary targets of opportunity for those seeking change.[28] A steel subcommittee, under the chairmanship of Senator Edward Martin (R-Pennsylvania) was set up to investigate "steel supply and distribution problems affecting smaller manufacturers and users"; by mid-1947 this body became the primary political forum in which the industry's capacity-expansion controversy was argued.[29]

One of the major postwar problems facing smaller firms (and, indeed, some larger ones as well) was provision of an adequate supply of steel. Not only was there strong domestic demand, but also foreign nations—many contending with war-damaged or war-demolished mills—were anxious to obtain American steel. The United States possessed three-fifths of worldwide capacity in 1947.[30] The wartime peak of 89.6 million tons of steel ingot output achieved in 1944, however, fell back to 79.7 million tons in 1945 and then dropped even further to 66.6 million tons in 1946 (see Table 2.1). There were several reasons for this reduction. For one, over 3.5 million tons of worn-out capacity were retired as soon as defense needs ended. For another, labor strife was beginning to crimp severely the industry's production schedules (as it did in other key sectors, too). By 1947, however, steel ingot output rebounded to 84.9 million tons, and the utilization rate jumped to 93%.[31] Yet, net capacity had not been increased by the industry even though demand was reaching historic highs. In the eyes of some critics, the steelmakers were attempting to raise profits through "a program of planned scarcity."[32] Regardless of rumors, the tangible effects were that smaller steel fabricators and users were being hurt by their inability to obtain sufficient product; as their complaints against the large integrated producers mounted, the small firms became a focus of governmental attention.

Table 2.1 American Steel Capacity and Production, 1940–60

Year	*Net Capacity (000 net tons)*[a]	*Total Production (000 net tons)*[a]	*Capacity Utilization (%)*
1940	81,619	66,983	82.1
1941	85,158	82,839	97.3
1942	88,887	86,032	96.8
1943	90,589	88,837	98.1
1944	93,854	89,642	95.5
1945	95,505	79,702	83.5
1946	91,891	66,603	72.5
1947	91,241	84,894	93.0
1948	94,233	88,640	94.1
1949	96,121	77,978	81.1
1950	99,983	96,836	96.9
1951	104,230	105,200	100.9
1952	108,588	93,168	85.8
1953	117,547	111,610	94.9
1954	124,330	88,312	71.0
1955	125,828	117,036	93.0
1956	128,363	115,216	89.9
1957	133,459	112,715	84.5
1958	140,743	85,255	60.6
1959	147,634	93,446	63.3
1960	148,571	99,282	66.8

[a]Ingots and steel for castings.

Source: AISI, *Annual Statistical Report* (various years).

The point of departure for this controversy was the so-called gray market in steel that had developed after the war. Due to vigorous demand and short supply, both semifinished and finished steel could be resold by original purchasers at a premium. The effect on many small users, such as fabricators, was financially debilitating; they found themselves either priced out of the market or unable to obtain the material at any cost, thus threatening the continuation of their normal operations. The loud pleas for assistance from this group found a ready ear in Washington. But somewhat typically, the large steelmakers at first refused to acknowledge that any such gray market even existed. Testifying before Martin's steel subcommittee in early June 1947, Executive Vice President Joseph L. Block of Inland Steel noted that although "efforts have been made to locate the fantastic tonnage offered [in the gray market], . . . it has long been apparent that most of this is just empty talk engaged in by speculators and charlatans."[33]

By adopting such a position, steel leaders perhaps believed the entire capacity controversy might be defused. If so, they were soon disabused of their beliefs through extensive testimony brought out at the congressional hearings. As a result, the industry agreed that the resale market was, indeed, a matter deserving correction.[34] But by the time the industry began to cooperate in eliminating the problem, the damage had already been done. In the opinion of the government, the press, and much of the public, the credibility of the large integrated firms and their concern for the public interest were suspect.[35]

By the middle of 1947 it was apparent to steel leaders that a more coherent public explanation had to be made to substantiate their no-growth position. The Senate Small Business Committee Hearings (headed by conservative Kenneth Wherry were calling increased attention to the capacity problem; one result was that this enhanced the standing of more radical political elements who were demanding governmental control of the mills.[36] The task of public explanation was at first assumed by Wilfred Sykes, president of Inland Steel. In May 1947 Sykes had presented a paper to the annual AISI convention that projected a stagnant domestic consumption trend for steel well into the future.[37] By 1950, he calculated, ingot demand would be on the order of 76.4 million tons and by 1955 only 78.5 million tons, compared to the actual 1946 output of 66.6 million tons. The basic logic behind these estimates involved an extrapolation from per capita steel production between 1911 and 1940. From 1911 through 1920, Sykes noted, U.S. per capita consumption (derived from production data) averaged 666 pounds; between 1921 and 1940, the average was 668 pounds, reflecting the maturation of the industry. Reasoning that the trend after 1945 would at best demonstrate only a slight advance over these prewar levels, perhaps 700 pounds per capita, Sykes found no justification for mill expansion; rated capacity already stood at 91.9 million tons at the end of 1946, enough to provide 1300 pounds of steel per capita. To accommodate potential peak demands in his projections, he utilized the point before 1940 when per capita consumption was at its highest level: that was 1929, when it reached 978 pounds. Factoring in a small increase to account for exports,[38] Sykes arrived at the conclusions shown in Table 2.2.[39]

It was clear from Sykes's premises that any further expansion would be illogical folly and financial intemperance. As Sykes noted to a reporter in June 1947:

> These figures would seem to indicate that no expansion in ingot capacity is required in the near future although, of course, there may be additional plants built either to replace existing uneconomical units or to satisfy some special needs.[40]

Table 2.2 Sykes's Estimation of Capacity Expansion

		Maximum per Capita Demand (lb per Annum)			
Year	*Projected Population*	*Domestic (A)*	*Export (B)*	*Total (A + B)*	*Ingot-Capacity Required (million tons)*
1950	143,896,000	978	83.5	1061.5	76.4
1955	148,186,000	978	81.0	1059.0	78.5

One June 19 Sykes appeared before the Senate's steel subcommittee to present his findings. He essentially made the same arguments stated earlier to the AISI gathering and concluded:

> I believe we are going to have real relief by the end of this year [1947]. I can see signs now in our own business of a let up of demand, for one reason or another.[41]

Yet the legislators wished to hear the other side of the controversy as well. Accordingly, Louis H. Bean was asked to appear. Though by now Bean had left the Bureau of the Budget and joined the Office of the Secretary of Agriculture, he decided to "look at the problem since this fitted in with [his] interest in postwar full employment problems."[42] In subsequent testimony Bean utilized the same historical data drawn on by Sykes, only he approached the matter from a different vantage: How much steel, he asked, would be needed to sustain a full-employment economy? Industry, on the other hand, asked: How much steel was actually consumed in the past? Thus the industry projection of stagnant demand was flawed, according to Bean, because its answer to the latter question was far-too-heavily weighted by the diminished output of the depression decade:

> At this point, it may be well to point out how some persons in the steel industry arrive at the conclusion that for years to come we will not need more than 80,000,000 tons of steel. In the first place, they are convinced on the basis of boom and bust experience, and the character of demand for their product, that steel production cannot be maintained year in and year out at a peak level. In the second place, they are conditioned by the experience of the 1920's and 1930's and are inclined to assume that some time in the near future we shall experience again a depression as severe as that of the 1930s. . . . In the third place, the use of unlike averages, for a 10- and 20-year period—as Mr. Sykes does when he cites the average per capita domestic demand for 1911 to 1920 as 666 pounds, and for 1921 to 1940 as 668 pounds—tends to hide the really significant upward trend in prosperity years, and the upward trend that has characterized and apparently still characterizes the growth of the steel industry. Finally, they conclude that there is bound to be a falling off in demand, and having found corroboration of a stable and lower than present demand in the misleading averages, it is assumed that the most optimistic assumption with regard to future maximum demand is that it will not exceed the maximum reached in 1929. It is on the basis of the 1929 figures now 18 years old that Mr. Sykes and others conclude that demand for the expected population of 1950 and 1955 will be about 76,000,000 and 78,000,000 tons respectively.[43]

Bean thus rejected the industry argument as unrealistic. Instead, he held:

> In the record of steel production from 1900 to 1947 . . . , we are concerned primarily with the trend of production in prosperity years,

> since the question we are dealing with is steel requirements for full employment. . . .
>
> On the basis of this rising trend in per capita consumption [evident between 1901 and 1929], we could have anticipated about 1,200 pounds or more in 1941. The actual figure turned out to be 1,243 pounds. For 1947, the rising trend suggests about 1,300 pounds compared with an actual production for the year of something less than 1,200. For 1950, if we are to have full employment, we would, on the basis of this experience, need around 1,400 pounds per person, and with a total population in that year of 148,000,000 total production would need to be 100,000,000 tons or more.[44]

The industry, of course, rejected Bean's analysis. Walter S. Tower of the AISI testified before the steel subcommittee some weeks later, noting that "present capacities, plus those now planned for completion over the next year, should meet every expectable demand in the near future."[45] He was followed in September by the heads of twelve of the larger firms; in a prepared joint statement, Ben Moreel, chairman of Jones & Laughlin Steel, declared:

> [Our] important program of expansion and improvement of steel facilities demonstrates that faith of steel companies in continued high peacetime demand. Together with record-breaking peacetime steel production this year, it is a complete refutation of statements that steel companies are holding back both production and expansion. . . .
>
> Meanwhile, analysis of all the facts clearly shows that where there are steel shortages, they are largely due to abnormal conditions.[46]

But protestations of innocence before congressional committees would not be enough to end the debate, as industry leaders now well realized. The capacity issue had become a major news story to the media, and the producers were forced to undertake countermeasures to persuade a suspicious public that their refusal to expand was not due to a clandestine plot to boost profits, as some had claimed.[47] Walter Reuther, head of the United Auto Workers of America (UAWA), wrote Truman in August 1947:

> Instead of increasing its output to bring the supply up to the demand during this critical period, the steel industry is trying to bring demand down to supply by raising prices and forcing customers out of the market.
>
> Its refusal to take the necessary steps towards maximum utilization of present capacity has meant short work weeks and recurrent layoffs in the automobile and other industries throughout the last year. Its refusal to expand capacity means mass unemployment in future years.[48]

Reuther reiterated his views before the congressional investigators and the United Steelworkers of America (USWA) also endorsed the call for more production, urging a government-backed output program if necessary.[49] Other published rumors even accused the industry of holding back investment because the steelmakers, too, believed all the horror stories that animated the Keynesian nightmares: that a new depression was imminent.[50]

The industry began to step up sharply its public relations program in order to repudiate the clouds of suspicion surrounding its motives. What followed was a blitz of official news releases, advertisements in leading newspapers and magazines, nationwide radio addresses, and informational plants to cooperative columnists—all designed to reassure the public and government that the steel firms had the situation well in hand and that there was no need for the government to intervene.[51] Hill & Knowlton, the chief public relations counsel to the AISI, made preparations for a massive lobbying campaign to individual members of Congress as well as a grass roots campaign directed to the public at large.[52] In addition, leading interests at U.S. Steel believed it would be "highly advisable for the Steel people to demolish [Louis H.] Bean's 'projections' of consumer demand."[53] To this end, a staff economist from the U.S. Steel Corporation, Bradford B. Smith, was directed to develop a lengthy rebuttal to Bean's analysis, which subsequently was published and widely disseminated through the AISI in early 1948.[54] Other executives from U.S. Steel also began to speak out publicly on the issue, labeling the expansion demands of critics as "fantastic" at best and generally rejecting any need to build capacity beyond the prudent levels established by industry experts.[55]

Yet still the problem would not recede from the headlines. Bean, at the invitation of Senator James Murray (D-Montana)—who was serving on the Wherry Committee and who had previously clashed with steel industry interests on several issues—prepared a third-round rebuttal to Bradford B. Smith's analysis:

> Members of the industry have presented testimony that individually they are expanding but that the industry as a whole is faced with a demand much below the present and with the prospect of a deep and prolonged depression. It is probably not unfair to say that the [AISI] and the industry statisticians appear to be feeding unwarranted pessimism to their superiors that does not square with the better judgment of the practical-minded leaders in the industry who are spending consumer's money in advertising their plans for expansion.[56]

This, of course, served only to elicit a further round of response in the deepening controversy.[57] There was also considerable debate on the question in the national media, as columnists and public affairs groups eagerly stepped forth with commentary, analysis, and recommendations.[58] The trade journal *Iron Age,* relying perhaps more on hope than accuracy, predicted in late 1948 that "the extreme steel shortage boogy man is bound to die a quiet death in

1949."[59] But such notions were naive; indeed, if anything, the issue escalated in 1949. Truman himself was receiving a variety of unsolicited proposals to resolve the problem, some of them quite bizarre.[60] A number of governmental agencies undertook official studies of the matter. The Department of Interior became involved in response to a request from the president that the department survey the nation's natural resource reserves to determine if foreign aid commitments could be met through existing stocks. The biggest question mark, reported Interior Secretary Julius Krug, was steel—presently "inadequate," he said, "to meet the demands placed on it."[61] An assistant secretary of the interior, C. Girard Davidson, went even further, urging the government to build 10 million tons of capacity if the industry would not.[62] Several defense-related agencies chose to investigate the steel-supply situation to gauge any potentially adverse impact on national security needs.[63]

The Commerce Department, too, became involved, but it chose to stay clear of any policy recommendations. This was due to Secretary Charles Sawyer's opposition to public intervention in the economy.[64] Nevertheless, the department did have a legitimate interest in costs, and its Office of Domestic Commerce prepared a highly authoritative analysis of potential outlays if 10 million tons of additional capacity were to be built (the minimum amount of increase necessary to full-employment demand, according to the critics).[65] Deducing a construction cost figure of $250 per ton for new integrated capacity, a total of some $2.5 billion would be required to do the job—an amount, U.S. Steel's Bradford B. Smith noted, greater than the entire steel industry's book value of plant and equipment in 1947.[66]

Yet nearly all the studies undertaken by the Truman administration seemed to support conclusions reached by the Senate's steel subcommittee. Issuing a final report in February 1949 on the industry's supply and distribution problems, the subcommittee found that "the steel industry in the United States does not have adequate capacity to produce the quantities of basic iron and steel that the country needs at this time."[67] Identifying the problem was one matter; resolving it was another. "No one appears to know, . . ." the report went on, "just how much steel capacity we should have to provide a free market to all comers."[68] Such indecision, however, did not characterize a lengthy study on the subject prepared by the Council of Economic Advisors (CEA) for interagency purposes and made available in March 1949. Finding that "analysis of the longer-run requirements for iron and steel under conditions of maximum production and employment indicates that substantial expansion of facilities and materials supply will be needed,"[69] the report went on to forecast a level of from 103 to 107 million tons of steel ingot output necessary in 1954 and from 112 to 120 million tons by 1958.[70] While acknowledging that the industry was expanding, the CEA pointed out that present plans for capacity growth would not meet these projected needs.[71] Pressure for action was building on the steelmakers. Nearly every month brought forth some new investigation or study that seemed to indicate unjustifiable recalcitrance by the industry. Something, obviously, would have to be done.

III

The postwar capacity issue—and the heated debate that it generated—masked more fundamental problems that would have to be addressed before any progress in meeting the nation's long-term demand for steel could be made. In many respects the steelmakers had legitimate reasons for their cautious approach to expansion. The industrial situation in Europe would first have to be resolved, especially in regard to Germany's steel sector, the leading international force in this commodity prior to the war. Second, if the domestic industry were to expand, it was unclear how this could be financed.

Indeed, this investment issue and related questions of costs, prices, and profits often attracted front-page headlines that stemmed from repeated disagreements between the industry and governmental officials over steel company financial performance. According to the capacity projections made by Louis H. Bean in 1947 (which nearly all the involved public policymakers accepted as valid), the industry would have to add an additional 10 million tons of steel ingot-making facilities by 1950 to meet expected demand. Using the Commerce Department estimate of a $250-per-ton expansion cost, this would (as noted) require a steel industry investment of some $2.5 billion. The industry claimed that it could not afford to undertake a new round of capacity investment given prevailing steel prices, whereas the Truman administration consistently asserted that current industry profits were more than adequate (see Tables 2.3, 2.4, and 2.5 for pertinent data).

There were a number of important technical considerations underlying the industry's caution toward expansion. These involve more than mere statistical extrapolations from past data, which, in fact, worked only to obscure the problem. A fully-integrated steel manufacturer requires a balance of facilities. These producers must have the capability both to process a variety of raw materials into the ingots, billets, and slabs of raw steel, and to convert the raw steel to so-called finished product. The finished product is then further processed into consumable items; this process is undertaken either by other firms, such as steel fabricators or end-users (like automobile manufacturers), or by the major integrated steel companies themselves who choose to participate in end-use markets such as bridge building.[72]

Yet, in the face of this complex degree of vertical integration and value additivity that characterized the major producers, the expansion debate was usually described only in terms of company melting capacity—that is, the firm's capability to process the raw material into raw steel (this is the capacity shown in Table 2.1). Any increase in the tonnage of this factor alone, however, would not automatically lead to greater steel availability in the economy unless comparable expansion were to occur in the capacity of finishing mills. Consequently, at least in the 1945–49 interval, much of the industry's new fixed investment went toward balancing its overall steelmaking capability, a function exacerbated by the transition in end-use markets brought about by the termination of the war and the rise in demand for peacetime products (different products require different configurations of mill layout). Thus the

Table 2.3 American Steel Industry Financial Data, 1945–52[a]

Year	*Revenue*	*Before-Tax Earnings*	*After-Tax Earnings*	*Dividends*	*Depreciation*	*Retained Earnings*	*New[b] Investment*	*Long-Term Debt*	*Stockholders Equity*	*Debt/Equity Ratio*
1945	$ 5,921	$ 303	$184	$138	$339	$ 46	$ 115	$ 485	$3,620	13.4%
1946	4,812	396	265	147	169	117	365	544	3,712	14.7
1947	6,705	694	412	184	239	228	554	605	3,927	15.4
1948	8,119	929	540	205	302	335	642	649	4,566	14.2
1949	7,436	910	533	222	278	306	483	681	4,885	13.9
1950	9,535	1,544	767	312	327	455	505	763	5,458	14.0
1951	11,845	1,961	682	312	374	371	1,051	1,030	6,038	17.1
1952	10,858	1,024	541	316	450	225	1,298	1,447	6,373	22.7

[a]All figures $000,000.

[b]New investment in plant and equipment.

Source: AISI, *Annual Statistical Report* (various years).

Table 2.4 Steel Prices, Wholesale Prices, and Consumer Prices, 1940–60

Year	*Basic Steel Prices*	*WPI*[a] *(all commodities)*	*WPI*[a] *(all but farm & food commodities)*	*CPI*[b] *(all items)*	*CPI*[b] *(all items but food)*
1940[c]	100.0	100.0	100.0	100.0	100.0
1941	100.4	111.2	107.2	105.0	102.9
1942	100.6	125.6	115.0	116.4	110.1
1943	100.7	131.1	116.7	123.5	113.1
1944	100.7	132.3	118.5	125.5	117.4
1945	103.1	134.6	120.0	128.4	120.2
1946	112.1	154.0	131.8	139.2	125.4
1947	131.4	188.6	160.4	159.4	137.0
1948	150.0	204.3	174.1	171.6	146.8
1949	162.6	194.1	170.5	169.9	148.4
1950	171.2	201.8	176.8	171.6	150.1
1951	184.6	224.7	195.1	185.3	159.7
1952	188.6	218.4	190.6	189.5	163.5
1953	203.6	215.5	191.9	191.0	166.7
1954	212.7	215.9	192.8	191.7	167.7
1955	222.9	216.6	197.0	191.2	168.2
1956	241.4	223.7	205.7	194.0	171.2
1957	264.6	230.1	211.4	200.7	176.9
1958	273.8	233.3	212.1	206.2	180.8
1959	278.4	233.9	215.8	208.0	184.3
1960	278.0	234.1	216.0	211.2	187.3

[a]Wholesale Price Index.

[b]Consumer Price Index.

[c]All indexes: 1940 = 100.

Source: Bureau of Labor Statistics. (Reproduced in U.S. Congress, JEC, 88th Cong., 1st sess., Hearings, *Steel Prices, Unit Costs, Profits, and Foreign Competition* [Washington, DC: GPO, 1963], p. 123.)

industry's capital investment figures often did not directly translate into an expansion in steel ingot capacity. U.S. Steel, for example, had expansion and modernization projects worth $950 million underway in early 1949. If all of this were to be committed to melting facilities, it would probably add 3.8 million tons of new capacity (using the Commerce Department's cost estimate of $250 per ton); instead, the particular configuration of projects then underway at U.S. Steel, though costing a total of $950 million, yielded only 600,000 net tons of new ingot-producing capacity. The bulk of this new investment was devoted to finishing mills.[73]

Other technical considerations also entered into the expansion-investment decision. One of the more important problems clouding the issue was the adequacy of the raw material supply necessary to achieve the projected steel-output goals. Government planners readily acknowledged that there were deficiencies in this respect. Of the more important raw materials used in steelmaking, the only one that seemed without major problems was coal,

Table 2.5 Year-to-year Percentage Changes in Steel Prices, Wages, WPI, and CPI, 1940–60

Year	*Basic[a] Steel Prices*	*Hourly[b] Steel Wages*	*WPI[c]*	*CPI[c]*
1940	+ 0.1%	+ 1.1%	+ 2.0%	+ 0.8%
1941	+ 0.4	+ 11.8	+ 11.2	+ 5.0
1942	+ 0.2	+ 10.1	+ 13.0	+ 10.8
1943	+ 0.1	+ 6.9	+ 4.4	+ 6.2
1944	0.0	+ 7.4	+ 0.9	+ 1.6
1945	+ 2.4	+ 2.3	+ 1.8	+ 2.3
1946	+ 8.7	+ 7.4	+ 14.4	+ 8.5
1947	+ 17.2	+ 11.3	+ 22.5	+ 14.5
1948	+ 14.2	+ 7.4	+ 8.3	+ 7.6
1949	+ 8.4	+ 4.4	− 5.0	− 1.0
1950	+ 5.3	+ 8.8	+ 3.9	+ 1.0
1951	+ 7.8	+ 10.8	+ 11.3	+ 8.0
1952	+ 2.2	+ 9.5	− 2.8	+ 2.3
1953	+ 8.0	+ 5.4	− 1.3	+ 0.8
1954	+ 4.5	+ 3.0	+ 0.2	+ 0.3
1955	+ 4.8	+ 8.4	+ 0.4	+ 0.3
1956	+ 8.3	+ 8.5	+ 3.3	+ 1.5
1957	+ 9.6	+ 8.9	+ 2.9	+ 3.5
1958	+ 3.5	+ 9.2	+ 1.4	+ 2.7
1959	+ 1.7	+ 8.1	+ 0.3	+ 0.9
1960	− 0.1	+ 0.6	+ 0.1	+ 1.5

[a]Basic steel prices, 1940 = 100.

[b]Hourly steel wages based on the total employment costs per hour for wage employees in the basic steel industry. These costs include—in addition to straight-time hourly earnings—premium pay for overtime and night work, and all fringes (i.e., paid vacations, paid holidays, jury duty pay, severance pay, supplemental unemployment benefits [SUB], insurance, pensions, and social security payments made by employers).

[c]Wholesale Price Index and Consumer Price Index, 1940 = 100.

Sources: Basic steel prices based on the price index compiled by the Bureau of Labor Statistics. Hourly steel wages from AISI, *Annual Statistical Report* (various years). WPI and CPI, U.S. Bureau of Labor Statistics, *Handbook of Labor Statistics* (Washington, D.C.: GPO, various years).

from which coke, a basic steelmaking input, was derived. In other inputs, many difficulties existed. High-grade iron ore deposits in the United States, concentrated in the Great Lakes Region, were nearing depletion. Vast deposits of rich ores had been recently discovered in South America, but plans to mine these and ship them to mainland mills were still highly uncertain.[74] Other required steelmaking minerals, especially chromite and manganese, were suffering from an even bleaker outlook because domestic sources were limited and, in the case of manganese, the Soviet Union had been a major supplier in the past.[75] Finally, scrap availability was perhaps the most significant input question mark, at least in the short run. Scrap is a vital ingredient in electric and open-hearth process steelmaking (which together accounted for over 94% of domestic capacity in 1947).[76] But the huge wartime and postwar steel demand had depleted most stockpiles of domestic scrap; although governmental officials devoted great energy and effort to finding scrap supplies abroad, they were not always successful.[77] As a consequence, there were no guarantees of sufficient scrap supply to meet demands and there was a strong possibility that such shortages would lead to exorbitantly high scrap prices (see Table 2.6).

Another problem concerned transportation costs and mill locations. By 1949 the freight charges for moving a ton of ore from the Mesabi Range in upper Minnesota to Pittsburgh, for example, exceeded the value of the ore itself.[78] Plans for the Saint Lawrence Seaway, which would have a significant impact on all transportation calculations in the industry, were still in the discussion stage. Meanwhile, termination of the basing-point pricing system—which seemed imminent in the late 1940s following several court decisions—would further complicate delivery problems. This factor was even more vexing because fundamental demand patterns in the industry were also shifting. The western regions of the country were significantly increasing their steel consumption, thus upsetting a traditional reliance on mills located in western Pennsylvania, eastern Ohio, and the Great Lakes area, which had hitherto been financially protected from just such geographical demand shifts by the basing-point system.[79] Producers would now have to decide where to locate any new mills in consideration of these changing and somewhat uncertain raw material, transportation, and demand variables. Moreover, the expanding producer would have to correctly forecast the type of product that would be demanded from any mill finally constructed, since a given combination of steelmaking facilities could not be easily converted from one kind of product configuration to another without considerable expense. If this were not enough, national defense factors would also need attention: Mill planners had to ask if the location of a new mill might be inordinately vulnerable to enemy attack (a frequently heard question in the age of Cold War buildup).[80]

A final problem that would complicate the expansion decision involved changing technology in steelmaking. In the late 1940s two significant innovations were under worldwide discussion. One concerned the use of continuous casting methods whereby raw steel could be immediately converted into usable shapes without first having to be processed as ingots, billets, or slabs and then cooled, reheated, and further processed into the bars, plates, tubes, and so on, that constituted finished steel. Savings in energy and labor costs could be substantial once this innovation was perfected.[81] The other major technological change involved the use of oxygen in steelmaking. This had been under discussion in the United States in one form or another since 1923 when the federal Bureau of Mines appointed a committee to study improvements in metallurgical processes that could be obtained through the use of oxygenated air.[82] But the heavy investment in open-hearth technology that had already been made by large domestic producers dictated, at least by their logic, that oxygenation experiments be limited to applications within the existing process. Innovations in this direction had begun in earnest in 1945, but by 1949 results were still inconclusive. If anything, previous forecasts of

Table 2.6 Price Indexes of Selected Steelmaking Materials, 1947–60

Year	*Industrial Crude Materials*	*Iron Ore*	*Coke*	*Pig Iron*	*Iron & Steel Scrap*	*Ferro-manganese*	*Industrial Power*
1947[a]	92.9	88.6	84.2	83.6	96.9	90.6	99.0
1948	108.5	96.5	104.3	102.7	122.3	98.2	100.3
1949	98.6	114.9	111.5	113.6	80.8	111.2	100.7
1950	109.9	123.0	116.0	116.2	104.6	113.6	100.4
1951	120.8	132.3	124.0	128.5	118.8	121.2	100.6
1952	109.3	137.6	124.7	131.1	114.2	132.0	101.0
1953	108.5	153.8	132.0	136.6	103.1	148.9	101.3
1954	103.3	157.7	132.4	138.4	79.8	147.6	101.7
1955	113.4	160.4	135.2	141.4	104.6	143.2	102.4
1956	120.0	173.0	149.7	149.9	132.5	164.0	103.2
1957	118.3	181.7	161.7	160.1	116.9	189.4	104.5
1958	113.7	177.1	161.9	163.0	93.8	183.8	105.6
1959	120.0	169.9	169.8	163.0	100.2	183.8	106.0
1960	115.3	171.0	170.4	163.0	82.9	166.6	107.7

[a]Indexes 1947–49 = 100.

Source: U.S. Congress, JEC, 88th Cong., 1st sess. Hearings, *Steel Prices, Unit Costs, Profits, and Foreign Competition* (Washington, DC: GPO, 1963), pp. 124–125.

greater output through such new technologies had proven much too optimistic.[83] As a consequence, steelmakers were still unsure of oxygen process payoffs.

In Europe, however, oxygenation experiments were being conducted not only in conjunction with open-hearth furnaces, but also with a new conversion technology utilizing a basic oxygen furnace (BOF); the first such experimental furnace was installed in Austria in 1947.[84] Much of the motivation for this BOF process could be found in the fact that it required significantly less scrap consumption than existing technologies; this input, of course, was in critically short supply in Europe at the time, just as it was in the United States. But in spite of these initial advances abroad, there were still numerous questions surrounding the utility of the innovation for the conditions that prevailed in the domestic industry in the late 1940s. All factors considered, the future extraordinary gains that could be derived from the revolutionary breakthrough of BOF were effectively hidden in 1949.[85]

In addition to these technical uncertainties concerning the availability of raw materials, the location of mill sites, and forthcoming technological changes, another serious problem plagued industry planners. This involved economic constraints. Given existing price schedules in the industry, it was difficult if not impossible to justify long-term investments or raise capital for steelmaking facilities, at least on purely financial grounds. Many administration officials, of course, tended to deemphasize this concern. The CEA, for example, noted:

> In both 1947 and 1948, profits of the steel industry after taxes were at record levels. The increase in profits from 1947 to 1948 was about 35 percent. Moreover, profits in the fourth quarter of 1948 were far above the average rate for the whole year, reflecting price increases in the third quarter and continued high rates of operation.[86]

As a result, the CEA concluded, the industry should be able to expand without public assistance. But on a comparative basis, the picture was not so optimistic. In 1947 fifty iron and steel firms showed a combined return on investment of 11.3%, placing them forty-second among forty-five manufacturing groups surveyed; in 1948 another study of thirty iron and steel firms found their combined return of 12.6% to be the lowest of the fourteen groups that were surveyed.[87] Although public planners may have had legitimate reason to complain about high profits in general, to industry executives it seemed to make little sense to select only their industry and impugn its investment decisions without reference to other competing users of investment funds.

Moreover, the inflation of costs involved in building new capacity had far outpaced recent price increases in the industry.[88] As a consequence, steel produced from new investment would have to command a price far higher than that from existing capital if market returns were to be obtained.[89] In spite of the large volume of self-serving rhetoric that surrounded these issues, the data seem clearly to indicate that steel was underpriced throughout much of the late 1940s and early 1950s and that the producers would have trouble independently financing the proposed new capacity additions. Between 1947 and 1952, the unweighted after-tax return on net assets for the steel industry as a whole averaged 12.2%, compared to an average for other leading manufacturing industries of 15.6%—a not-altogether unhealthy performance (see Table 2.7). Yet, there is a bias in these figures insofar as they are based on historical costs, which in the steel industry with its older, long-lived assets were, of course, substantially lower than replacement costs. When the latter factor is taken into consideration, returns are diminished. After-tax profits per ton of steel shipments in the 1947–52 interval averaged only $8.55; assuming a per ton investment cost of $250 and a utilization rate of 92% (actual figures), then the return on investment for new capacity would average only 3.14%. If before-tax figures are used, the return still averages only 6.4%—and this assumes a constant capacity utilization of 92% into the future (see Table 2.8 for calculations).

What this data would seem to argue for was either less restraint by governmental officials over price increases in steel or the provision of some form of state assistance to spur investment (by accelerated amortization, federal

Table 2.7 Average Return on Net Assets: Steel Versus Other Leading Manufacturing Industries, 1940–60

	After-tax Average Return on Net Assets				
Year	*Steel*	*Other Industries*	*Difference: Steel Versus Other Industries*	*Steel Rank*	*Total Industries Surveyed*
1940	8.5%	10.3%	− 18%	32	45
1941	9.6	12.4	− 23	40	44
1942	6.5	10.1	− 36	45	45
1943	5.6	9.9	− 43	43	44
1944	5.2	9.8	− 47	44	45
1945	5.0	9.1	− 45	44	45
1946	7.5	12.1	− 38	41	45
1947	11.3	17.0	− 33	42	45
1948	14.0	18.9	− 26	38	45
1949	11.5	13.8	− 17	24	45
1950	15.3	17.1	− 10	28	45
1951	12.3	14.4	− 15	25	46
1952	8.8	12.3	− 28	35	46
1953	11.6	12.5	− 7	21	46
1954	9.4	12.4	− 24	32	46
1955	15.2	15.0	+ 1	14	41
1956	13.9	13.9	0	17	41
1957	13.2	12.8	+ 3	17	41
1958	8.2	9.8	− 16	27	41
1959	8.4	11.7	− 28	35	41
1960	7.8	10.6	− 26	29	41

Source: First National City Bank (New York), "Monthly Economic Letter" (various years, April issue).

Table 2.8 Steel Industry Return on Investment, 1947–52

Year	*Net Profit (000,000)*	*Net Shipments (000 tons)*
1947	$412	63,057
1948	540	65,973
1949	533	58,104
1950	767	72,232
1951	682	78,929
1952	541	68,004

Total after-tax profits: $3,475,000,000 in 1947–52.
Total net shipments: 406,299,000 tons in 1947–52.
Therefore, profits ÷ shipments = $8.55 per ton.

New mill construction costs = $250 per ton of capacity. (But assuming a 92% capacity utilization, then $250 ÷ 0.92 = an adjusted construction cost of $272 per ton of capacity.) Therefore, $8.55 ÷ $272 = 3.14% return on investment, *ceteris paribus,* between 1947–52.

Before-tax profits between 1947–52 were $7,062,000,000 in the steel industry. Therefore, profits (before taxes) ÷ shipments = $17.38 per ton on a before-tax basis. Assuming a similar adjusted construction cost of $272 per ton of capacity, then $17.38 ÷ $272 = 6.39% return on investment, before taxes for 1947–52.

Source: Data used in calculations are from AISI, *Annual Statistical Report* (various years).

loans, or the like).[90] If steel were as critical to national economic objectives as the public policymakers claimed, then such interventionist programs would appear justified. Yet up until 1950 neither outcome would prevail; instead, there was only a continuing panorama of politically charged hearings, accusations, and counteraccusations, none of which seemed designed to bring the disputants together to actually achieve mutually beneficial goals. The result (as Chapter 3 will detail) would be a deepening of the already considerable schism that existed between the federal government and an industry that many perceived as vital to national economic well-being.

3

Truman and the Steelmakers, 1945–48

The steel capacity controversy detailed in Chapter 2 rose to prominence amid clashes over several other issues that appeared to threaten—at least in the view of management—the ability of the private firm to control its own economic affairs. These issues included product-pricing decisions, wages and benefits paid to labor, the basing-point distribution system, strategies for vertical integration of the firm, and tariff and export regulations. In addition, larger questions of monopoly and corporate power often attached to these more particularized issues, thereby extending the dimensions of the business–government debate in steel and deepening the sense of suspicion and hostility that often animated the contestants. The steelmakers' primary fear was that the government would expand its economic role and demand that the firms alter their traditional private behavior in one way or another. At heart, these were ideological fears; they represented the accumulated beliefs and attitudes of company leaders who since 1901 had felt that their decisions should be unconstrained by external concerns.

The unexpected elevation of Harry Truman to the presidency in April 1945 created uncertainty in many quarters. Among businesspeople the primary question revolved around New Deal economic and social policies: Would Truman continue to push for change as his predecessor had or would he opt for less ambitious goals? Truman's prior record offered little guidance on these matters.[1] Nevertheless, most business leaders probably agreed with the sentiments of Myron C. Taylor, a board member and former chairman of the United States Steel Corporation (USS), who wrote to a friend shortly after Roosevelt's death: "I feel again the necessity of laying aside all preconceived ideas regarding the [new] President and getting behind him and helping him because he is our President."[2] Yet such an attitude seems more reflective of the honeymoon period that is accorded most new public officeholders; the real test would be a function of time. And as Truman's years in the White House passed by, the steelmakers' unhappiness with his administration grew. "Let it be noted," stated Walter Tower of the American Iron and Steel Institute in the spring of 1945, commenting on steel's role in the war, "that the complete record of what steel men have done is an imperishable monument to the ability and efficiency of *private* management in industry."[3] This right to managerial independence by private firms would become a basic theme among industry leaders for the remainder of the Truman presidency. According to industry perceptions, some groups within the government would be satisfied with nothing less than a nationalization of the mills. It thus became their duty to resist.

The public agencies that entered into the steel–government controversy from 1945 to 1952 is, indeed, impressive. These included not only the White House directly, but also the Council of Economic Advisors (CEA); the Federal Trade Commission (FTC); the Departments of Justice, Labor, Commerce, Interior, and Agriculture; several Defense Department–related agencies; the Economic Cooperation Administration (ECA)—administrator of the Marshall Plan; and a number of specially empowered congressional committees or interagency bodies investigating one aspect or another of steel industry activity. One may safely conclude that the Truman years represent the most sustained era of volatility in the steel industry's long history of unhappy relations with its government, and it unfortunately propelled the producers into an oftentimes rigid posture of defensiveness that would prove detrimental to their long-term interests.

I

In 1945 the United States was rapidly shifting toward Keynesian economic policy, at least implicitly. This was obviously an important factor in ensuing business–government divisiveness throughout key industries. That this development had actually been underway for some time did not matter; what caused objections to be raised now was that public leaders were consciously calling attention to the need for government to undertake interventionist policies to spur economic growth. Steel spokespersons, of course, had long been resistant to any governmental initiatives that did not conform to their own conception of priorities, thus they were not keen to embrace any public calls for private compliance. The many controls imposed by government during the war, mainly on prices and profits, were by 1945 under attack on several fronts. Walter Tower at that year's AISI convention said:

> To the great credit of members of the steel industry, they appear to be of one mind in urging that with Germany defeated, production and

> use of any steel not needed for war programs shall not be subject to further controls. They argue, and with sound logic, that any other policy means needless hardships. . . .
>
> For steel it seems obvious that removal of controls as promptly as possible will greatly aid progress through reconversion, which in the aggregate is bound to be troublesome under the very best circumstances. It is encouraging that the first steps towards return of industrial freedom are already being taken. But there are still many things to be done to remove all the elements of a fast-growing threat of permanent regimentation.[4]

Economic planners in the administration were, in fact, equally concerned about the need to implement decontrol. But the main emphasis of their attention was on the effect this would have on unemployment; thus they devoted their primary efforts to preventing a recession.[5] Most planners believed they would have perhaps a full year to develop careful programs for reconversion; the sudden deceleration of the war in the Pacific following Hiroshima, however, meant that this comfortable lead time was now foreclosed. As a result, thousands of contracts were hastily canceled and tens of thousands of workers found their hours cut or their jobs eliminated, creating confusion throughout the nation and among Truman's domestic advisors.[6] Yet what became clear as the months went by was that no recession would develop; rather, prices began to climb steeply. As this phenomenon spread, labor demanded that its wage packet go up likewise—not only to compensate for higher prices, but also to make up for the reduced hours it now faced. In consequence, by the end of October 1945 the president was cautioning against both "wide unemployment" and "runaway inflation."[7] Yet even this was soon overwhelmed by the rapidity of postwar change: The administration abandoned for the time being its inherited full-employment program that only months before had been declared as basic economic policy.[8] It now became apparent to most observers that inflation alone would emerge as the administration's dominant economic problem.

The effects of higher wages would be the instrumental factor in the inflationary spiral that followed the war's end.[9] Organized labor, with some degree of justification, felt that its sacrifices had been greater than those of both management and shareholders during the war.[10] Consequently, it vowed to make up for the pent up imbalance through steep wage demands on employers; autoworkers, for example, called for an immediate 30% pay hike.[11] Although the resulting management–labor clashes did not lead to the physical violence so common to the struggles of the 1930s—testimony to the growing institutionalization of labor as a legitimate political force—they nevertheless did cause acute pain and suffering to the national economic welfare. Almost five thousand strikes were called in 1946 alone, putting more than 4.5 million workers out of work at one time or another and costing the nation some 116 million man-days of lost employment (by far the highest total ever recorded).[12] Although a 113-day autoworkers' strike was perhaps the most expensive stoppage, national shutdowns in the coal mines and on the railroads were equally calamitous (resulting in both cases in federal seizures of property in order to maintain essential services).[13] Not surprisingly, steelworkers, too, became involved in this spreading web of labor dispute.

The confrontation in steel can be traced (as it could for many of the 1946 strikes) to events that occurred in late 1945. In August of that year the industry applied to the Office of Price Administration (OPA) for authority to raise prices by $7.00 a ton (the OPA, the federal price-control agency set up during the war emergency, was still operating in 1945). The OPA, however, first cut the steelmakers' request down to $2.00 and then rejected it altogether, citing fears that similar pleas might soon follow from other sectors, which, of course, would only add to national inflation.[14] Meanwhile, the industry contract with the United Steelworkers of America (USWA) was coming up for renegotiation. Benjamin Fairless, president of U.S. Steel, abruptly announced that no wage increases could be granted without a prior agreement with the OPA that price hikes would be approved—a direct rebuff to administration stabilization policies that called for a clear distinction between wage and price issues.[15] With workers already on strike in the auto industry, steelworkers foresaw intransigence ahead; in late November 1945 they voted to authorize a walkout when their contract expired in January. This prompted the president to announce that a fact-finding board would be appointed to study matters, though this did little to ameliorate the parties.[16]

As negotiations became more intense in early January what became most apparent was the internal disarray then prevailing in Truman's administration. John W. Snyder, a conservative ex-banker who was head of the Office of War Mobilization and Reconversion (and a close personal friend of the president ever since their service together in World War I), apparently chose to cut through all channels and personally promise Fairless that he would get a price increase averaging perhaps $4.00 a ton from the OPA. The industry, however, sensing a retreat on the administration's anti-inflation policy, held out for more. The union, assuming that prices would probably go up anyway, then stiffened its own resolve and demanded a raise of twenty-five cents per hour. At this juncture Truman felt compelled to call the principal adversaries—Fairless of U.S. Steel and Philip Murray, head of the USWA as well as the Congress of Industrial Organizations (CIO)—into the White House for a conference; both the president and his advisors believed that a strike would be disastrous for the economy because the 750,000 steelworkers comprised the biggest union in the country and steel was in strong demand. After some informal bargaining, a tentative agreement was reached at this conference, providing the workers with a raise of nineteen and one-half cents per hour and the employers an implicit promise of at least $4.00 a ton in price relief (and probably more).[17]

But Fairless could not hold up his end of the deal. The rest of the industry balked, reviving memories of the great Steel Workers Organizing Committee (SWOC) confrontations of the mid-1930s when U.S. Steel was also spurned by its rivals over recognition of union demands.[18] The reaction in the White

House was one of humiliation and embarrassment at this rejection of presidental conciliation because the press had avidly been following and reporting on every move in the negotiations. In a final attempt to avoid a strike (and after rejecting arguments to seize the industry), Truman called the parties back to the White House for more talks, and privately offered his own compromise solution to the impasse. Fairless, however, rejected the president's plan as too expensive. On January 20, 1946, steelworkers struck the industry in the largest single walkout in American history.[19]

In this dispute, one of Truman's first major domestic tests as president, his performance was marked by ambivalence, ineptitude, and indecision. He seemed unable to choose between conflicting goals of holding the inflationary line in the national interest (thus resisting both labor and industry demands) or of championing a healthy pay raise for labor, perhaps the most powerful single interest group in the New Deal coalition. Truman wrote to his mother and sister in late January:

> Things seem to be going the wrong way here in labor matters, but I am hopeful of an ultimate settlement. The steel strike is the worst. . . .
>
> The steel people and General Motors I am sure would like to break the unions and the unions would like to break them so they probably will fight a while and then settle so both will lose in the long run only the man in the street will pay the bill.
>
> Big money has too much power and so have big unions—both are riding to a fall because I like neither.[20]

Yet, faced with the alternatives of abandoning all economic controls, caving in on his anti-inflation program on a piecemeal basis, or exerting a strong hand through seizure of the mills by executive edict, Truman ultimately chose the most expeditious middle route (though this came only after a protracted period of publicly scrutinized indecision).[21] In mid-February he finally allowed the OPA to authorize a $5.00 per ton rise in steel prices, after which the industry and the union agreed to a wage increase of eighteeen and one-half cents per hour (the amount privately suggested by Truman in his last meeting with the parties before the walkout). The strike was ended on the fifteenth, and the workers were back on the job by the eighteenth.[22]

But by now the damage had been done. Not only did the strike contribute significantly to the 16 million tons of lost steel output in 1946 (at a time when this commodity was in critically short supply), but more important, it created a permanent break (or "bulge" as Truman euphemized) in the government's attempt to hold down inflation during the period of reconversion. By giving in to big labor and big business, the president perhaps implicitly signaled to other powerful interest groups that compromises on the government's position could be obtained. As the anti-inflation program was washed away in a flood of strikes, wage hikes, and price increases in the months that followed, most observers tended to place blame squarely on the president's ineptitude and indecisive leadership.[23]

Following this initial incidence of interaction, Truman's relationship with the steel industry would grow progressively more strained, ending eventually in a bitter federal seizure of the mills in 1952. Perhaps the president harbored an abiding sense of suspicion toward the industry because of its rejection of his proposals for labor conciliation in January 1946. In any case, this was hardly a propitious start to a relationship already historically characterized by mutual doubt; for the next seven years, there was little the industry would do that did not evoke some reply, usually critical, from the White House. For their own part, steel leaders allowed few opportunities to pass without publicly denouncing one action or another of those in government.[24]

II

The end of formal economic controls was one result of Truman's handling of the steel-strike negotiations. Early in the administration, decontrol was a question that many governmental officials had treated with caution, fearing that economic chaos might follow abrupt relaxations and perhaps lead not only to rising inflation, but even to renewed depression.[25] These views initially prevailed over business calls for immediate termination of controls; in June 1945, Congress therefore enacted a one-year extension of the OPA. In the following year the administration tried to obtain a further extension, but with the war ended, both public and politically partisan opposition increased. Controls legislation that finally passed Congress was greatly altered from the president's request—so much so, in fact, that he vetoed it, stating it gave him only "a choice between inflation with a statute and inflation without one."[26] After intense political maneuvering all too characteristic of the reconversion period, a revised OPA extension law was finally passed and reluctantly signed. But only weeks later, in November 1946 and again owing to partisan politics, Truman suddenly ended all existing price restraints in the nation. Any subsequent increases would be the Republicans' problem, the president concluded, since they were the ones so ardently demanding a lifting of controls; now he would give them their wish.[27] In the next four months, in fact, consumer prices jumped 6%.[28] The steel industry, however, would not obtain the managerial freedom over price that Truman's action would otherwise seem to permit.

As the nation began to operate without formal controls for the first time since 1942, the administration adopted a practice that had often been used with success by President Roosevelt. This was jawboning, whereby the White House brought highly publicized pressures to bear on important economic interests in an effort to have them voluntarily reduce their demands (a practice that would become increasingly popular in future administrations as well). Faced with the fact that his prior stabilization policies were in trouble, Truman realized that new anti-inflationary measures would have to be taken or else his public standing could decline even further. Accordingly, the administration began to turn to both the media and private exhortation to make its point.[29] Yet, for the most part, this revised approach met the same fate as previous efforts. As the months passed with one White House crisis after another capturing headlines, Democratic party leaders grew increasingly disenchanted with the president. This disaffection was sharply exacerbated by the firing of Secretary of Commerce Henry Wallace, still the favorite of the liberal wing of the party, in September 1946. As the congressional by-elections approached, the most noteworthy fact of the president's involvement was his near total absence from the campaign; Democratic candidates seemed to shun any guilt-by association that his presence might project.[30] Yet even this proved insufficient to stem the tide of voter disapproval. When the composition of the eightieth Congress was decided in November, it fell under Republican control for the first time since 1930.

It was at this low point of his personal standing that Truman finally began to undergo one of the most remarkable revivals in American political history. Shaken by the breadth and depth of critical reaction to his past record of leadership, both he and his advisors now realized that the White House itself was in serious danger of being lost in 1948 unless the past eighteen months of turmoil were checked. Consequently, both a new strategy as well as sense of purpose began to emerge, providing direction and conviction previously unseen in the president's behavior. Out of this change would eventually come the Fair Deal, Truman's amalgamation of his basic midwestern roots with a more sophisticated eastern liberalism carried over from the New Deal era. The latter influence would clearly predominate; certainly, this was the case in the transformed Truman's attitude toward big business. Indeed, much of the president's prior ambivalence surrounding the allocation of blame for the economy's inflationary tendencies would now be cast aside; in its place, the large monopolistic corporations would be identified as the culprit. Although, surprisingly, prior New Deal allegations of administered pricing did not arise, there was little room for doubt about where the administration placed blame for inflation. For the steel industry in particular, long the symbol to many of unrepentant corporate power, the effects of this political revitalization of New Deal tensions would be unpleasant.

Truman launched his new domestic strategy in earnest early in 1947. In February the Justice Department filed an antitrust suit against USS when that firm announced it would acquire a small West Coast steel fabricator.[31] This was followed over the next several months by the filing of twenty two more such complaints against a wide variety of alleged monopolists.[32] In March and April the president began a renewed attack on inflation by publicly urging price cuts by large corporations. Addressing a group of newspaper editors on April 21, Truman stated:

> There is one sure formula for bringing on a recession or a depression: that is to maintain excessively high prices. . . .
>
> Our private enterprise system now has the responsibility for price. . . . Private enterprise must display the leadership to make our free economy work by arresting this trend.[33]

The steel industry was singled out as a basic sector to monitor in this regard, and there were reports in the press that the administration had privately implored the chairman of U.S. Steel, Irving S. Olds, to lower his firm's prices.[34] But the company chose not to comply. When the coal miners' union signed a generous new contract in mid-1947, Olds felt it would inevitably lead to higher steel prices (coal-derivative coke was a significant input to overall steelmaking costs).[35] The administration, however, disagreed; the CEA, the new White House advisory agency created by the recently passed Employment Act of 1946, issued its first midyear economic report soon afterward and stated that steel firms should be comfortably able to absorb higher coal costs without raising prices.[36] Truman himself made a public plea for restraint, saying any price increase "would be a serious blow to our economy."[37]

The president's jawboning approach was having some effect on U.S. Steel. The corporation still felt that price hikes were justified, but it was reluctant to make any moves under the glare of the president's public spotlight.[38] The need for delicacy, however, was soon removed: In the last week of July 1947 several other integrated producers raised finished steel prices across the board by about 7%.[39] Claiming it could no longer be a "goody-goody and hold out on everybody else" within the industry, U.S. Steel insured that the increase would stick when it, too, joined in on August 1; it was soon followed by Bethlehem Steel, second-largest producer in the industry.[40] Olds and Eugene G. Grace (now chairman of Bethlehem) knew they would be on the spot by defying the president's wishes for reductions, yet constantly increasing costs, they maintained, made their action unavoidable.[41]

Although some battles may have been won by the firms, the war itself was being lost. The steelmakers—primarily U.S. Steel, public lightning rod of the industry—were becoming increasingly defensive about the role of national economic villain into which they were being cast through Truman's more aggressive political posture of 1947. Thomas W. Lamont, chairman of J. P. Morgan & Company and a USS director since 1928 complained:

> If there is one thing that [the government] ought to appreciate, it is that the U.S. Steel Corporation is the largest basic industry [*sic*] in the world, and aside from every other consideration it is of [such] vital importance to the welfare of the community that it should be kept sound, stable and liquid. If the Steel Corporation should even come up against it, it would be a calamity for the whole country. I should think these people at Washington would want to be applauding every measure taken to keep it as a leader in sound and conservative management.[42]

But Lamont's wishes in this matter would go unheeded. The ability of the president to dominate the levers of public opinion-making machinery was yielding dividends for the first time since Roosevelt's death, and the industry's own public relations response seemed feeble in comparison. When U.S. Steel had lost its PR-oriented chairman, Edward R. Stettinius, Jr., to governmental service at the beginning of the war, it had replaced him with the affable and even relatively liberal Irving S. Olds, a former Wall Street lawyer.[43] Although Olds seemed to grasp the strategic necessity of maintaining a sophisticated corporate public relations function to counter a more influential external environment, he was unable to convince all of his colleagues of this need. Benjamin F. Fairless, the corporation's president and chief operating officer, basically remained an old fashioned steel master. Though certainly not in any manner as unreconstructed as, for example, Tom Girdler of Republic Steel—his first mentor in the business—Fairless still believed that values more characteristic of generations past should and could shape the industry's posture vis-à-vis government.[44] Powerful directors of the firm, including Morgan's chairman Lamont, seemed to side more with Fairless's view than with Olds's. Lamont urged Fairless to speak out "in short, simple sentences," to tell the nation exactly what was what in steel and why.[45] Though this approach contained a certain romantic quality that might appeal to many in the nation, it was overmatched against the more powerful campaign that Truman was able to launch from his commanding position in the White House.

Yet few if any of U.S. Steel's top officers believed that problems with Washington were minor and hence easily overcome. Quite the contrary, they began to see the administration's attack as the chief obstacle to implementation of their own domestic strategy for competitive success.[46] The huge estimates of future steel demand forecast by governmental planners were, steel leaders believed, still hopelessly too optimistic. Nevertheless, they became seriously concerned about their ability to finance the new capital investments that they thought would be necessary.[47] In 1946 the industry's important capacity-utilization rate dropped to only 72.5%, owing largely to labor disputes that shut down the mills for lengthy periods, but in 1947 it rebounded back to 93%. Profits, too, were substantial in the latter year as industry net income topped $400 million, the highest total since 1930. Earnings for U.S. Steel alone were $127.1 million net in 1947; total cash flow was substantially higher.[48] But the costs involved in expanding steel-production capacity were enormous: Amounting to less than $100 per ton prior to the war, they rose to over $300 per ton by the late 1940s.[49] As far back as 1945, U.S. Steel was rumored to be considering construction of a giant new integrated steel mill on the East Coast; moreover, the firm also felt it had a substantial amount of modernization work to complete on existing installations if it were to remain competitive with its smaller rivals, who continued to chip away at its once overwhelming market-share lead.[50] What was needed, officials of the corporation concluded, was more investment money, and they felt that could come from only one source: higher prices.[51] As we have noted in Chapter 2, there was some truth to these conclusions.

The administration's focus on steel in its reinvigorated anti-inflation program was quickly eliminating the price option as a means to increase cash flows. With some justification, industry leaders believed they were being forced to mitigate price adjustments while not only their costs went unchecked (especially labor), but also while other industries jacked up prices at a far faster rate[52] (see Tables 2.4 and 2.5 for comparative data). "One has to remember," grumbled senior U.S. Steel board member Lamont, "that the Adminstration that has been in power since 1933 has constantly encouraged labor's demand, and . . . it is not possible for any single corporation to combat [such] unjust demands."[53] Although not an unbiased charge, this clearly conveyed industry attitudes as it groped for some response to the price dilemma.

It should be noted, however, that both sides—business and government—probably felt constrained by the bounds under which they were forced to debate. What the combined force of historical tradition and contemporary political necessity presented was a rather unappealing choice between two extremes: private off-the-record discussions—never well received nor legitimated in a pluralistic democracy—or highly politicized and publicized hearings—where a pressurized atmosphere often prevailed and partisan political considerations usually dominated proceedings. Nowhere, it seemed, could be found a more hospitable environment to resolve the complexities of what, in fact, amounted to formulation of a national industrial policy. Private groups—for example, the Business Council and the Committee for Economic Development (CED), which otherwise might have served this function—were becoming increasingly polarized by the political chaos that surrounded Truman's early years in office.[54] The CEA, led by its chairman Edwin G. Nourse, met privately with U.S. Steel officials in September in another attempt to get them to hold down prices. But the danger that plagued such tactics was that secrecy would be compromised and positions perhaps distorted. In this instance, a press leak resulted in a syndicated national columnist claiming that the CEA "virtually went down on its knees" trying to persuade the firm to hold the line. The results, of course, were a hasty public denial by an embarrassed Nourse, a foreclosure of this route as a means to discussion, and a further strain to the participants in an already tense situation.[55]

As positions became locked and flexibility for maneuver diminished, it was perhaps inevitable that the conflict would be elevated to the level of an acrimonious public spectacle. The industry, perceiving itself increasingly isolated in the eyes of public opinion, responded to the government with growing anger. The new political initiatives launched by Truman in early 1947 were obviously working, at least in steel. Not only were private pricing decisions subjected to greater scrutiny by the public, but another area in which industry interests had long exercised autonomy was also brought on to the agenda of business–government confrontation. This was the basing-point system of steel distribution.

III

In August of 1947 the FTC filed suit against the AISI and (eventually) 101 individual steel firms, seeking once again to force them to terminate their traditional method of assigning precalculated freight charges to steel deliveries.[56] The FTC had originally enjoined this practice as far back as 1921 when the agency filed a successful suit against the United States Steel Corporation demanding the system be abandoned.[57] The order to terminate Pittsburgh Plus was handed down in 1924; in its unofficial acceptance of this finding, however, U.S. Steel said it would comply only "in so far [*sic*] as it is practicable to do so." What the firm and the rest of the industry then proceeded to do was substitute a multiple basing-point system in place of Pittsburgh Plus, a relatively minor modification that did little to alter the ultimate effect of the practice. This unofficial state of partial compliance remained intact through 1938. In that year new federal legislation was enacted that made all FTC decisions (including past ones) final unless petitions for review were filed within sixty days. Attorneys for U.S. Steel immmediately asked for just such a review of the 1924 decision. By the time any substantive judicial resources could be committed to the case, however, pressures for stable wartime production had overtaken the steel industry and several other basing-point suits involving different parties (namely, the Cement Institute case and the Salt Producers Association case) were already quite well advanced within the courts. Consequently, U.S. Steel and the FTC agreed to stay proceedings on their suit until these other cases were resolved.[58]

The upshot of this legal maneuvering was that although the FTC would eventually win its case in the Supreme Court (in 1948), agency officials made a tactical decision to prosecute the aforementioned 1947 AISI complaint in order to terminate once and for all any variations of a basing-point scheme that the steel firms might employ. Another consideration, of course, was that this new case would provide a hedge if the Supreme Court went against them (as the Court of Appeals had in the Cement Institute proceedings). The new complaint was, therefore, "one of the biggest and best prepared since the [FTC] decided" to go after the steelmakers, at least according to industry observers. Moreover, these same sources saw the specter of political motivation behind the government's action.[59] This was because the suit was filed only two weeks after industry price adjustments in late July, which were made despite Truman's plea for restraint. Insofar as such observations were believed, they would do nothing to improve soured relations between the involved parties.

In addition to these new public threats to the industry's independence of decisionmaking in both price and distribution, steelmakers were still fighting the capacity issue. By early fall 1947 expansion had become a headline topic through the efforts of the Senate committee investigating small business concern with steel availability (noted in Chapter 2).[60] As a parade of unhappy witnesses marched into hearings held across the nation by the touring lawmakers, the larger firms in the industry were again drawn into the unwelcome role of avaricious predators unconcerned with the national interest.[61] And with repetition, industry spokespersons seemed incapable of any convincing denials. A J. P. Morgan & Company partner wrote to an associate:

> I don't know whether you will agree with me, but more and more I am impressed by the inadequacy of the intellectual aspects of defense

> that we Americans put up in defending our case,—whether it is this attack on the steel industry [over capacity expansion] or the Russian attack on America. The role that ideas play in history is far more important than our spokesmen seem to allow for. Unless we come to grips with the intellectuals that attack us, the rest of our case hangs in mid-air and is ineffectual. We are not an intellectual people and we are at a disadvantage in a conflict of ideas. We rely on superficial effects.[62]

Although these remarks were not necessarily applicable to all members of the nation's corporate elite, nevertheless, they seem highly appropriate to the major competitors in the steel industry. They were indeed at a distinct disadvantage in the conflict of ideas, probably for the simple reason that their philosophy of economic competition had long since been rejected by the vast majority of the American people. If anything, the observation of the Morgan partner clearly underscored why: The industry had been unable to articulate the specialized nature of competition that it faced (both domestically and abroad), and it had been paying the political price for that inadequacy since 1901. The "superficial effects" of public relations campaigns were insufficient weapons to overcome entrenched public beliefs on these matters.

But the government cannot be totally absolved of its own resort to "superficial effects" to win its case. Although Truman's campaign against mounting inflation was a legitimate exercise of presidential leadership in the national interest, one should still not dismiss the fact that the principal tactics used were political broadsides against large corporations (and not powerful unions, though this sector had in the past come in for criticism from the administration).[63] Moreover, the more cynical observers might also have pointed to the political opportunism that permeated most of the president's 1947 agenda: Having finally found his voice, he began to use it more and more to articulate slogans designed to reawaken the declining New Deal coalition. Truman needed issues to blunt not only the challenge from the right posed by the Republicans, but also those on the left personified by Henry Wallace, who almost immediately after his dismissal from the cabinet the previous September became an attractive candidate for the Democratic presidential nomination.[64]

In order to delineate the political dimensions of the inflation question—that is, to position the issue as a Republican problem and by association a consequence of unchecked corporate greed—Truman called for a special congressional session in October 1947.[65] This would be devoted to a discussion of interim European aid and domestic inflation. The debate surrounding the former issue would not be so polemical because in general many Republicans tentatively agreed with Truman's policy of aiding Europe as a means to contain Communist expansion. But on the domestic front there was a fractious disparity of positions between the two parties. Truman's apparent strategy was to present Congress with a New Deal–oriented program for the control of inflation so sweeping "as to be absolutely unpalatable to the Republican majority."[66] In mid-November he unveiled a ten-point anti-inflation proposal that provided for extensive federal controls over economic activity. As planned, it forced the Republican-controlled Congress to go on the defensive: Unless some substantive counterpolicies could be enacted, Truman would be able to confidently proclaim to the nation in the coming election year that the detested wave of high prices was the fault of a greedy big-business sector and its Republican puppets in Congress. Realizing their dilemma, GOP leaders tried vainly to fashion some coherent response. But they were caught within their own discord over domestic economic policy because more conservative elements in the party would not agree to any attempts to institute new economic thinking into public policymaking. In the end, only three of Truman's ten points resulted in legislation, all diluted considerably from what he originally proposed; the one action affecting the steel industry was a program of voluntary allocation of scarce commodities, to be administered by a generally friendly Commerce Department.[67] The president thus won his political skirmish with the Republican Congress, declaring on December 28, as the special session ended, that he felt "deep disappointment" at "such feeble steps toward the control of inflation."[68]

It should be noted that this voluntary allocation plan resulted in the reestablishment of a mechanism that had been tried in the past to facilitate business–government cooperation in America. The Commerce Department, now under the guidance of W. Averell Harriman (named as Henry Wallace's replacement in the fall of 1946), chose to utilize a network of industry advisory committees to implement the allocation of scarce commodities to priority users. Originally created by Bernard Baruch's War Industries Board in World War I, the *ad hoc* committee system consisted of private-sector executives, organized along specific industry lines, who could meet as a unit with governmental officials to thrash out problems, develop agendas for action, and in general oversee a program of cooperation between the two parties.[69] As such, these committees theoretically could provide the forum long missing between business and government to effect real progress in sector-specific settings.

The steel committee was the first to be created by the Commerce Department's Office of Industry Cooperation in January 1948. Initially, fifteen leaders of the industry met with Secretary Harriman, and it was expected that steel would set the pattern for the many others that would follow.[70] Twenty-eight industry advisory committees were soon formed, and a total of fifty-three meetings held. Of the nineteen actual allocation agreements reached by these various committees, all pertained to steel or pig iron.[71]

Although providing a potentially powerful tool to resolve some of the nation's escalating business–government conflicts, the industry's advisory-committee system unfortunately never took hold. The reasons had to do with objections raised by Justice Department and other federal officials as to the legality or propriety of committee activities (in view of antitrust strictures) as well as protests from many smaller firms who felt they were being unfairly excluded from important deliberations. In the face of such objections, few top governmental leaders were willing to promote this cooperative system as a solution to the growing business–government stalemate.[72] Truman, the champion of the little man and a staunch supporter of small business, certainly was not about to do so. Thus, lacking support, the program terminated along with the Voluntary Agreements Act in the fall of 1949. Confrontation, not cooperation, would remain the order of the day.

Meanwhile, it was against the backdrop of intense politicization of domestic economic issues that the steelmakers entered 1948 and the election campaigning that promised only more of the same confrontational tactics. "If the [steel] industry felt that it was being mistreated . . . over the pushing around it has received from the federal government since the war's end," declared the trade journal *Iron Age,* "it hadn't seen anything until the dawn of 1948."[73] This prophecy of discord soon turned to fact after U.S. Steel announced a 10% rise in the price of semifinished steel in mid-February, citing continued higher costs.[74] Yet this time even some Republicans felt compelled to register disfavor, testimony to the low level to which the industry had sunk in public opinion as well as the increasing untenableness of its political position. The congressional Joint Economic Committee, chaired by conservative Senator Robert Taft (R-Ohio), voted to conduct a hearing on the price hike. Joint Economic Committee member Senator Ralph Flanders (R-Vermont), stated:

> In an industry as fundamental as steel, there are public questions involved as well as questions of private business policy. Any rise in prices at this time and under these conditions has to be defended on public grounds as well as for business reasons.
>
> . . . The steel industry apparently does not yet realize the nature of the public interest in their business policies.[75]

Truman also immediately seized on the opportunity presented by Big Steel's price hike, ordering investigations by the CEA, the Commerce Department, and even the Federal Bureau of Investigation. The latter swooped down on sixteen firms to seek data that might indicate conspiracy.[76] Both Commerce and the CEA quickly produced studies of the situation, which the president made available to the press in early March (along with the FBI report, which found no conspiratorial intentions).[77] The Commerce report was mostly an impartial collection of statistics with accompanying text; for those who chose to examine the numbers in detail, they would seem to have justified U.S. Steel's price rise at least on a comparative basis to other industries (Figure 3.1). The CEA report, however, was more concerned with larger policy implications than simple economics:

> Fortunately, the public attention aroused by your [i.e., Truman's] call for an investigation of the rise in semi-finished steel prices has served notice on industry that any general price increase will have to be defended strongly before the bar of public opinion. Had it not been for this publicity, the rise in steel prices would undoubtedly have acted as an encouragement to others to proceed with price advances.[78]

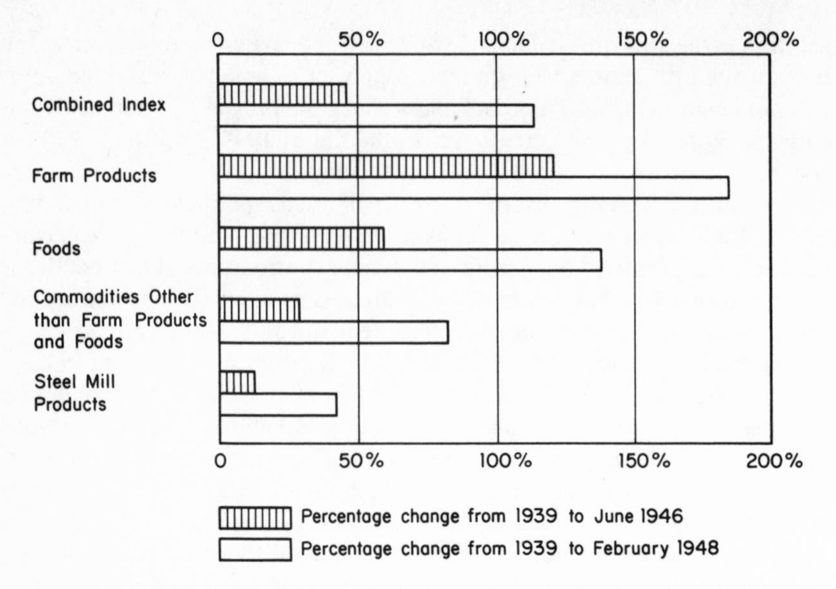

Figure 3.1 Steel price increases relative to other wholesale prices, 1939–1946 and 1948. *Source:* U.S. Department of Commerce, press release, March 12, 1948.

The JEC hearings were held on March 2, featuring appearances by officials representing Big Steel, Little Steel, and the smaller independents.[79] Most of the hard questioning came from Senator Joseph O'Mahoney (D-Wyoming), an avid New Dealer who had chaired the Temporary National Economic Committee (TNEC) investigations of monopolistic business practices ten years earlier and who in the past had often been critical of the behavior of the largest steel companies. The industry—especially U.S. Steel—unfortunately did not take this opportunity to improve its public image. *Iron Age* commented:

> Whether caused by unpreparedness or from other causes, U.S. Steel apparently failed to make a strong case for itself under questioning of the committee which centered its fire on the corporation, holding it to be the bellwether for the industry in general. The committee obviously was less impressed by efforts of Mr. Fairless than by those of Mr. Homer [Arthur B. Homer, president of Bethlehem Steel] who shot back figures without hesitation and used Presidential statements on the need for reasonable profits with telling effect.[80]

But congressional hearings—even those under Republican auspices—were hardly designed to provide a showcase for corporate image building, a fact not lost on U.S. Steel's Fairless who testified that "much of the criticism [of his firm's action had] been emotional or political in character."[81] Yet it was this very obdurateness of mind exhibited by steel leaders—an inability to sufficiently grasp the facts of political reality—that contributed further to the industry's deteriorating image (and it is worth noting that Bethlehem, the industry pioneer in modern public relations, was able to make a better presentation of its case than did its larger rival).[82] The hearings, although they had no immediate effect on steel pricing, nevertheless underscored both the increasing frustration of the producers' actions as well as their continued inability to manage the political environment. Truman and the government were not just winning the battle for control of public opinion, they were dominating it. The short-run social and political damage of this outcome was bad enough, but in a longer-term perspective it would work to produce even worse consequences, both for the steelmakers and the nation. The firms became more apprehensive about the intentions of Washington and as a result more cautious in their marketplace response.[83] As we shall see in following chapters, this timidity would soon prove disastrous, especially in the international environment, a sector where at least a modicum of business–government cooperation was mandatory to success.

Still recovering from the scrutiny of the JEC hearings, the industry was dealt another blow in April 1948 when the Supreme Court overturned a lower-court decision and upheld the FTC in its rejection of the Cement Institute's basing-point system.[84] Shortly afterward, U.S. Steel reluctantly announced that it, too, would end its historic reliance on this means of equalizing freight charges on product deliveries. Steel spokesmen gloomily predicted impending chaos (and even some higher prices) as an FOB system of distribution was finally implemented throughout the industry.[85] In May 1948 Walter Tower stated:

> If this country wants steel in its modern efficiency it must accept bigness. If it rejects that bigness, it rejects the main prop of this country's economic strength and World influence.
>
> Yet here comes the highest court with an opinion which if carried over into this industry could mean only confusion and difficulty for both makers and users of steel. It could even put an end to the economies and efficiency which comes from bigness, and pull the props from under thousands of small users, who would face the tough choice of moving or folding up.[86]

But this was, of course, more tirade than truth, for the tough choices would be faced not by users, but by producers: Should they construct new mills nearer to customers in order to obtain competitive advantages? And the problem of competition in the industry, as Tower well knew, was that if one firm were to choose this expensive option, then others would be forced to follow. Not unexpectedly, steel leaders preferred that such important decisions be made according to their own timetable of need, not the government's.

Steel lobbyists thus mounted one last rearguard action before capitulating to the FTC. Working with representatives from other industries also forced to abandon the basing-point system because of the new legal precedents, steel interests obtained the introduction and eventual passage of congressional legislation that would permit an extension of a modified basing-point system.[87] Truman expressed outrage at this ploy; speaking before a campaign rally in Louisville, Kentucky, in late September 1948, he said:

> Four days after this [the Supreme Court Cement Institute] decision, the chairman of the U.S. Steel Corp. . . . gave his marching orders to the Republican Congress, stating that the decision would so upset marketing procedures that the matter "should receive the attention of Congress."
>
> Three weeks later, [Republican leaders] took steps to see that it did "receive the attention of Congress" by introducing a resolution authorizing an investigation into the effects of this decision. The Republican Congress obligingly appropriated $50,000 for this investigation which is now underway. . . . It is as if the inmates of a jail were asked to decide whether or not the laws under which they had been sentenced should be repealed. . . .
>
> If by any chance the Republicans do win in November, there can be little doubt that these and many other exemptions will become the law of the land, thus reducing the antititrust laws to a shambles and removing from the average citizen his last vestige of protection against the power and greed of private monopoly.[88]

This episode finally ended in defeat for the industry when President Truman vetoed the bill Congress had passed.[89] For the first time in the twentieth century, steelmakers would be forced to adopt a universal FOB method of delivery, and mill location would come to play a more important role in competition: This, no doubt, was one contributing factor in the huge expansion program that the industry finally entered into in the 1950s.

There would, of course, be more controversy between steel and the government before these changes got underway. The casual observer, in fact, might have concluded that there were no conciliatory breaks whatsoever in the long span of confrontation between these antagonists. At least the union had not gone out on strike in either 1947 or 1948, though this was largely due to generous new contracts that the USWA leadership was able to win from industry; raises of fifteen cents and thirteen cents per hour were obtained in successive years (amounting to 11.3% and 7.4% annual gains, respectively—see Table 2.5).[90] The continuing strong postwar demand for steel apparently was a factor in management's decision to accept higher wages as a necessary cost to keep the mills in operation. The capacity-utilization decline of 1946 jumped sharply back to higher levels in 1947 and went up even further in 1948 to 94.1%. But, as we have noted, the steel producers passed on at least part of these costs in higher prices, which rose over 45% between 1945 and 1948. With the 1948 presidential elections looming closer, Truman continued to emphasize the inflation issue—and the "unholy alliance" of combined big business and Republican intransigence to quell it—as a major plank in his campaign platform.[91]

The industry, meanwhile, made good on its prediction that an end to the basing-point system would cause prices to rise. The manner in which this occurred, however, was "one of the most unusual in steel history" according to *Iron Age,* traditional voice of the smaller producers.[92] In late April 1948 U.S. Steel said it would not grant steelworkers any pay hike in forthcoming contract negotiations, and it then proceeded to reduce some prices on finished product.[93] Some in the trade press applauded this as an "attempt to smash inflation."[94] But according to a subsequent FTC study, this was part of a less compelling strategy to eliminate competition from smaller unintegrated steel firms who tended to purchase most of their semifinished steel needs from the major producers:

> It is apparent that by raising prices of semifinished steel in February [of 1948] and by cutting the prices on the products made therefrom in May, U.S. Steel Corporation applied a double squeeze on the smaller semi-integrated and nonintegrated mills. The leadership of U.S. Steel was followed by the other large integrated companies in both instances, though apparently somewhat more reluctantly on the second occasion. These companies were themselves caught in the squeeze when the prices of finished steel products were cut.[95]

Regardless of intentions, the plan did not work very well. Most of the larger integrated firms refused to go along with Big Steel, perhaps realizing both that they were more vulnerable to union pressures if they renounced wage hikes and that they were generally less exposed to any pressing need to cater to public relations. In July U.S. Steel therefore reversed itself and raised finished prices by nearly 10%, citing the by-now forthcoming pay boost to steelworkers as its reason.[96] Although this elicited the usual reply from the White House—Truman was soon to comment that he thought steel prices were always too high—other observers noted that the industry was about the last one to boost wages and prices in this so-called third round of postwar inflation.[97] Moreover, this episode also reflects the increasing desperation of USS to obtain some means of increasing its cash flow in order to finance growth: It was willing to risk not only the enmity of the government, but also that of its industry colleagues.

IV

As November 1948 approached the steelmakers apparently had only a one-prong strategy for recovery from the depressed public standing into which they had sunk. This revolved around anticipation that Truman would soon be replaced. Such a presumption was certainly not limited to steel executives: "Anyone who has had any doubts about the election of Thomas E. Dewey as President of the United States has only to come to Washington to have these doubts dispelled immediately," wrote *Iron Age* in late October.[98] But the industry's hope for the installation of a more friendly administration was soon to be shattered in one of the most stunning electoral triumphs of modern American history. The revitalized Truman defeated Dewey and the Democrats swept both houses of Congress.

Any plans by the steelmakers for a quick reduction in business–government tensions were thus dealt a severe blow by the defeat of industry-supported candidates. The trade press commented:

> Any way you look at it, 1949 is going to be a big year in Washington for the metalworking industries.
>
> As far as the steel industry is concerned, you can bet your bottom dollar that President Truman and the new Congress are getting ready to roll up their sleeves, spit on their hands, and go to work.
>
> The working-over may be done with (a) an axe, (b) a broom, (c) both.
>
> There is a growing feeling of self-righteousness in the Administration and in the new Congress that license for an axe- or broom-wielding treatment was granted in the November elections. These forces are attempting to exploit to the fullest the "mandate from the people" argument.[99]

The agenda of fear that the industry saw ahead included price controls, compulsory allocation of steel to designated customers, governmental operation of some new mills, price cooling-off periods to combat inflation, taxes on excess profits, antitrust suits, and prolabor legislation (reversing, perhaps, the controversial Taft-Hartley Act that the previous Congress had passed over Truman's veto in 1947).[100] When asked by reporters to comment on a statement by a Democratic congressman that "business had nothing to fear" from the results of the elections, the president replied: "Has business had anything to fear in the past three and one-half years?"[101] There is no record of steel industry reaction to Truman's statement, though surely its leaders would have disagreed, some vehemently.

But whatever the industry response may have been, it no doubt changed for the worse in early January 1949, shortly after Truman delivered his State of the Union address to Congress. "Every segment of our population and every individual has a right to expect from our government a fair deal," said the president in enunciating the theme of his new term.[102] He then went on to list a series of legislative objectives that would please New Deal advocates the nation over, including such measures as a more equitable tax structure, repeal of Taft-Hartley, minimum hourly wage laws, a new farm program, national medical insurance, protection of civil rights, federal aid to education, and other such social and economic goals long cherished by the party's liberals. Near the end of his message, Truman addressed the needs of industry:

> So far as business is concerned, it should plan for steady, vigorous expansion, seeking always to increase its output, lower its prices and

> avoid the vices of monopoly and restriction. So long as business does this, it will have the help and encouragement of the Government.[103]

To combat inflation, the president outlined an eight-point program of action, including credit controls, export controls, allocation of key materials in short supply, and standby authority for price controls over basic commodities; finally, he concluded, Congress should

> authorize an immediate study of the adequacy of production facilities for materials in critically short supply, such as steel; and, if found necessary, to authorize Government loans for the expansion of production facilities to relieve such shortages, and furthermore to authorize the construction of such facilities directly if action by private industry fails to meet our needs.[104]

The president's message not only elevated the capacity-expansion question back to the top of the list in the industry's continuing conflict with government, but it also aroused new fears in some quarters. *Iron Age* editorialized, "In his message on the State of the Union the President served notice on the steel industry that it was in danger of nationalization."[105] Irving S. Olds, U.S. Steel chairman, stated that "while [nationalization] may seem . . . a remote possibility, I should not be surprised to learn that the eventual nationalization of the steel industry along proposed British lines is in the minds of some of our critics."[106] Olds's reference to England concerned the Iron and Steel Bill that was introduced in Parliament in 1946 following the Labour party's election victories; it would nationalize the steel industry in that country. That Olds's concern was more than the usual business executive overreaction to any perceived threat from the state was confirmed the following November when the bill became British law.[107]

Yet, others in American industry reacted with less agitation to the president's State of the Union remarks. "He just repeated what he's said a dozen times before," commented one.[108] Political rhetoric notwithstanding, there was one aspect of the message that the steelworkers did not automatically dismiss:

> If President Truman with congressional support should immediately cause the passage of a proposed bill for accelerated depreciation, the President likely would find an expansion by industry exactly as took place during the war when accelerated depreciation caused an expenditure of $6.5 billion.[109]

Truman encouraged this less controversial approach to expansion when, several days after his congressional speech, he clarified his statement: He meant that the government should take "progressive steps" to insure sufficient steel supply, not the immediate construction of mills by government; federal loans should be made available to industry first.[110] The underlying political nature of this "nationalization" statement—as opposed to any real policy proposals—was made even clearer in subsequent months. Both the CEA and JEC, in reporting on the State of the Union legislative recommendations, would not give an endorsement to point eight (even though the CEA was directly under the president's control and the JEC was chaired by Senator O'Mahoney [D-Montana] and dominated by Democrats).[111] Then, in May 1949 Secretary of Commerce Charles Sawyer addressed the annual meeting of the AISI in New York and told his audience:

> No government official ever will or can run a business as well as a businessman can do it. It is not the intention of the Government to run your business. Contrary to any rumor you may have heard, the Government never intended to take over the steel business of the United States, nor even considered that as a possibility.[112]

If any final evidence of Truman's real intentions in this matter were needed, it came in June when the Economic Expansion Bill—recommending among other steps massive construction of federal plants in critical industrial sectors—was introduced by Wright Patman (D-Texas) in the House and James Murray (D-Montana) in the Senate; the latter had long given his endorsement to plans for government-owned steel plants. This measure never received administration backing and, consequently, never got out of committee in either congressional house.[113] Truman's threats to build government-owned mills were thus little more than political bluff, words designed to evoke a response by industry that could not be achieved by any better means. The real message here, however, is not the courage of elected officials in facing up to big business, but rather the incredibly barren arsenal of public tools to fashion national competitive policies. Blame in this regard would have to be equally borne by both business and governmental leaders.

Meanwhile, the level of relations between these adversaries would continue to be fraught with uncertainty and political intrigue. The year 1949 would not pass without still another industrywide strike by the steelworkers' union nor without the by-now ritual price hike and subsequent congressional hearings.[114] A continuing issue that would also not dissolve peacefully was that of capacity in the industry—not surprising, of course, given its treatment by the president in his January message.

The steelmakers continued to loudly resist governmental calls for expansion while demanding approval of their own privately devised schedules for growth made according to their own calculations of demand. Thus Chairman Ernest T. Weir of National Steel would proclaim in January 1949, shortly after Truman's speech, "It is obvious that the huge immediate need for steel is abnormal and temporary. It would be foolish and damaging to the United States to build permanent capacity in proportion to this passing situation."[115] Yet, just days prior to Weir's outburst, the AISI had issued a press release stating that "steelmaking capacity of the steel industry, at 96,000,000 tons per year, is now greater than ever before in war or peace, and further large expansion is planned during the next two years."[116] At the same time the institute was broadcasting its news, U.S. Steel was reassuring the nation with details of its $650 million in expenditures since V-J Day for additional steelmaking facilities. "So long as the nation's need for steel continues to grow," the firm announced, "United States Steel intends to do its part to supply its customers, large and small, old and new."[117]

The surprising election of Harry Truman in 1948 thus widened the divisions between the major steelmakers and public policymakers. Choosing to run on a platform that emphasized the economic perils deriving from large and powerful corporations, Truman was obviously in no position to pursue cooperative policies with these firms in the achievement of national objectives. In response companies like U.S. Steel retreated further into the defensive postures that historically had characterized much of their reaction to political authority. The outcome of such behavior would soon prove detrimental to both the firms and the nation's long-term economic interests.

4

Domestic Steel and the New International Economic Order

The public debate surrounding the expansion of steel capacity received considerable attention in the late 1940s. This is not surprising given the critical role played by steel in the nation's economy. Yet throughout the discussion one topic never seemed to attract much attention: The power of the international environment to shape the domestic industry's choice of strategic action. Not all, of course, were oblivious to this potential; some company managers openly alluded to European reconstruction as a potential threat to industry stability.[1] But when the various governmental studies were made of the capacity question, relatively little public notice was given to a consideration of how foreign factors might influence domestic options. The Senate's Small Business Steel Subcommittee, for example, focused primarily on export license violations when it considered the foreign market, whereas other governmental groups interested in steel pondered means of remanding foreign scrap left in war zones back to the United States to meet domestic requirements.[2] For the most part, however, postwar problems of international trade in steel were left almost entirely in the hands of agencies and commissions devoted to the restoration of war-torn economies, not with groups seeking to establish any long-term industrial policy or strategy for American producers.[3]

The domestic steelmakers themselves did relatively little to call public attention to the implications of international trade for their future well-being. This appears surprising in view of prewar industry behavior regarding foreign competition. Surely such concerns were afloat in industry circles and perhaps even paramount in the minds of some steel leaders. Indeed, the history of steel's political stance from the 1920s onward forces attention to how international variables help explain the industry's holding pattern toward postwar domestic expansion—a strategy (as we have seen) to which it clung tenaciously despite strong public opposition. The present chapter will consider these changes in the international environment of steelmaking.

I

Any probe of the intersection between domestic and international activity in steel in the postwar period leads immediately to a most compelling question: How does one account for the huge increase in domestic capacity that finally occurred in the United States after 1950? As we noted, the industry's resistance to expansion was strident from 1940 onward. Yet the 1950s reveal an impressive record of growth (see Table 4.1). Moreover, the rate of capacity increase in the nation's coastal zones was considerably higher than in inland locations (with Texas, New Jersey–Delaware–Maryland, and California, respectively, ranking first through third of all states in rate of growth).

As noted previously, domestic steel producers paid close attention to expansion programs of foreign rivals in planning their own growth. This was due primarily to competitive fears: Foreign producers often enjoyed comparative advantage in labor as well as certain other cost factors. This competition was most likely to occur in coastal American markets, which were more vulnerable than inland areas that enjoyed natural protective barriers because of expensive overland transportation costs. In addition, protective tariffs had been lowered after enactment of reciprocal trade treaties in 1934; this trend continued after the war, thus further exposing the domestic industry to potential inroads from cheaper imports. Indeed, these fears of international competition had been so great that they caused the large domestic steelmakers not only to join a world cartel during the late 1930s, but also to resist expansion at the outset of World War II when it seemed clear that American national security might be endangered without more steel.

Given this prior history of global interdependence in steel-expansion decisions, one naturally must raise questions when evaluating the decision to increase capacity after 1950. What uncertainties existed in the international environment of the postwar years that might have contributed to the domestic industry's initial reluctance to expand? And what eventually occurred in

Table 4.1 Capacity Growth

	1950 (ingot tons)	*1955 (ingot tons)*	*1960 (ingot tons)*	*1950–55 Increase*	*1950–60 Increase*
Total United States	99,982,650	128,363,090	148,570,970	28.4%	48.6%
Coastal states	18,725,980	27,410,420	32,229,870	46.4%	72.1%

Source: AISI, *Annual Statistical Report* (various years). See Table 4.2 for details.

Table 4.2 Expansion of Steel Capacity by Geographical District, 1950–60

District (by state)	*Net Capacity as of* Jan. 1, 1950 (net tons)	Jan. 1, 1956 (net tons)	Jan. 1, 1960 (net tons)	1950–60 Change (%)
MA, RI, CT[a]	498,280	683,200	177,000	(64.5)
NY[a]	4,763,900	6,914,080	7,420,010	55.8
PA	29,313,830	35,012,150	37,321,130	27.3
NJ, MD, DC[a]	5,755,580	7,094,300	10,930,260[b]	89.9
VA, WV, KY, TN	4,032,220	4,766,740	5,324,190	32.0
GA, AL, FL[a]	3,736,590	5,385,020	5,832,020	56.1
OH	19,749,240	24,835,660	28,318,880	43.4
IN	11,262,840	15,637,500	18,440,510	63.7
IL	8,411,980	10,643,200	12,794,400	52.1
MI, MN	5,456,460	7,733,420	8,957,000	64.2
MS, MO, CO, OK	1,674,000	2,187,000	2,825,000	68.9
TX[a]	862,320	1,824,350	2,524,080	192.7
UT	1,440,000	1,937,000	2,300,000	59.7
WA, OR[a]	459,920	514,000	551,000	19.8
CA[a]	2,649,390	3,195,470	4,795,500	81.0
TOTAL	100,066,550	128,363,090	148,510,980	↓
			1950–55 Increase =	28.3%
			1956–60 Increase =	48.4%
COASTAL STATES	18,725,980	27,410,420	32,229,870	↓
			1950–55 Increase =	46.4%
			1956–60 Increase =	72.1%

[a]Coastal states.

[b]U.S. Steel's Fairless Works capacity of 1.8 million net tons is shown under New Jersey rather than Pennsylvania because of its tidewater location.

Source: AISI, *Annual Statistical Report* (various years).

this arena that might have influenced the expansion that finally did get underway in the early 1950s—even in regions considered highly vulnerable to foreign competition and even though tariff walls continued to be lowered?

As the war came to an end, an obviously important issue that should have concerned public policymakers and industry officials alike was the future trend of foreign trade in steel.[4] Between 1941 and 1945 domestic producers had continued to enjoy sizable export markets (still booming from the high demand experienced in the years 1937 to 1940 when Europe prepared for war).[5] There was uncertainty, however, about the postwar years. *Iron Age* noted:

> It is believed in some circles that by the time domestic demand is taken care of, foreign countries which are busily reconstructing their steel plants will be in a position to furnish steel in quantity which will be beyond current expectations.[6]

Yet, only one week later, the following appeared in the same journal:

> Steel plant construction . . . will require 5 to 10 years for completion and even then plant capacity will hardly exceed each nation's own immediate needs.[7]

To help clear up some of these questions, Congress in early 1946 requested the U.S. Tariff Commission to undertake an analysis of how the war's aftermath would affect America's international competitive position. The commission's survey covered a number of industries. Regarding steel exports, the report noted:

> The competitive strength of the European continent, probably for a considerable period, will remain weaker than it was before the war. . . .
>
> During this later postwar period, as during the years immediately following the war, the United States will probably be the only important source of capital for international investments. Whatever expansion of transportation, mining, and manufacturing facilities takes place throughout the larger part of the world will have to be financed mainly by outside capital, which will have to be supplied by the United States. . . . The ability of foreign countries to buy iron and steel, as well as other commodities, from the United States during the long-term postwar period will be materially affected by the policy of the United States Government and of the nationals of this country as to investments abroad.[8]

On imports, the following conclusions were drawn:

> During the later postwar period the prospects as to imports of tonnage steel into the United States are more problematical. Rehabilitation of the iron and steel industries in continental Europe and the installation of new plants and equipment may increase their efficiency so that they may be able to compete more effectively in the United States markets than during the early years following the war.[9]

The key variable to foreign trade in steel, the commission said, would be the role played by Germany. This was a fact clearly recognized throughout the industry, as noted by the trade press:

> Possibly the most important corollary to all this confusion [in Europe] is the fact that the interdependence of the western European countries before the war was so great that until a policy is laid down for the future of industrial Germany, none of the industrial countries can make any definite plans for the rebuilding of their shattered economies. France cannot know whether she will get Ruhr coal or be able

> to sell ore; Sweden does not know if she can sell her iron ore; Belgium and Luxembourg must face the coal procurement problem; Great Britain cannot determine whether her eager industrialists will be able to capture Germany's markets.[10]

And most important for the United States, its domestic producers could also not formulate their future plans until the German question was resolved one way or another.

It would not be easy for Allied policymakers to do this. Indeed, the problem lay at the very heart of emerging global issues that would dominate the international political agenda for the next decade and a half.[11] Briefly, the debate revolved around two contradictory Allied objectives: the implementation of measures to insure that Germany would be incapable of fomenting more destruction in the future and the implementation of measures to block the spread of Communist expansion into Western Europe. There was considerable dissension within both the Roosevelt and Truman administrations on the German issue, divided along lines of corrective measures (disarmament, denazification, and decartelization) versus constructive measures (restoring the German economy and integrating the country back into the European system).[12] Henry J. Morgenthau, Jr., Roosevelt's treasury secretary, had long advocated the corrective option, proclaiming that the Ruhr district be stripped back to the point "that it can not in the foreseeable future be an industrial area."[13] The Morgenthau Plan, as it became known, seemed initially to be favored by President Roosevelt, who allowed an early postwar planning conference to conclude: "This programme for eliminating the war-making industries in the Ruhr and in the Saar is looking forward to converting Germany into a country primarily agricultural and pastoral in character."[14]

But Morgenthau's views soon came into sharp conflict with others who favored the constructive approach. The most vocal of these proponents, at least initially, were American business interests and their allies in the State Department. Both Roosevelt's last secretary of state, Edward R. Stettinius, Jr., as well as Henry Stimson, former secretary of state (and then secretary of war), voiced strong opposition to the Morgenthau Plan. They were joined in this by many others who occupied positions of influence within the government.[15] When President Truman assumed office in early 1945, he inherited this unresolved conflict over Germany's future. But a complicating factor seemed to bind his hands: Agreements reached by England, Russia, and the United States at the Yalta Conference in February 1945 (under Roosevelt) had already established tentative guidelines for German reparations—guidelines that, if implemented, would move Germany toward the pastoralization option. Among other things, these called for eventual dismantling and distribution to the Allies of much of Germany's productive machinery and equipment, a step already underway by Russia in its zone of influence in the occupied Axis territory. According to some accounts, Truman was at first reluctant to renege on what he believed were fixed accords; consequently, less than one month after Roosevelt's death he approved a military directive for the European occupational forces (known as JCS 1067) that contained important elements of the Morgenthau Plan.[16]

Truman and his advisors, however, had second thoughts on this approach. In its place, and after due consideration, there began to emerge a repudiation of the Yalta agreements.[17] It was noted that Russia could not be trusted to abide by prior terms, that Stalin's intentions were clearly expansionist in nature, and that unless Germany were given some means to support itself, its economic problems would ultimately cause a huge drain on the taxpayers of America, who would be forced to support the welfare needs of the ravaged population. The president seemed favorably disposed to these latter persuasions (writing later in his memoirs, in fact, that Morgenthau's ideas "could starve Germany to death").[18] Consequently, at the Potsdam Conference of July 1945, a follow-up to the Yalta talks, Truman began to back away from earlier commitments on German reparations and, indeed, seemed to take a new hard-line position toward all dealings with the Soviets.[19] As the American policy toward Germany's economic future thus began to change, the fate of that country's devastated iron and steelmakers—her leading prewar industrial sector—would inevitably rise in importance as a subject of international discussion and conflict.

The immediate effect of this change in policy was that the initial efforts at dismantlement of the German economy began to drag.[20] Truman left all reparations activities to the jurisdiction of the military government installed in Germany after the victory. The head of the occupation forces, Major General Lucius D. Clay, later wrote:

> Fortunately the provisions of JCS/1067 were in some respects general in nature, so that the degree of application was left to the judgment of the military governor and some of its more drastic economic and financial provisions were tempered by the agreements reached in Potsdam.[21]

Working under Clay as head of the Economic Division in the American zone was Brigadier General William H. Draper, Jr., who in civilian life was a vice president in the investment banking firm of Dillon, Read and Company.[22] Draper was a vigorous supporter of the new policies to rebuild German industry that began to filter out of the confusion following Potsdam; Germany, he stated to Clay, would necessarily have to reenter world markets and in so doing it "has got to have the opportunity to have efficient industrial organizations; and where that requires sizable industry or plants, that should be permitted."[23] Working with Draper as head of his division's Steel Section was Rufus Wysor, president of Republic Steel, who was knowledgeable about the organization of German industry prior to the war (Republic had been a key American member of the International Steel Cartel [ISC] organized by the Germans in the 1930s).[24] Other leaders of the American steel industry also favored German reconstruction. Bethelehem Steel's Eugene G. Grace said in reference to himself and other leading steel executives:

> When we saw the end of the war approaching [and were asked] what our ideas were in respect to the rehabilitation of the German steel industry [w]e said by all means rehabilitate it to a substantial production. It was said they will go to war again. But that is not necessary. You can produce steel in Germany but control its application.[25]

Attitudes such as these were soon able to overcome the objections of many officers in the Decartelization Branch of Draper's Economic Division who continued to believe in the corrective approach to the German problem. The latter's zeal was so strong, in fact, that they were labeled "extremists" by General Clay, "sincere but determined to break up German industry into small units regardless of their economic efficiency."[26] This extremist view was not to prevail.

Proponents of Germany's reindustrialization—not only in steel but in other vital sectors as well—would thus win out over those favoring pastoralization. The underlying motives behind the American steel industry's support of this policy are not precisely clear, however. Some believed the reason was rooted in a desire by the producers to renew the disbanded ISC.

II

The trade press was rife with speculation that a new cartel was being formed. *Iron Age,* traditional voice of the smaller firms in the iron and steel industry, noted as early as the spring of 1946 that (in regard to European producers):

> It would be difficult to prove in court today that there is in existence a single marketing agreement in regard to a steel product line, but it is not too difficult to show logically that in all probability the agreements are being drawn up. Some seem likely to have been completed.[27]

Iron Age's correspondent—a highly informed source—went on to conclude, "1946 is in fact an ideal year to draw up a steel cartel. I am personally convinced that the latter is true" (i.e., that such an agreement had been reached).[28] Such stories would continue to surface through the early 1950s, and on several occasions there was vague corroboration of rumors from official U.S. governmental sources.[29] But few details were ever mentioned.

There can be little doubt, however, that leaders of the American steel industry did favor U.S. reconstruction aid to their European rivals. This attitude was evident not only in the early policy debates over Germany's future, but also when the United States embarked on a new program of massive foreign aid under the auspices of the Marshall Plan. By the end of 1946 much of the government's official policy toward the future of Europe had shifted to a position of assistance; as Truman's Secretary of State James F. Byrnes stated late in that year:

> We want to assist in European reconstruction because we believe that European prosperity will contribute to world prosperity and world peace. That is not dollar diplomacy. That is not imperialism. That is justice and fair play.[30]

Closely linked to this foreign policy, however, was the obvious—but officially unstated—desire to check Soviet aggression by rebuilding Europe in the image of American capitalism. The Marshall Plan, unveiled in the early summer of 1947, provided the mechanism whereby the United States could undertake such a program without having to work through any international bodies or with any former allies who might thwart the ideological objectives of reconstruction.[31] The steel industry generally supported this development and concurred with its political goals.

To a large extent the European Recovery Program was formulated and implemented by officials from the largest business and financial firms in the United States.[32] Although strong opposition might have existed between this group and the Truman administration over domestic economic policy, its members were essentially united behind the government's pursuit of international objectives. Steel industry support was evident when representatives from the United States Steel Corporation (USS), Armco, Republic, Inland, and Bethlehem—comprising the largest of the integrated producers—either gave testimonial endorsement or actual time and effort to promoting Marshall Plan goals. Most observers perceived the aid program to be a boon to domestic producers:

> Historically Europe has not been a large market for American steel, but the accumulation of demand through the war years, plus the retarded reconstruction of most European countries has left the Continent in a position where only America can ship the large tonnages of steel required to fill pipelines and set the wheels of industry in motion.[33]

Others, however, were not so sure about these prospects.[34] One stated that the ability of the United States to remain the world's leading exporter was problematic at best, since not only Europe, but many other nations (formerly importers of steel) were also building new capacity to meet internal needs—for example, Brazil, Chile, Argentina, India, the Netherlands, South Africa, and Australia.[35] The tight steel market in the United States certainly was not helping American producers to hold on to export customers; by 1947 domestic consumers were exerting political pressures for export curbs. Moreover, the strong worldwide demand was driving up prices in the export market, thus inducing some nations to build new capacity just to serve such outlets.[36] The Truman administration was in somewhat of a quandary over how to deal with these foreign needs because it wanted not only to calm the complaints of domestic consumers, but also to insure that Europe would obtain sufficient steel to meet recovery goals: "Maximum [steel] production," wrote Truman in December 1947, "is essential to curb inflation and to support our foreign aid programs as well as the domestic economy."[37] In response to these conflicting demands, the State Department proceeded to revise downward its estimates of European consumption while Congress authorized the Department of Commerce to initiate an export-control program to take effect in the first quarter of 1948.[38] The effort was successful: Iron and steel exports dropped by 30% that year.[39] This was seen as a positive step by those administration officials who did not wish the industry to become too dependent on export markets lest they destabilize overall domestic economic goals; steel shortages, it must be kept in mind, were perceived as a serious threat to the domestic economy at this time.[40]

But fears of an enlarged domestic export sector were being checked by more than governmental controls. One steel engineer returned from a tour of European mills in 1948 and announced that the reconstruction and utilization of new technological improvements that he had observed would result in continental exports that would soon affect American coastal markets.[41] Thomas W. Lamont, a board member of U.S. Steel, inquired whether foreign steel might again reappear in domestic markets owing to the altered conditions of the postwar economy, but he was assured by Benjamin Fairless, president of USS, that it would be "quite awhile before that takes place again" since not only were there persistent shortages abroad, but foreign steel was selling at a 20% premium over U.S. prices.[42] Yet not all indicators were as optimistic as Fairless might have wished his board to believe. Indeed, by 1949 there was justifiable room for doubt about the future ability of the United States to meet foreign competition in steel. With this country beginning to provide billions of dollars in aid, Europe rebuilt much faster than anyone expected, and her industrial output increased accordingly (see Table 4.3).

Nowhere was this more apparent than in Germany. "From the utter prostration of 1945," remarked the London *Economist,* "the German steel industry has worked its way up in 1949 to a position where it is the fifth largest producer in the world."[43] Crude output totaled 9 million tons in 1949 compared to only 2.5 million tons in 1946.[44] American resistance could be credited for much of this redevelopment in steel, along with the vigorous efforts of the Germans themselves to restore their past economic strength.[45] Not only did the United States supply funds, but it also intervened forcefully in the continuing dissension over how the German steel industry should be reorganized. France and England along with Russia were opposed to reestablishment of the large combines that characterized Germany's prewar industrial structure. France and the Soviets were worried about future security if Germany reemerged with a capability in armaments production, whereas Great Britain was concerned with threats to world export markets that a rebuilt

Table 4.3 International Production of Crude Steel[a]

Year	*United States*	*United Kingdom*	*Germany*[b]	*France*	*Italy*	*Benelux*	*(ECSC)*[c]	*Japan*	*Russia*	*ROW*[d]	*World Total*
1936	53.5	13.2	20.7	7.4	2.2	5.7	—	5.5	18.0	12.0	138.2
1939	52.8	14.8	29.6	9.4	2.5	5.2	—	7.4	20.7	9.8	152.2
1945	79.7	13.2	1.7	1.8	0.4	1.2	—	2.3	13.8	11.1	125.2
1946	66.6	14.2	3.3	4.9	1.3	3.9	—	0.6	14.3	12.5	121.6
1947	84.9	14.3	4.4	6.3	1.9	5.1	—	1.0	15.7	15.0	148.6
1948	88.6	16.7	7.9	8.0	2.3	7.0	—	1.9	20.5	15.3	168.2
1949	78.0	17.4	10.1	10.1	2.3	6.7	—	3.4	25.6	15.7	169.3
1950	96.8	18.3	13.4	9.5	2.6	6.9	—	5.3	29.8	24.5	207.1
1951	105.2	17.5	14.9	10.8	3.4	9.5	—	7.2	34.5	27.8	230.8
1952	93.2	18.4	17.4	12.0	3.9	9.6	—	7.7	38.6	31.9	232.7
1953	111.6	19.7	17.0	11.0	3.8	8.7	(43.6)	8.5	41.8	36.2	258.3
1954	88.3	20.7	19.2	11.7	4.6	9.6	(48.3)	8.5	45.0	38.1	245.7
1955	117.0	22.2	23.5	13.9	5.9	11.2	(58.2)	10.4	50.3	42.8	297.2
1956	115.2	23.1	25.6	14.8	6.5	12.0	(62.8)	12.2	52.9	48.5	310.8
1957	112.7	24.3	27.0	15.5	7.4	12.1	(66.1)	13.8	56.2	52.7	321.7
1958	85.3	21.9	25.1	16.1	6.9	11.9	(64.1)	13.0	60.9	57.8	298.9
1959	93.4	22.6	28.5	16.8	7.5	13.0	(69.8)	18.3	65.8	71.3	337.2
1960	99.3	27.2	37.6	19.1	9.1	14.7	(80.5)	24.4	72.1	78.1	381.6

[a]All figures 000 000 net tons.

[b]After 1945, Germany is West Germany.

[c]European Coal and Steel Community; these figures represent the total for West Germany, France, Italy, Belgium, Netherlands, and Luxembourg (as well as the Saar before it was absorbed into West Germany).

[d]ROW: rest of world.

Source: AISI, *Annual Statistical Report* (various years).

Germany would pose. As a result, there were persistent calls to dismantle Germany's steel mills, establish an international authority to control all Ruhr industry, and break up the vertically integrated iron and steel firms into small and independent units of production.[46]

The United States generally opposed these plans, as did the Germans themselves, though for somewhat different reasons. Revitalization of the Ruhr was seen as the heart of the Marshall Plan, and an increase in coal and steel output in the Bizonal Area (that portion of Germany under U.S. and British mandate after 1945) had the highest priority from American advisors.[47] As one indication of this intent, U.S. officials placed the individual who had been director of the ISC prior to the war in supervisory control of all Bizonal iron and steel operations.[48] Stating that British calls for further deconcentration went "too far"—the resulting companies would "never be able to compete on the world market"—U.S. policymakers concluded:

> We have decided upon the advice of the . . . experts who have studied the problem that a certain degree of vertical integration is necessary to put both the Ruhr coal mines and the steel plants into world competitive position.[49]

German industrial dismantling had been under attack by officials in the United States ever since an initial reparations program calling for the elimination of 1500 plants was agreed on by the Allies in early 1946.[50] As the goals of the Marshall Plan came into better focus, this number was reduced to 859 plants, yet even this level was still considered too high by some.[51] Accordingly, in early 1948, Truman appointed a Cabinet Technical Mission to review the entire dismantling situation, since this issue was causing increasing political strains not only with America's Western Allies, but also with Russia, which felt that agreements made earlier at Potsdam were now being forsaken just as agreements had been abandoned after Yalta.[52]

But this cabinet-level group would not reverse the growing sentiment to reindustrialize Germany. It now recommended that only 600 plants be dismantled and that at least 300 critical facilities be saved under any circumstance in order to insure European recovery; steel mills in particular were designated as essential because they represented the keystone to reindustrialization.[53] In consequence, steel plants were designated for further evaluation and review prior to dismantlement. The Economic Cooperation Administration (ECA)—the agency that administered the Marshall Plan—was assigned responsibility to undertake the study, and in November 1948 ECA Administrator Paul G. Hoffman (formerly president of the Studebaker Corporation) approved an Industrial Advisory Committee in steel to make an analysis and submit recommendations. The individual chosen to head the group was George M. Humphrey, president of M. A. Hanna Company, a traditional supplier of coal to American steelmakers and, as such, in close contact with domestic industry opinion. The Industrial Advisory Committee undertook a thorough study of the situation, yet it relied heavily on a previous study made in early 1948 by executives of USS—an investigation that strongly recommended retention of significant steel capacity in western Germany.[54] A total of 381 plants were reviewed in detail by the Humphrey group and ultimately 167 were recommended for retention; the committee's rationale for its findings was stated as follows:

> Although ours was primarily a business approach, we considered the problem with a view to laying the foundations for a lasting peace. We were guided by the desire of the American citizen to have all participating nations, including western Germany, reconstitute sound economies of their own at the earliest possible time, in order first to minimize and then terminate the necessity for continued American assistance.[55]

American intervention was thus an important factor in the maintenance of a sizable German iron and steel industry in the postwar years. Although this policy officially emanated from governmental agencies, it is clear that the leaders of the largest American steel firms had a major voice in shaping these recommendations; by no means was a revitalized German steel industry imposed on the domestic producers against their will.

What is not so clear is why these firms consented to assist in the reconstruction of their foreign rivals or what concessions they were able to obtain, if any, for their cooperation. But first it should be noted that American efforts were not completely successful in the prevention of decartelization in German industry. The large Vereinigte Stahlwerke (United Steelworks Union), established by merger in 1928 and producer of 40% of Germany's prewar crude steel output, was ordered broken up; the Krupp family, the second largest prewar crude steel interest, was deprived of much of its former iron and steel properties.[56] Indeed, decartelization of iron and steel was one of the more symbolic goals of those who would impose American-style antitrust statutes on postwar Germany; at least in theory these forces succeeded in gaining their ends. The twelve large combines that controlled the prewar industry were separated into twenty-eight different firms, with fabrication facilities placed apart from crude steelmaking; moreover, none of the new companies was allowed to control more than 75% of its own coal requirements (as coal ownership was thought to have been the prime variable in the earlier power of the cartel).[57]

Although proponents of reduced firm size and vertical disintegration were temporarily victorious, limits on total industry output were overturned. The level of steel production imposed by the Allies, initially set at 5.8 million tons of raw steel per year (to be derived from a maximum of 7.5 million tons of capacity), was raised to 10.7 million tons of output at the urging of American officials.[58] Even this latter figure proved flexible because steel was vital to European reconstruction, and Germany's natural factor endowments had always made her the most prolific producer on the continent.[59] Yet there were still some disappointments to those who favored rapid reindustrialization.

The French were able to prevail in the establishment of an international authority to oversee Ruhr production, a move to which the Germans were bitterly opposed.[60] On the other hand, English calls to nationalize German iron and steel interests (as was occurring in Great Britain) were rejected. American influence was able to persuade the Western Allies to allow a future German government the freedom to determine for itself questions of ownership as well as industrial organization within its national boundaries.[61] After the Western Zones were granted authority to become the Federal Republic of Germany in 1949, the new government chose to repatriate many of the mills to their former owners and to allow the return of former executives to their prior jobs in the industry. Within a few years the earlier pattern of vertically integrated firms began to reappear as seven large combines emerged to dominate the industry.[62] Not surprisingly, necessity demanded that economic efficiency prevail over political ideology. Only a few years previous, in fact, it would have been considered ludicrous to even suggest that the steel masters who supplied Nazi Germany would soon be back in control. Yet they were—a turn of events owing to the influence of their colleagues in American iron and steel firms as much as to anyone else.

III

With the outlines of Germany's long-term steelmaking posture finally beginning to take shape, it became somewhat easier for the Allies to plan for European reconstruction as a whole. This task continued to fall within the purview of the ECA, which worked closely with the Organization for European Economic Cooperation (OEEC), a supranational structure set up in 1948 by the sixteen nations receiving aid from the U.S. Economic Recovery Program.[63] In March 1949 the ECA issued a prospectus on the rehabilitation program for European iron and steel through the fiscal year 1952–53. The report, which took into account that rapid expansion would create problems, stated:

> It is estimated that the combined program will provide a finished steel production of about 44.3 million tons within the participating countries in 1952–53 as compared with the 1935–38 average of 31.5 million tons. In this estimate it is assumed that the Bizone [Germany] will reach a level of 7.7 million tons by 1952–53, which compares with 1935–38 average production of 10.4 million tons.
>
> . . . Gross exports of finished steel from participating countries are estimated at 10 million tons for 1952–53, while gross imports are estimated at about 4 million tons. The net exportable surplus to non-participating countries is estimated at 6.6 million tons as compared with actual prewar exports of about 3 million tons a year.
>
> The total investment to be made in production facilities for iron and steel as submitted to the Steel Committee of OEEC is estimated at the equivalent of about $3.1 billion. . . .

> It is believed that the program as submitted to OEEC, if carried to completion, will result in a substantial excess capacity in the steel industry of Western Europe especially in the case of flat products where the major modernization and expansion effort is concentrated. The rationalization of these programs seems to be a problem of first priority for the Steel Committee of OEEC.[64]

These plans for a resumption of European exports would, of course, excite comment in the United States. Eugene G. Grace, head of Bethlehem Steel, expressed worries that American foreign aid would eventually result in cheap imports into the United States and layoffs of American workers; "We are industrializing the whole world," he said, "[and] deindustrializing the United States."[65] Ernest T. Weir of National Steel said it was now time "to begin scaling down financial aid, with a view to closing the purse strings entirely by 1952 as scheduled."[66] The OEEC, on the other hand, requested a seven-to-eight-year extension of the iron and steel aid program to allow participating countries time for more orderly development and a better rationalization of their industrial structure.[67]

The implications of these requests raised some concern. "Rationalization," after all, had been a code word for the restructuring and subsequent cartelization of the German iron and steel industry during the late 1920s and early 1930s.[68] With the industry seemingly in recovery and, indeed, with all of Europe in the middle of an apparent glut of new mill construction that would lead to overcapacity, the odds seemed favorable that there would soon be a sharp increase in competitive pressures. This fact was readily observable as early as 1949. In April of that year ECA Administrator Paul G. Hoffman remarked that "European production is increasing. European exports of steel will offer heavy competition in all parts of the world and will help in cutting U.S. aid by building up countries abroad."[69] But only several weeks prior to this, his agency had issued its iron and steel prospectus (mentioned earlier) in which it warned continental producers that they "may find themselves priced out of the export market unless appropriate measures are taken."[70] The question for many, no doubt, was what these "appropriate measures" might entail.

Was this a signal that the Europeans desired resumption of a steel export cartel? English producers were convinced that Germany perceived the OEEC as a precursor to a new cartel—grounds for worry, they felt, as the new Labour government seemed to be pulling back from participation in the move toward economic cooperation underway on the continent (and, hence, might not allow English membership in such arrangements).[71] There were also continuing French–German tensions over the latter's economic revitalization; French fears about Germany's industrial strength remained high, as did France's desire for Ruhr coal, and thus any ideas about revival of the steel cartel would have to undergo careful and even delicate consideration before agreements could be reached. In the United States, meanwhile, the red flag of recartelization was still being waved about by some as grounds for official investigations and inquiries.[72] Indeed, in a 1950 congressional probe of alleged monopolistic practices in the U.S. steel industry, a direct accusation was made by one witness, James S. Martin (a former member of the Decartelization Branch of General Clay's military government in Germany), that a new cartel had already been formed in Europe. Moreover, it was stated by Martin, the American firms that had previously joined the ISC "are now participating in the establishment of the same kind of unofficial economic foreign policy for the United States."[73] The reason for this contemptible behavior, he claimed, was that the participating firms

> were not interested so much in expanding American export sales as they were in limiting imports into the United States, so that the American steel industry internally could be better organized and protected against foreign competition.[74]

No evidence of a smoking gun variety could be produced to support these charges, and representatives of U.S. Steel (chief target of the attack) vehemently denied all such allegations before members of the congressional investigating committee.[75]

Yet something seemed afloat in Europe at this time. The overcapacity situation that was rapidly developing on the continent was having the expected results; one source noted that the Europeans "seem to be lost without a cartel. They are just floundering around trying to sell the tonnage that their nationalistic spirit determines they should produce."[76] By early 1950, export prices were falling drastically throughout the world because of excess output in Europe.[77] It is difficult to believe that American steel interests would not be involved in, let alone unaware of, any developments designed to rationalize such behavior, though this apparently was the case.[78] Nevertheless, the undercurrents of rumor suddenly emerged into sharper outline in May 1950 when Robert Schuman, the French minister of foreign affairs, dramatically announced to a shocked world that

> the French Government proposes that the entire French–German production of coal and steel be placed under a common High Authority, in an organization open to the participation of the other countries of Europe.[79]

What this would soon lead to was the formation of the European Coal and Steel Community (ECSC), a bold and innovative organization designed to achieve the fundamental goal of European unity that hitherto had seemed impossible to bring about, that is, structural integration of Germany into the mainstream of Western European economic activity.[80] Schuman's proposed plan would consist of only six European nations: besides France and Germany, the two principals, there would be Belgium, Holland, Luxembourg, and Italy; as its steel producers had earlier predicted, England would not choose admission.[81] The treaty formalizing the ECSC was not signed until

April 1951, and the group did not actually begin coordinated operations until February 1953. Yet these were relatively minor delays, for the organization was truly a breakthrough of historic proportions. Indeed, the ECSC would serve as the mainspring to formation of the European Economic Community (EEC) in 1957, the Common Market that would (hoped its backers) lead to an eventual United States of Europe capable of significantly altering the prevailing balance of power in world political and economic affairs.

Aside from these grander implications, however, there were also other considerations for the involved parties concerning the mutual self-interest of the steelmakers in both Europe and America. Although the latter voiced public indifference toward the Schuman Plan when it was first unveiled—"The production facilities and resources are there, and it won't make much difference to us in the long run . . . whether they are operated as a pool or separately," said one[82]—there was probably a much more profound reaction at work under the surface. Were the Europeans able to coordinate their involvement in world steel markets, the repercussions for American producers (and domestic American markets) would undoubtedly be significant.

IV

Thus far we have centered the discussion of international steel on European reconstruction. Another area of concern that needs mention is non-European international development in the industry. We have already noted in passing that many nations that previously were importers of iron and steel products began to construct their own production facilities after the war. In some cases this was simply an extension of programs underway since the 1920s and 1930s. In others, it reflected new fears that steel self-sufficiency would be a necessary condition of the tense postwar world in which hostilities might break out at any time.[83] Whatever the motivation, both world capacity and world production grew rapidly after 1945 (see Table 4.3).

In one nation in particular, however, American steelmakers did not show too much interest toward this potential threat to traditional export markets. This was Japan. Indeed, from 1945 through the late 1950s, the domestic industry paid scant attention to Japanese recovery, especially compared with the intensity of American surveillance of European developments in iron and steel. Much of this, of course, can be explained by history: Japan had never been a critical factor in world iron and steel before, and it certainly did not appear ready to assume such a role after the destruction imposed by World War II.[84] Moreover, in the past the Japanese had self-consciously insulated much of their industry from foreign entanglements.[85] As the U.S. Tariff Commission noted in its 1946 survey of future trends in the international steel trade, "producers in . . . Japan were not members of any of the [prewar] cartels or syndicates."[86] Finally, when the West did give attention to emerging postwar steel-producing nations, the primary focus was on the Soviet Union, whose influence in world markets appeared far more portentous than that of any other region.[87] For all of these reasons, there were few indications that would lead one to predict that in less than twenty years the Japanese would become the third-leading manufacturer of steel in the world.[88]

This is not to say, however, that Japan's industry escaped all notice in the United States. Immediately after surrender there was confusion in Allied policy toward the defeated empire. Similar to the debate surrounding the German question, some American officials advocated a heavy dose of deindustrialization and reparations as a form of punishment for Japanese war crimes as well as insurance against any future Japanese aggression. An internationally composed Far Eastern Commission was established in Washington in 1945 to oversee postwar control of Japan, and Edwin J. Pauley (a California oil entrepreneur with a background in political fund-raising for the Democratic party) was named as the U.S. reparations commissioner to the group.[89] Pauley recommended that three-fourths of Japan's iron and steel machinery be dismantled for reparations and that the industry be limited to only 2.25 million tons of raw steel output per year (compared to a wartime peak of 7.65 million tons of output in 1943).[90]

But pastoralization was not to become policy in Japan any more than it had in Germany. The Potsdam Conference had called for "eventual Japanese participation in world trade relations."[91] With the rapid escalation of Cold War tensions, however, it became apparent that an industrially robust Japan under Western influence could provide a major bulwark against further Communist expansion in Asia.[92] By the end of 1948 those attitudes became policy and with it Japanese economic development was given top priority.[93] But for Japan's iron and steelmaking interests there appeared to be formidable problems to the implementation of this strategy. Most of the rapid growth enjoyed by this sector during the 1930s was the direct result of Japan's military buildup; few peacetime markets seemed capable of exerting comparable demand. Moreover, the country was dependent on extremely heavy importation of raw materials in order to produce steel. Prior raw material sources located in Manchuria and Korea were cut short after the war, whereas vital scrap supplies were often earmarked for diversion to the United States to meet the latter's own pressing need for this commodity. Thus even though most of Japan's steelmaking capacity had been left unscathed from the war (82% of open-hearth melting capacity and 87% of rolling capacity survived), the outlook for the future was not bright. Only 155,000 tons of steel were produced in the last quarter of 1945 and 557,000 tons in all of 1946.[94]

Nevertheless, the effort to rebuild was begun. Occupied Japan was effectively under the control of General Douglas MacArthur, Supreme Commander, Allied Powers (SCAP), who was an ardent supporter of the program to restore productive capacity to prewar levels.[95] An intricate system of cooperation between the government and industry soon emerged to provide reconstruction and development assistance. In October 1946 a Steel Technology Committee issued a seminal report: It outlined a future program of large integrated steelworks that would utilize the most modern and efficient technology available and that would be located on sites best able to take advantage of transportation-cost savings.[96] The powerful Economic Stabilization Board, a cabinet-level agency, gave the industry the highest priority in achieving its production goals, especially through preferential treatment in raw material quotas.[97] Beginning in 1946 the state-owned Reconstruction Finance Bank began providing assistance, financing more than half the industry's capital requirements through 1948.[98] The United States, meanwhile, provided not only direct funding for industrial rehabilitation, but also oversaw the allocation of scarce raw materials from Western sources (especially iron ore).[99] When SCAP requested technical assistance for the industry in late 1947, engineers and other specialists from USS were sent to offer guidance and advice on plant modernization.[100] In 1950 a new Iron and Steelmaking Technology Committee was established through the joint efforts of the Ministry of International Trade and Industry (MITI), the Iron and Steel Institute of Japan, and the Japan Iron and Steel Federation; the committee was vested with responsibility for insuring that the industry's expansion goals would be obtained in the most efficient manner possible.[101] Given the nation's comparative disadvantage in raw materials, it was important that such planning and coordination take place. Further cooperation between business and the state was evident in August 1950 when Japan adopted its First Rationalization Program in iron and steel; this plan, which was in effect through 1955, was designed to modernize existing facilities in the industry, improve efficiency, and lower costs.[102] It proved highly successful.

American steel interests were well aware of these Japanese developments. Yet they never perceived them as a threat to their own domestic stability. *Iron Age* commented in early 1949:

> Recovery in Japan continues with less interference and controversy than in any other country affected by the war. Under the present program she is not intended to become a factor in the world trade market. But her program, instead, aims at self sufficiency and sufficient increase in steel production to take care of increases in domestic steel demand in the future.[103]

All factors considered, the domestic American industry's attitude toward Japan seemed justified. Certainly there was nothing on the horizon in the late 1940s to indicate any serious future competition from that quarter. By 1950 SCAP ordered the largest Japanese steel firms deconcentrated (in conformance with American-style antitrust laws imposed on Japan in 1947); not long afterward, both American economic aid and Japan's own price subsidies to the industry were terminated.[104] Europe, and not Japan, would remain the focus of American concern in evolving patterns of global competition in steel throughout the 1950s.

Thus, as the new decade began to emerge, many of the international uncertainties confronting American steel producers shortly after the war's end began to resolve themselves. The German question, perhaps the most confusing part of the puzzle, now seemed less opaque, as did patterns of steel redevelopment in other parts of the globe. To be sure, however, not all doubts were erased: The true intentions of the ECSC remained to be seen. Yet, on the whole, it appeared that earlier American fears about the international implications of domestic capacity expansion would begin to dissipate. As such, the industry was poised for growth.

5

The Triumph of Conflict: Truman and Steel, 1949–52

The atmosphere of distrust between the large steel producers and the Truman administration did not improve after the president's electoral triumph in November 1948. As we noted, his January 1949 State of the Union message—with its oblique reference to federally controlled steel mills—only exacerbated tensions between these two groups. Nationalization and creeping socialism became pervasive subjects for discussion at industry gatherings. Amid this situation, prospects for improved relations only worsened. By 1952 they would finally erupt into one of the most bitter business–government confrontations of the postwar era when Truman unsuccessfully attempted to seize the industry during a labor dispute.

The steelmakers' problems with organized labor were often at the root of business–government conflict, since labor had been perhaps the key supporting interest behind Truman's unexpected victory. Even though the president had on a number of occasions voiced irritation with big labor's policies during his first term in office, he consistently sided with the United Steelworkers (USWA) in its confrontations with management after 1948. As one result, the industry experienced rising labor costs throughout the period, while attempts to generate necessary capacity-expansion funds through higher prices almost invariably met with stiff governmental resistance. This inevitably had an adverse impact on the industry's ability to expand and modernize its mills in the most cost-efficient manner possible, which in turn led to deterioration of domestic steel's global competitiveness.

I

The capacity-expansion debate, perhaps the most divisive issue between the steelmakers and the administration during Truman's first term, seemed to dissipate somewhat in the spring of 1949. At that time a mild recession caused demand to ease and prior gray market distribution problems to evaporate.[1] But labor troubles soon emerged to create new distractions for any improved industry–government cooperation. In May the USWA announced it wanted to reopen contract talks in order to secure not only higher wages, but also pension rights for its members. The companies, the union insisted, would have to provide complete funding for this new benefit; labor would make no contribution.[2] These demands, of course, were rejected by the industry. In the "present business decline," declared Benjamin Fairless, president of U.S. Steel, "nothing could be more damaging to the economy" than to grant the union claim.[3] As intransigence on both sides hardened, it appeared that the second national steel walkout since the end of the war might occur by mid-July.

The administration did not accept this prospect lightly. Although the pressure to increase steel output was perhaps not as overriding as it had been at the beginning of the year, the president and his advisors still believed national security might be threatened by a shutdown.[4] But how to mediate the labor situation was another matter. Only days before the scheduled strike date, the Senate had rejected an administration-backed proposal to repeal the antilabor Taft-Hartley Act and replace it with a milder bill.[5] Organized labor, of course, had strongly supported the president in his 1948 victory; indeed, one of Truman's campaign pledges to this group had been an end to Taft-Hartley. Given these political considerations, he could hardly be expected to invoke the law now to halt the impending strike, no matter how critical the situation.

What Truman settled for was appointment of a fact-finding board to investigate the dispute and make nonbinding recommendations.[6] After initial reluctance, the firms agreed to participate in this exercise; the union, meanwhile, had immediately accepted the plan and granted a sixty-day reprieve to the strike. Subsequent hearings before the Steel Industry Board, however, proved long, tedious, and often bitter to all sides. The industry, led by the United States Steel Corporation (USS), refused to accept any of labor's demands. Industry representatives, in fact, challenged the legitimacy of the board itself, portraying it as the portent of a "new social order" that meant only further erosion of traditional economic liberties.[7] They also argued that recent wage increases had already outstripped productivity gains (see Table 5.1).[8] Pensions, moreover, were deemed appropriate only to the "improvident [and] incurable spendthrifts" and were also labeled as "socialistic" in nature. Besides, the firms said, adequate welfare programs were already in place. But the union countered with its own cadre of experts. Labor productivity far outpaced wage gains, they testified; also, steel company profits were going through the roof. And in perhaps the most skillful of their rebuttals, it was pointed out that the executives of the industry enjoyed the prospect of

Table 5.1 Employment Costs and Output per Man-hour: Steel and the Private Nonagricultural Economy

	Employment Cost per Unit of Output		*Employment Cost per Man-hour*		*Output per Man-hour*	
Year	*Steel*	*Pvt. Nonagr. Economy*	*Steel*	*Pvt. Nonagr. Economy*	*Steel*	*Pvt. Nonagr. Economy*
1940[a]	100.0	100.0	100.0	100.0	100.0	100.0
1941	99.0	107.9	110.2	111.4	111.3	103.1
1947	143.3	177.2	169.2	200.7	118.0	113.2
1948	153.2	188.9	182.6	216.2	119.2	114.3
1949	161.4	186.3	194.0	219.9	120.2	117.9
1950	155.6	187.5	209.0	234.3	134.4	124.8
1951	170.7	201.9	230.6	262.4	135.0	129.9
1952	188.3	210.3	254.0	280.1	134.9	133.0
1953	190.8	216.3	268.6	297.0	140.8	137.3
1954	207.0	218.5	282.2	307.0	136.3	140.3
1955	194.5	217.3	303.5	319.6	156.1	146.9
1956	209.7	230.5	328.4	338.4	156.6	146.8
1957	230.9	237.0	358.5	356.5	155.3	150.3
1958	261.2	240.6	387.0	364.9	148.1	151.5
1959	254.9	243.3	421.6	385.6	165.4	158.3
1960	264.7	249.3	419.5	398.9	158.5	159.8

[a]All indexes: 1940 = 100.

Sources: Bureau of Labor Statistics; U.S. Department of Commerce; American Iron and Steel Institute. (Reproduced in U.S. Congress, JEC, 88th Cong., 1st sess., Hearings, *Steel Prices, Unit Costs, Profits, and Foreign Competition* [Washington, DC: GPO, 1963], p. 94.)

huge pension benefits upon retirement. "Where does socialism start?" pointedly inquired Philip Murray, president of the USWA.[9]

On September 10 the fact-finding board submitted its report.[10] It rejected the union's wage-boost arguments but found the company's rejection of benefit demands also to be wanting; it recommended company-funded pensions. The industry was surprised at this turn of events, with *Iron Age* perhaps best expressing the producers' sentiments when it editorialized that the report was "neither as bad as steel people expected nor as good as the union had hoped."[11] It appeared that governmental intervention into the industry's affairs would not result in further compromise of the steelmakers' financial strength vis-à-vis its labor force, as most industry spokespersons had feared.

But this initial bout of satisfaction soon soured. By September 15 Fairless rejected the board's findings. The imposition of pension costs would be too great a burden for the firms to bear, he felt.[12] With negotiations thus at an impasse, and Truman without "any more aces up his sleeve," the strike began on September 30.[13] It lasted a full month, until Bethlehem Steel broke the deadlock and negotiated a new contract along the lines recommended by the Steel Industry Board's report. Within several weeks all of the other producers caved in (with U.S. Steel among the last to sign).[14] By mid-November, the mills were back to normal operations.

The strike did nothing to improve the industry's relations with either the public or the administration. Most press commentary characterized the board's recommendations as eminently fair, which indeed they seemed to be.[15] Soon after their publication, a wide range of other industries whose own labor contracts were also in negotiation settled according to the pension plan outlined by the neutral panel.[16] The steel industry was thus left with only a reinforcement of its negative image as an atavistic throwback to the nineteenth century, stubbornly refusing to recognize the realities of the day regardless of human compassion or national consequence. Not unexpectedly, steel's revived public relations program—recently expanded in scope to meet the demands of the capacity issue—was tarnished in the eyes of those whom it was intended to impress.[17]

One may only surmise the motivation behind U.S. Steel's rejection of the Steel Industry Board's labor bargain.[18] Perhaps the easiest conclusion is that this represented the firm's usual reflex resistance to any incursions of external authority into its "freedom" to do as it pleased. U.S. Steel and indeed the entire industry had never been well disposed to accept organized labor as a legitimate interest whose rights were to be taken seriously. Thus pay raises, pensions, or any other benefits for workers were to be fought with vigor before they were agreed to by management.

But there were obviously other less ideological factors at work that solidified the firm's resolve. For one, despite their own claims to the contrary, it was apparent that postwar steel demand was exhibiting a strength far beyond earlier expectations and that expansion would be necessary; this would not be inexpensive. For another, European steelmakers were rebuilding faster than anyone thought they could (even though many questions about Germany's iron and steel industry still remained unresolved in 1949). Thus, if the United States were to increase its own capacity to meet enlarged domestic demand as well as remain competitive with potential rivals, it would need to do two things: increase cash flows to fund new investment and hold down wages to minimize cross-national cost disparities. Seen in this perspective, the steelmakers' intransigence appears, if not more palatable, at least more logical. Indeed, the attempt to hold down labor charges—accounting for 35% of domestic steelmaking costs in 1948—might even be observed as a long-term strategic necessity were the industry to remain a viable factor in global markets of the future (including its own).[19] In combination with the industry's traditional antilabor posture, this mounting threat of renewed international competition demanded that rising labor costs be checked. Although U.S. Steel—the most important domestic firm participating in the international market—was unsuccessful in this attempt, it nevertheless seemed to be making an effort to hold the line.

In mid-November 1949, *Iron Age* rhetorically asked: "Will rising labor and raw materials and services costs force a general increase in steel selling prices?" The journal's answer was, "Probably not this year."[20] It was wrong. Exactly one month later, U.S. Steel announced a raise in domestic base prices and so-called extras averaging $4.00 per ton, a raise intended, the producer said, to offset higher operating costs.[21] In addition, the firm reduced export prices—a confirmation of its competitive strategy in the face of increased uncertainty in the international market. But when the rest of the industry followed U.S. Steel's lead and raised their own prices, some members of Congress became alarmed. The Democratic-controlled Joint Economic Committee (JEC), now under the chairmanship of Senator Joseph O'Mahoney (D-Wyoming), correctly observed:

> The steel industry is basic to the American economy. Decisions made by steel executives on production, on expansion of facilities, and on prices have a profound effect on the entire Nation. Indeed, the strategic importance of steel makes it essential that steel management exercise a high degree of statesmanship in its policy judgments.[22]

Questioning that statesmanship, the JEC instituted a formal investigation of the price hike.

Not everyone in government approved of this decision. "The steel industry's primary mistake," said one federal economist, "is that it did not raise prices as much as producers of other basic commodities. If it had done so during the early post-war period, none of this furor would have ever been raised."[23] In both 1946 and 1947 the wholesale price index (WPI) had increased considerably faster than steel prices, whereas in 1948 the situation was reversed (see Table 2.5). But 1949 was a year of recession for the American economy and hence also a period of declining wholesale prices; steel's price hike that year was decidedly out of step with overall economic trends. As a result, the JEC called four days of investigative hearings in late January 1950 in which some six hundred pages of testimony were taken from a variety of witnesses.[24]

In late March 1950 the JEC issued its report. The political nature of the document seemed obvious: The eight Democrats on the committee concurred in the findings, whereas the six Republicans issued a condemnation in the attached minority report.[25] "The conclusion," stated the majority, "is clear: increased steel prices for United States consumption were possible only because competitive conditions in the steel industry were lacking."[26] Four recommendations were then made: (1) the Federal Trade Commission (FTC) should systematically collect data on costs, prices, and profits from the steelmakers; (2) producers should file advance notification of price increases with the government, which would then hold hearings prior to actual implementation of such proposed hikes; (3) a new study should be undertaken by the government to find ways to increase competition in steel; and (4) it should be determined if the steel industry be regulated like a public utility. This last point was most controversial; the JEC report reads here as follows:

> A study should be authorized to examine the extent to which the steel industry has developed technological and economic similarity to public utilities and has acquired such strategic importance in war, peace,

> and in the maintenance of high-level employment as to become uniquely affected with a public interest in order that the Congress may determine what, if any, legislation should be adopted for the preservation of competition.[27]

This last recommendation was justified by the legislators through reference both to remarks made by Judge Gary in 1911 that the industry and government should set prices jointly and to comments in 1938 by Myron C. Taylor, U.S. Steel's then-chairman, that USS was a "national institution" whose affairs "cannot be successfully managed solely and restrictedly as a commercial enterprise."[28] The accuracy of these statements could not be denied—indeed the concept of steel as a utility seemed deserving of thought—but the political context in which they were paraded forth for display served only to draw further attention to the rising tide of discontent between the large steel producers and government.

There obviously was a need for industry and government to undertake closer ties, not only because of domestic capacity-expansion problems, but also because of the changing context of the international steel trade. The industry's ability to meet demands for mill modernization and growth was seriously strained by stagnating financial performance, which made it difficult for steelmakers to raise sufficient amounts of investment capital through traditional channels. Some form of public assistance appeared justified at this time. Yet for members of Congress to suggest that the industry be treated as a public utility seemed to serve no purpose other than to fuel an already politically charged atmosphere. As there was clearly no support for this proposal in the full Congress, the primary intent of its backers served only to increase the companies' bitterness and distrust about the true plans of the incumbent Democrats for their industry's future.[29] Soon after the JEC report was issued, USS sent President Truman a copy of a recent speech made by its president, Benjamin Fairless, noting that he might wish to peruse its contents; it was titled "Guilty Before Trial."[30]

The publication of the JEC report coincided with yet another congressional probe of the industry. Beginning in July 1949 a House Subcommittee on Study of Monopoly Power began hearings into problems of industrial concentration in the United States and possible remedies that might be applied through revisions of the antitrust law; this became known as the Celler Committee in honor of its chairman, Representative Emanuel Celler (D-New York). On April 17, 1950, the committee turned its attention from generalities to specifics; the first industry to be questioned in respect to monopolistic practice was steel. Over the next three and one-half weeks, a wide range of issues in this industry were explored as numerous witnesses were called to testify.[31]

The testimony carried charges that large firms in the steel industry not only collaborated illegally with the International Steel Cartel (ISC) before the war, but that they were also involved in attempts to revive this discredited association in the late 1940s.[32] Officers of USS (it will be recalled from Chapter 4) vigorously denied these allegations (though it appears that their congressional interrogators were not wholly persuaded of their innocence). As we have indicated, there was some reason to believe that a new cartel might be advantageous to the domestic producers in the postwar era. Curiously, these particular hearings did not lead to further investigation of this subject.[33]

Instead, the congressional committee was most concerned with firm size. There was little doubt that steel firms were inordinately big; of the fifty largest American manufacturing firms operating in 1948, seven were integrated iron and steelmakers.[34] When U.S. Steel had announced its price hike the previous December, Representative Celler termed it an act of "social irresponsibility," and further noted that this "heedless action boldly points the need for a re-examination of the prerogatives of bigness."[35] Celler's hearings were designed to explore this question in detail. The industry, in fact, was convinced that Celler had already reached conclusions of his own on the matter—namely, that the producers, especially U.S. Steel, should be deconcentrated under the powers of the antitrust laws. A number of witnesses gave testimony to this effect, particularly academic economists called in by Celler to present expert opinion. For example, George Stigler, then at Columbia University, stated:

> The real question is, Do we have enough competition in the industry to dispense with the necessity for further social controls? And my answer is that the forces of competition in the steel industry are not now sufficiently strong to justify us to leave the industry alone.[36]

The industry was properly fearful of the consequences of this challenge. Far more than the Temporary National Economic Committee (TNEC) investigations of the late 1930s, the present inquiry seemed liable to accomplish what Justice Department lawyers could not do in their celebrated 1911 case against the Steel Trust. In response, U.S. Steel mounted a vigorous counterattack before the Celler Committee. President Benjamin Fairless and a phalanx of his top associates spent three full days responding to heated questioning, presenting their own rationale for the firm's extraordinary size. Fairless began:

> United States Steel Corporation is successful; it is profitable; it is efficient, and it is a large enterprise. These are the simple facts and I am proud of them.
>
> United States Steel is large because it has had to be. . . .
>
> But within our government today there are powerful agencies and groups which hold that all this is wrong. They hold that there is something inherently vicious in bigness, and growth, and success. . . .
>
> To my way of thinking, the advocates of these theories are the most dangerous reactionaries of this twentieth century. By dismembering business, they would turn back the clock to the "horseless buggy" days of 50 years or more ago, and would try to squeeze a modern,

> dynamic, efficient America once more into the puny production patterns of its industrial childhood. Or, by subjecting American productive enterprise to the deadening hand of political regulation, they would borrow from the Old World the dismal economic philosophies that have led most of Europe to desolation and despair. They would substitute governmental regulation for competition and political pressures for customer control.
>
> With the unsound and dangerous notions of such reactionaries, I emphatically disagree.[37]

What followed was a spirited defense that surprised many observers given the corporation's rather unimpressive showing in prior political public forums.[38] Company officials went to some length to demonstrate that they were not a monopoly: Detailed charts and exhibits were presented; more outside experts were called in; and representatives from other steel firms were summoned to provide testimony.

In the end the industry was victorious, as no Justice Department dissolution suit was filled.[39] Yet it appears that perhaps additional factors had influenced governmental attitudes besides mere tactics of aggressive counterattack. What seems probable is that the same critical forces that had saved the corporation from judicial challenge in both 1915 and in the late 1930s were again at work. This, of course, involved national security threats posed by war. It was probably deemed by public officials that potential temporary supply cutbacks stemming from industrial reorganization—were the steel industry broken up—would be too great a risk to endure given existing international political conditions (the Korean War broke out on June 25, 1950, while the congressional committee was considering what action to take based on the hearings).[40] This same logic, it might be noted, would soon terminate the FTC's attempt to stop American participation in the international petroleum cartel.[41]

Although Celler eventually was able to persuade Congress to make some adjustments to the Clayton Act, he could not generate sufficient enthusiasm for the imposition of major structural changes in the steel industry. In many respects his congressional hearings mark a turning point in the industry's dismal record of postwar episodes with government. The combination of a more aggressive posture by industry executives and a rapidly changing international political environment would create public attitudes that were far less hostile to big business. Steel in particular was able to parlay this change into tangible advantage. Suspicious questions concerning an international cartel soon slipped from prominence, as did general public uneasiness over industry price changes, alleged monopolistic practices, and charges of irresponsible corporate power (though some members of Congress, to be sure, maintained their doubts on these issues).[42] In the November 1950 by-elections, Republicans picked up twenty-eight additional seats in the House and five in the Senate, substantially reducing what had been decisive Democratic margins in both bodies. Yet just when progress toward more cooperative business–government relations in steel could and should have been made, the two sides were pulling further apart. Within eighteen months the industry would score a dramatic victory over Truman in one of the most publicized business–government clashes of the era.

II

Before turning to these events, however, it is necessary to take note of another industry development of importance: the capacity-expansion issue. Among the witnesses before the Celler Committee was Louis H. Bean, the Department of Agriculture economist so prominent in the capacity debates of 1947. He was again called to testify on this subject.

> It is probably sufficient . . . to suggest that if by 1960 per capita consumption should continue to expand to, say, three-fourths of a ton per person, and if total population should increase at the rate of only 1 percent per year, say to 165,000,000, we would need a total production of around 125,000,000 tons per year and a capacity to provide the necessary "elbow room" of at least 135,000,000 tons. . . .
>
> MR. LEVI: Then, you suggest that in this expansion there would seem to be ample room for new steel companies. How many are you assuming there might be room for?
>
> MR. BEAN: That, again, is a matter of assumptions you want to lay down. If you want new firms of a capacity of a million tons, then obviously there would be room for 20 or 25 additional companies [given Bean's estimate of a shortfall of some 25 million tons of capacity by 1960].[43]

By January 1950 annual ingot steel capacity had grown to over 99 million tons. This achievement would seem to have fulfilled earlier projections set by the industry's critics (calling for approximately 100 million tons by 1950). Yet continuing high demand, exacerbated by military needs stemming from the outbreak of war in Korea, now rendered this earlier target inoperative. As shortages developed in some product lines, governmental planners as well as elected officials began to worry once more about capacity. The House Small Business Committee began an investigation into general market conditions in the industry.[44] In response, steel firms began announcing new plans for more expansion; in June 1950 it was stated that 4 million tons would be added, then 6 million, and by October over 9 million tons of new construction had been scheduled.[45] In a news release issued on October 2, Secretary of Commerce Charles Sawyer applauded the industry's efforts, calling them "an encouraging indication of the willingness of industry to forge ahead," and he especially noted the many difficulties that such rapid growth would have to overcome. Only several weeks later, however, the head of the National Security Resources Board (NSRB), W. Stuart Symington, said, "There now appears to be a national shortage of steel," and this was more than just "temporary." It was also implied that an industry goal of perhaps 130 million tons of annual capacity now seemed appropriate.[46]

The industry was thus again involved in expansion controversies. This time, however, government officials would provide more than just verbal encouragement (or threats) to promote growth. The outbreak of hostilities in Korea was a decided factor in this change. It prompted a temporary run on consumer goods as the public feared commodity shortages similar to what had occurred in World War II; this behavior naturally tended to drive up prices. But the administration was reluctant to declare that a full mobilization was required to meet the Korean situation, and it was skeptical about the imposition of any direct economic controls to mitigate the short-run inflationary impact. As a consequence, Truman decided to combat the problem through a form of supply-side economics. The Revenue Act of 1950 and the Defense Production Act of 1950, both passed in the early fall, were designed to stimulate supply and thus reduce hoarding and price inflation.[47] Contained in the measures was language permitting accelerated depreciation for new investment in plant and equipment related to defense needs. The steelmakers were generally pleased, of course, as for many years they had advocated revisions to the tax depreciation laws as one means to help them fund new capacity expansion.[48] Truman designated Symington's NSRB to administer the new program; similar to World War II practice, the agency would provide certificates of necessity for approved projects, allowing a five-year write-off of qualifying investments rather than the twenty-year term under existing tax law. In addition, the Reconstruction Finance Corporation (RFC), a federal lending agency, announced it would provide low-cost funding for any holders of these certificates.[49] These new policies were enthusiastically accepted by the steelmakers even though they did not precisely meet all their hopes (industry executives had previously called for depreciation deductions based on replacement costs rather than traditional original-cost practices).[50]

The immediate effect of these new certificates of necessity and associated RFC loans was a further increase in the projections of capacity expansion. Industry sources soon began to talk about 120 million tons or more to be available by the end of 1952.[51] There was a new and interesting aspect to this development, however. Many of the certificates were awarded to entrepreneurs who promised to build independent new mills in locales never-before considered by the industry. Thus a group from Comanche, Iowa, petitioned for $100–$150 million to start up the North American Steel Company; Howard Baker, Jr. (son of a Congressman and later to achieve political standing himself), asked for $10 million for Tennessee Steel to be built in Verdun, Tennessee, apparently an ideal industrial site in that state. Yet another group desired to erect a million-ton integrated steelworks in the deserts of California, alleging that they had access to a secret source of iron ore in the area that so far had escaped detection by both industry and governmental geologists.[52]

But not all the plans of these entrepreneurs were so fanciful. The Lone Star Steel Company received a $75 million government loan that allowed the firm to establish integrated operations in Dallas, Texas, and a major new mill was approved for both certification and financial subsidy near New London, Connecticut.

This latter project—to cost $250 million and to have a capacity of approximately 1 million tons of flat rolled product per year—was to be the first integrated mill in the New England region; its construction was regarded as a serious competitive threat by existing firms in the industry.[53] By the end of January 1951 a total of 15.7 million tons of ingot capacity had been granted certificates of necessity by the NSRB and applications for another 8 million tons were pending; the construction costs for approved projects alone totaled over $1.4 billion.[54]

The government thus appeared to obtain both of its goals for the steel industry: more capacity and more competition. This development would obviously have a significant impact on the competitive strategies of existing producers. With the announcement of new entrants springing up all around them (as it must have seemed), they would have to act fast if they, too, were to receive a share of the certificates and thus maintain the existing competitive balance in the industry.[55] U.S. Steel actually got a head start in this when in May 1950 it confirmed long-circulating rumors that it would build a large new integrated mill on the Delaware River near Trenton, New Jersey.[56] This would mark the corporation's first major assault into the eastern region then dominated by Bethlehem Steel (which had a large mill at Sparrows Point near Baltimore). The move was motivated by a number of factors, however. Postwar inflation had driven up transportation costs to the level that the Sparrows Point mill enjoyed a $4.00–$5.00 per ton advantage in major East Coast markets over steel shipped from Pittsburgh; even if the basing-point system were to be revived (and Truman did not veto the new legislation aimed at renewing this system until June 1950), the cost difference was still too much for Pittsburgh-based suppliers to absorb given the thin margins on which finished steel was now sold.[57] Consequently, U.S. Steel decided to move directly into the east to protect its ties to customers there and to utilize newly discovered Venezuelan ore holdings to supply the projected Fairless Works (as the plant was to be called, in honor of the firm's president). Moreover, observers noted that this facility, located on a waterway with easy access to the Atlantic, would also allow the company to supply booming West Coast markets. This was because the relative lack of natural resources for steelmaking on the West Coast required that any mill constructed there absorb transportation costs necessary to haul in raw materials; on a comparative cost basis, U.S. Steel could ship the finished product from the East Coast and still remain competitive.[58]

The decision of U.S. Steel concerning this new facility triggered a reaction that would surprise those who did not believe the industry to be competitively structured. Although firm behavior obviously did not resemble perfect competition, major industry entrants were nevertheless unwilling to allow the leader to capture presumed new growth markets unchallenged. Consequently, Bethlehem Steel announced a $30-million expansion of its Sparrows Point mill, adding some 900,000 tons of capacity and National Steel soon followed with plans to penetrate the east through a new 1.75-million-ton plant to be located only 35 miles from U.S. Steel's proposed site for the Fairless Works.[59] With expansion by other firms planned not only in the mid-Atlantic area but elsewhere too, the industry's competitive pattern was undergoing considerable alteration from its traditional configuration.[60] Indeed, the steelmakers were embarking upon the greatest growth spurt in their history.

As the industry moved aggressively to increase capacity, leaders of the largest firms—particularly U.S. Steel—began to conduct themselves more forcefully in dealing with congressional investigators. This new stance was clearly evident at the annual AISI convention in May 1950:

> There is not the slightest doubt as to the ultimate objectives of those who are leading the attack [against the industry]. They give devout and impassioned lip service to free enterprise. But in the same breath they propose measures which would lead inevitably to nationalization and socialization of the steel industry.
>
> We cannot turn back this attack with defensive measures alone. Defense cannot win the war of propaganda and public opinion any more than it won the war of bullets and bombs. The only way we can frustrate our attackers is by an even more vigorous counter-attack. If we wait until a bill to nationalize the steel industry is actually introduced in Congress, it will be too late. . . .
>
> The thing this industry needs more than anything else is men who know how to talk, men who can dramatize the issues in this fight and take them to the public. We know how to make steel and we usually know how to sell it. But our whole future depends on our ability to demonstrate to the public that we know how to run this industry under private management far better than the government could possibly run it. . . .
>
> We are not going to win this fight by moderate, pacifist tactics. We're going to have to start slugging. And the minute we do, I am sure we will find more and more of the American people in our corner cheering us on.[61]

This vigorous antistate rhetoric would soon be put to the test, for the exigencies of war were forcing the establishment of stricter public controls over prices and wages than had previously been thought necessary. Accordingly, the steelmakers and the administration found themselves on a collision course within the oft-contested arena of private managerial prerogative and public duty.

III

When the Korean War broke out in June 1950, the general assumption was that it would be a relatively short-lived conflict. This belief had contributed to the administration's decision not to seek direct domestic economic controls through the revenue and production acts passed in September 1950. At first, American and Allied forces seemed to be able to make easy progress against their North Korean opposition; but in October a fateful decision to cross the 38th parallel brought Communist China into the conflict on the side of North Korea. When the Chinese surged into battle during November, it soon became apparent that a much longer war was inevitable. As one result, panic buying developed and prices again shot up; for another, vast quantities of basic commodities would obviously now be required to meet enlarged defense demands.[62] To mitigate the inflationary impact this would have on the economy, Truman on December 15 imposed direct wage and price controls over a variety of critical materials and products related to military need (a power he had already been granted standby authority to use).[63] In addition, the president reorganized the recently established temporary stabilization agencies to better meet the new challenge. An Office of Defense Mobilization (ODM) was created to take charge of all aspects of the emergency: The existing Economic Stabilization Agency (ESA)—with its twin Wage Stabilization Board (WSB) and Office of Price Stabilization (OPS)—would now report to the ODM rather than to the president as before. Following creation of yet another agency, the Defense Production Administration (DPA) in January 1951, the overall federal bureaucracy for dealing with the Korean emergency at the beginning of 1951 is shown in Figure 5.1.[64]

The steelmakers, meanwhile, had granted their labor force a 10% pay adjustment in December 1950 before the new controls took effect; moreover, they also hiked prices that same month, adding 7.7% to the composite price level of finished steel. Although some concern with this was expressed in Washington, none of the congressional challenges that had accompanied the prior year's price change were heard (no doubt because of worries over the military situation abroad).[65] 1950 had been a record year for the industry: Aggregate revenue, profits, production, and shipments hit all-time highs. Korean War demand had clearly injected a "shot in the arm" to the domestic steelmakers' performance.[66] This also carried over to foreign producers; U.S.

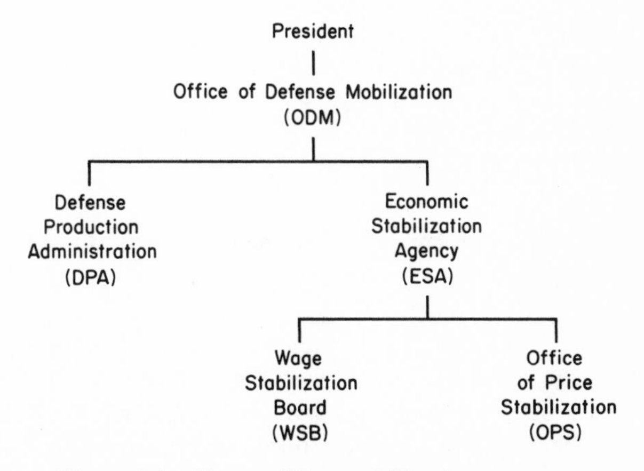

Figure 5.1 Korean War mobilization structure.

imports of steel mill products jumped over 300% from 1949 levels, and the Japanese steel industry in particular was able to profit from the upsurge of defense orders from America.[67]

Demand continued strong through 1951 as the war intensified. Indeed, annual ingot output for the first time in history exceeded 100 million tons, and capacity utilization actually was above 100% for the year as a whole. Revenues again set a new record, but industry aggregate net income declined by 11% as steelmaking costs (especially raw materials) continued to rise despite federal controls (see Table 2.6).[68] As the year progressed, however, the USWA began to campaign for another large pay hike; the union's contract would expire on December 31, and the workers appeared determined to stay ahead of the war-induced price spiral.[69]

By October 1951 the White House began to express concern over the impending steel negotiations. John R. Steelman, assistant to the president and for many years chief White House spokesperson for labor issues, warned the president of "signs of hysteria" on this matter; "the burning question," he continued, "seems to be what the Government is going to do on wages and prices."[70] There had been increasing dissatisfaction on the part of organized labor with the controls implemented the previous December. In fact, many segments of the nation were beginning to resent these economic impositions. A formal declaration of war had never been made in the Korean conflict, and throughout 1951 Truman's personal popularity was sinking as a progressively restless citizenry was voicing its disappointment with the apparent military stalemate.[71] The president's standing was complicated by other factors as well: Senator Joseph McCarthy (R-Wisconsin) was finding "communists" in all branches of government; numerous charges of corruption and scandal were being leveled at the administration (and in some cases proven); and in general the electorate seemed to be tiring of twenty years of Democratic rule.[72] Congress, too, began to reflect the mood of frustration as it failed to legislate much of Truman's Fair Deal program. Political uncertainties generated by the prospect of a presidential election in the following year seemed only to add to these tensions.[73]

Amid this backdrop of deteriorating domestic and foreign conditions, Truman approached the steel problem. His primary concern was that the mills continue to run full-blast in order to supply the war effort.[74] After a visit from USWA officials in mid-October, he seemed assured of this objective: "They want no work stoppage," he informed Steelman, and "they [are] ready to bargain and . . . want to do the right thing. I believe them."[75] Yet what the "right thing" translated to in terms of a labor contract was (in the words of the trade press) "the most formidable [package] ever tossed at the steel industry": a union shop, numerous improvements in benefits, a guaranteed annual wage, and a "substantial" pay raise.[76]

Although the union's initial demands were no doubt padded to provide negotiating leverage, there was nevertheless a widespread belief that the workers would make a tough stand before agreeing to any compromises; President Fairless of U.S. Steel certainly seemed to think so, commenting that the bargaining "will probably have to be decided . . . in Washington."[77] On the very day of this remark, the Council of Economic Advisors (CEA), in fact, submitted a memorandum to the president expressing that agency's concern "that the forthcoming steel negotiations may again, as last year, result in excessive wage and price increases, and ring the bell for another round of inflation."[78] Consequently, the memo went on to recommend:

> The price line in steel is to be firmly held. There should be no wage settlement of a size necessitating or resulting in general price increases in the steel industry. Current and prospective steel profits are high enough to absorb such wage increases as are desirable and to attract the new capital needed for the steel industry's expansion program.[79]

To back this, the agency's staff economists included a brief analysis purporting to show that the producers could easily absorb a pay hike of fifteen cents for their workers (the expected minimum demand) and perhaps even up to twenty-two cents per hour. Moreover, the CEA soon urged that the administration specifically inject the needs of the national interest into the negotiations before the two sides agreed between themselves that high wages justified high prices and vice versa.[80]

But the president did not follow this advice. Instead, he chose to heed Steelman's counsel that the parties be left alone and that existing mechanisms be allowed an opportunity to operate before the White House voiced concern. But this policy soon foundered when the employers, concerned about the adequacy of investment funds, refused to bargain until they had some assurances of a compensating price hike from the stabilization controllers.[81] As the strike deadline of January 1, 1952, approached, Truman asked for both an extension of the walkout date and utilization of the WSB by the disputants. "It is of the utmost importance to prevent an interruption in the production of steel . . . ," he stated, and "continuous production of this industry is essential in order to meet urgent demand for steel—steel for weapons, for factories, for highways and hospitals and schools."[82] The stay was thus granted. Industry leaders perhaps recollected that in 1946 White House intervention eventually brought desired price increases under somewhat similar conditions of mandatory national economic controls; the union, meanwhile, could look back to federal intervention in 1949 that resulted in the companies' acquiescence to a noncontributory pension plan. Perhaps both sides thus perceived grounds for gain through the White House.

The WSB and its sister agency, the OPS, spent significant time and effort on the steel case (only one of many such mediation problems before them). When conclusions were finally reached, the majority of the tripartite WSB felt them fair and adequate to both sides.[83] They called for a wage increase of about fifteen cents per hour and a compensating $2.00–$3.00 per ton price hike for the industry. The producers, however, rejected this when word of the proposals leaked prematurely; if fifteen cents were to be granted, they countered, then $6.00–$9.00 per ton more would be necessary in price relief.[84]

Nevertheless, on March 20, 1952, the WSB held to its convictions and publicly announced the formal recommendations: an eighteen-month contract to ultimately provide seventeen and one-half cents per hour in raises (phased in over three steps); additional benefits that would bring the total compensation package up to about twenty-six cents per hour over the eighteen-month span; adoption of the union shop clause; and recommendations for further bargaining between the parties on some of the other issues (such as a guaranteed wage).[85] All of these "moderate and just" proposals, stated Nathan Feinsinger, chairman of the WSB, would allow the steelworkers merely to "catch up to increases already granted . . . in major segments of American industry."[86] (It might be noted that at the end of 1951 the gross average hourly earnings of steelworkers were 22% higher than the all-manufacturing industry average).[87] The producers, however, were discomfited with the WSB proposals once they became public and promptly raised their own demands for price relief to $12 a ton if the board's wage figures were to be met.[88]

The USWA immediately accepted the WSB recommendations and set a new strike date of midnight, April 8. Yet at this point the negotiations seemed to be headed for trouble. According to many outside observers, the package appeared strongly in favor of labor—acknowledged as perhaps the key constituency in Truman's 1948 victory and assumed to fulfill an equally important role in the upcoming campaign. On the other hand, the proposals had strong defenders in the White House: The adjustments were within bounds when compared to other sectors, it was argued by Steelman, and the CEA said the employers should easily be able to absorb higher labor costs even if no price hike were granted.[89] Moreover, to give in to the demands of the industry would wreck the entire stabilization program and open the gates to runaway inflation (a theme backed by officials of the stabilization agencies, who perhaps recalled a similar outcome after the 1946 steel–labor capitulation by the administration).[90] While considering these arguments, Truman was also receiving constant input from his Defense Department advisors: They pleaded that under no circumstances should the steel supply lines to Korea be cut, for to do so "would be a national tragedy . . . that we could not possibly stand."[91] This same argument had animated the president's actions ever since the steel talks had begun the previous fall, and he obviously felt no less disposed toward such counsel now.

While the parties were considering their next move, another complication developed. On March 28 Charles E. Wilson, a Republican and president of General Electric Corporation, resigned from his public service post as head of the ODM, the government's top agency for coordinating the domestic response to the Korean emergency. Wilson apparently believed he had Truman's backing to raise the previous OPS offer of a $3.00-per-ton price hike to the steelmakers, and he tried to do this in order to secure industry approval of the WSB recommendations. However, he incautiously remarked to press reporters his belief that the board's wage package was excessive. When this appeared in the media the following day, the unions immediately objected, as did stabilization agency officials. Truman then backed down from his previous commitments to the ODM chief; feeling his authority compromised, Wilson resigned.[92] In the eyes of public opinion, this was viewed as yet another indication of the increasing confusion in the White House; liberals happily accepted the resignation, but others did not, concluding it was another victory for labor provided by an antibusiness administration.[93] The controversy stemming from this incident compelled Truman to publicly demonstrate his nonpartisan desires to keep the mills operating. On March 29 he announced his (privately long-held) intentions not to seek presidential reelection in the fall.[94]

With negotiations getting nowhere and the strike deadline now only days away, the administration began to consider its options—under constant pressure from the Defense Department for no letup in the production of steel.[95] The alternatives bore a resemblance to those facing Truman in his 1946 encounter with these same parties: give in to the companies' price demands and possibly undermine the stabilization program, or seize the mills under one of several apparently legal maneuvers available to him.[96] Fred Vinson, chief justice of the Supreme Court, had privately counseled the president that such a seizure would be upheld by the courts.[97] There was, however, a further choice: invocation of the 1947 Taft-Hartley Act that called for an eighty-day cooling-off period while workers stayed on the job and negotiations continued. Truman, however, would not consider this latter option, just as he refused to do so during the 1949 steel impasse. The act's provisions, he later noted, "didn't serve in arbitration and settlement. The authors didn't want a means of settlement. The objective was destruction of labor unions. Read Hartley's book."[98] Though he publicly defended his refusal to use the measure on less emotional grounds, it is clear that Truman was ideologically opposed to this legislation—which had become law only over his veto—and under no circumstances would it receive his endorsement.

This left only the other alternatives if the mills were to continue to operate. In a final effort to compromise, the head of the OPS secretly offered the companies a $4.50 per ton price hike on April 3, but this was met with a counterproposal asking for $5.50–$6.00. Steelman, who had now replaced Wilson at ODM and obviously was speaking with the authority of the president, said no.[99] Consequently, at White House meetings on Tuesday, April 8, in which Defense Department personnel expressed serious concern with the maintenance of steel production, it was decided to seize the industry.[100] At 10:30 P.M. that evening, President Truman went before a national radio and television audience to announce, "By virtue of the authority vested in me by the Constitution and laws of the United States . . . the Secretary of Commerce is hereby authorized and directed to take possession" of the steel industry.[101] The historic steel seizure of 1952 was underway.

This case has attracted a significant degree of attention over the years. However, most of the commentary and analysis tend to focus on either the Constitutional issues involved in the president's use of the inherent-powers doctrine or the role of the corporations in exerting their own power to dictate regardless of the public interest.[102] Indeed, the stern behavior of the firms was a question that puzzled contemporary participants in the drama: "Some day," wrote Roger Putnam, head of the ESA, "I hope we shall know the real motives behind the mysterious conduct of the steel industry in this tragic incident."[103] The most thorough analysis of the seizure case also expresses similar ambiguity. "Perhaps," writes legal historian Maeva Marcus, the companies rejected the last OPS offer because they mistakenly believed Taft-Hartley would soon be invoked. Or again, she notes later, maybe "the Union's demand for compulsory Union membership" was the cause. And finally, "the industry [may have] decided early in the dispute to wage a campaign for public opinion," for, in the words of one steel executive, "we had a responsibility that went beyond our industry, and a remarkable public relations opportunity."[104] Yet an unexplored avenue in this puzzle is the industry's perception of its diminishing competitive position in global markets: The firm's intransigence to compromise on higher labor costs may have been because of its concerns over rapidly modernizing foreign steelmakers.

Regardless, the legal battle—which quickly became the focus of attention—needs no further elaboration here given its lengthy treatment elsewhere.[105] The industry's public relations assault, designed to show the country that Truman was clearly unjustified in his actions, was vigorously implemented under leadership of the American Iron and Steel Institute (AISI) and its outside PR agency: Considerable sums of money were spent on this campaign, and it appears to have played an important role in the seizure's outcome.[106] Renewed bargaining, meanwhile, got underway immediately following the seizure, and after a slow start, it appeared to be making some headway.[107] According to Steelman's recollection of events, there was no enmity between Ben Fairless of U.S. Steel (who bargained for the industry) and Philip Murray of the USWA; industry appeals to mass opinion for repudiation of Truman's "sad mistake" came primarily from representatives of the other firms, whom Fairless was under pressure to appease in order to maintain his industrywide coalition. In fact, Steelman said both U.S. Steel and union negotiators also had "no antagonism towards seizure," knowing it had been used before as a settlement technique. The expensive industry PR campaign to discredit the president following the seizure was only an "afterthought," Steelman believed, and USS did not participate in it to the degree of the smaller firms.[108]

Private talks between Fairless and Murray were within a day or so of agreement when a preliminary ruling by the Supreme Court on May 3 ended hopes for a quick settlement. The ruling seemed to favor the industry's position; hence, no further progress was made until the final decision was handed down. This occurred on June 2, when by a six-to-three vote the Court ruled in favor of the steelmakers and thus returned the mills to their legal owners. That same afternoon, the USWA walked out on strike.[109]

The resultant work stoppage lasted a total of fifty-three days. It was marked by a series of negotiations, breakdowns, and new startups. The major stumbling block was the union shop clause of the WSB recommendations; on wages and other fringe benefits, there appeared to be general agreement.

Within a week of the Supreme Court's decision it also seemed clear that the administration would grant a price hike in the $5.00–$5.50 per ton range (which the companies found acceptable).[110] But the industry would not bend on the union shop issue. It is difficult to determine exactly why it was so adamant on this point, though obviously the long history of opposition to organized labor was a key factor. The potential labor-force rigidity that a union shop might cause was not that significant (Fairless had earlier indicated to Murray that he would be willing to concede on this issue within a year or so).[111] What perhaps motivated the steelmakers to hold out was the knowledge that they had already won the battle. With the courts having ruled in their favor on the seizure question, with public opinion moving to their side, and with the administration now willing to grant an acceptable price hike, they perhaps saw no reason to concede to the USWA on the closed shop issue.

The stalemate finally ended on June 24, shortly after the secretary of defense held a news conference in which he stated that the strike in steel was causing more damage than "the worst possible enemy bombing raid could have done" to the war effort in Korea.[112] An angry Truman then demanded reconciliation, and wage and price terms in line with the amounts discussed in the first round of talks some six months earlier were agreed to; a modified union shop clause was also installed whereby an employee could opt out of union membership after a given period of time had elapsed. Though he felt "convinced that it was wrong," the president nevertheless gave his approval to the industry price hike. The rest of the firms soon reached agreement along the same terms as the U.S. Steel–USWA pact, and by July 26 workers began returning to their jobs at the mills.[113]

It was readily apparent that the president was the principal loser in this highly publicized conflict between government and the steel industry. The union and the companies (especially U.S. Steel) seemed to mend their differences rather quickly and return to business as usual.[114] But Truman was left in a position of defeat; according to one scholar, this was "perhaps the most serious setback the Presidency has ever suffered" (an opinion voiced, however, before the Nixon administration took office).[115] The essential element of Truman's defense throughout the seizure and strike—his insistence that nonstop steel was vital to the war effort—seemed overstated to most observers.[116] Steelman, in fact, later claimed that the Defense Department had purposely inflated the military risks involved in a steel shortage and that this had caused Truman to perhaps place too much emphasis on avoiding a shutdown at any cost; the president himself agreed with this analysis.[117] At any rate, none of the dire military outcomes feared by Defense Department bureaucrats came to pass during the strike.

It is perhaps ironic that both industry and government approached the domestic labor problem from an international perspective: Truman wanted to insure steady supplies for Korea, and the steelmakers wanted to protect themselves against the increasing competitiveness of foreign producers. But neither side was ever able to articulate the importance of this foreign dimension well enough to reach some middle ground of compromise. What this implies as much as anything else was the critical lack of any substantive structures available to either the state or the industry that could have been utilized to resolve questions of national industrial policy. This was a shortcoming, we have noted, that plagued these particular adversaries throughout Truman's time in office.

In combination with the many other problems Truman experienced in 1952, his entire administration seemed tinged with failure as a result of the steel-seizure episode. Moreover, the public's mood of disenchantment carried over to the rest of the Democratic party as well. When voters went to the polls in November 1952, they chose to terminate the Democrats' twenty-year domination of the White House. Clearly, the industry was now in a position of triumph over Truman after seven years of often frustrating and sometimes even bitter relations. It would be left to other actors to determine if a less hostile course could be found for the future.

6

Eisenhower and Reconstruction of the International Steel Industry

The selection of Dwight D. Eisenhower as the nation's thirty-fourth president was greeted with enthusiasm by American steel producers. Among other things, they expected a reduction of governmental "interference" in industry affairs after twenty consecutive years of Democratic administrations. Arthur B. Homer, president of Bethlehem Steel, echoed the sentiments of the entire industry when he observed in early 1953:

> Now, as to Washington we probably have reason to believe that the election [of the Eisenhower administration] presented industry with a new opportunity for service in an atmosphere in which the relationships between business and Government will not be poisoned by mutual distrust.
>
> The change in Administrations in Washington means that we should and will establish new relationships with Government. It would seem, at the present time, that the basic economic philosophy of the Eisenhower regime is that the country should rely on free enterprise to continue prosperity and high living standards and that the government should act only as a stabilizer of desirable conditions—with a minimum of interference.[1]

With the Republicans' return to power, the steelmakers assumed they would again be able to turn full attention to traditional objectives: forging metal and earning profits. In many respects the industry was well positioned to do both. In the international arena an emerging sense of order seemed to be developing following the uncertainties of the immediate postwar years, when American producers were urged to undertake an expensive capacity-expansion program without any assurances of protection from potentially lower-cost foreign competitors. Whether this new optimism was the result of improved performance, due to the boom generated by the Korean War, or simply of renewed confidence stemming from an internal decision to finally meet offshore competition head-on, remains problematic. Nevertheless, the results were clear: During the early 1950s, American steel mills were modernized and expanded as if there would be no significant threat of foreign penetration into home markets within the foreseeable future.

With the international environment apparently stabilized for the time being, a domestic strategy for industry improvements could now be defined. It would consist of the following factors: operating costs (especially labor) would be carefully managed, and minimized wherever possible; continued modernization and prudent expansion of capacity would proceed as planned; prices would be adjusted upward as necessary to support the capital outlays needed for modernization and to provide adequate levels of profit; and the industry would pursue sound and cooperative relations with the new administration to insure tranquility in this previously bilious arena of interaction. As well, the industry would also reach out to improve its dealings with other external constituencies through an expanded public relations program.

Initially, the new strategy seemed to work exactly as planned. By 1955 record output in ingot production had been achieved, while net industry earnings in that year reached an all-time high of more than $1 billion. The index of finished steel prices rose 18.4% from 1953 to 1956 (while the nation's wholesale price index [WPI] was up only 3.8% in the same period). During the American Iron and Steel Institute's (AISI's) annual meeting of that year, the president of one major steel firm could state:

> When one thinks in terms of the hostility to business which for so many years was Washington's stock in trade, the change in attitude along the Potomac has been profound. There . . . is heartening evidence that we can and do have public servants who know what makes the whole economy tick; who are not afraid of raising a hand against regimentation which would stifle the operation of a free economy. . . .
>
> I believe we can be thankful for what we now have in Washington.[2]

A large-scale opinion survey conducted in 1955 found the public's overall estimation of the steel industry to be not only high, but that it was also regarded as the nation's most essential productive sector.[3]

This sanguine period of achievement, however, would prove shortlived. Domestic steel production peaked in 1955 at 117 million net tons. Ten years would pass before that level would again be reached. The annual growth rate of crude steel output, a healthy 5.8% from 1946 to 1955, dropped to only 1.4% during the 1956–70 interval. In short, the industry began to enter into economic decline during the latter 1950s.[4] Meanwhile, the large steelmakers' carefully crafted strategy of international stability and control over their domestic environment turned to ashes. Indeed, the exceptional performance of the early 1950s served only to temporarily mask what was in truth a flawed agenda from the outset. The industry's triumph over Truman in the 1952 steel seizure case would turn to defeat within a decade; by 1962 a new Democratic administration would publicly humiliate Big Steel leaders in a dramatic confrontation over prices and corporate power. It is between these two symbolic business–government clashes that one must sift for the evidence underlying the American steel industry's present state of decline.

I

The critical factor in the large steelmakers' projected strategy for success during the Eisenhower presidency was the maintenance of a stabilized situation in global trade relations. Fundamentally, this meant that cheaper foreign imports should not be allowed to penetrate domestic steel markets in any significant volume. As we have seen, it was the threat from international rivals that had long been decisive in shaping the industry's competitive strategy; it remained of the utmost importance that this arena be brought under control if management objectives were to be achieved. Following announcement of the new European Coal and Steel Community (ECSC) in May 1950—an event (as we saw) shrouded in both secrecy and potential intrigue—it appeared that the goal of renewed international stability in steel might finally be at hand. This attitude was reinforced in the spring of 1952 when major ECSC producers formed an export cartel. Moreover, the imminent seating of Republican majorities in both houses of the eighty-third Congress along with the new Republican president seemed to bode equally well for this objective of insulation from international trade turbulence: The GOP had traditionally been associated with not only iron and steel interests specifically, but also with protectionism for American industry in general. If for some reason the new entente that the ECSC seemed to have provided were broken, then the Republican administration could be expected to render political assistance in building a shield from lower-priced imports.

By Eisenhower's inauguration in January 1953, however, U.S. foreign trade policy was becoming somewhat blurred. The Reciprocal Trade Agreements Act (RTAA) of 1934, which had finally reversed an historic American endorsement of protectionism by installation of new bilateral trade agreements, was again under congressional attack. Although the original three-year act had been routinely extended throughout Roosevelt's term, its expiration date of June 1945 fell amid a flurry of controversy surrounding the future direction of American foreign policy.[5] Not only was a new president sitting in the White House, but as well the portent of an end to war—and with it a vastly altered regime of global political economy—caused congressional leaders to reflect anew on America's position in the world market.

Truman was clearly committed to the proposition that increased world trade was a necessary ingredient to peace. "A large volume of soundly based international trade is essential," he stated in 1946, "if we are to achieve prosperity in the United States, build a durable structure of world economy and attain our goal of world peace and prosperity."[6] In addition, the president viewed trade as a means of extending the influence of American capitalism throughout the world and thus countering the menace of Communist expansion.[7] His administration therefore engaged in intensive lobbying to persuade a skeptical Congress of the value of liberalized trade, which resulted in 1945 in another three-year extension of the RTAA.[8] Although the measure would ultimately receive further extensions in 1948, 1949, and 1951, each time the legislative process imposed new restraints on the basic principles of free trade embodied in the original act.

What the Truman administration was essentially seeking throughout this period was a conversion of bilateralism to multilateralism. The nation-by-nation and item-by-item nature of the bargaining process that was ushered in with RTAA in 1934 proved difficult to administer in the complexities of the postwar world. What was needed now, it was believed, was a more streamlined approach that would delineate fundamental parameters of trade for all participants rather than the itemized intricacies of the old method.[9] In order to institutionalize such a new regime, the State Department proposed creation of a multination International Trade Organization (ITO) to manage future negotiations. Though the ITO concept was eventually shelved in 1950, an interim mechanism was established in 1947 to provide administrative oversight to global trade matters. This was the General Agreement on Tariffs and Trade (GATT).[10] Although never formally approved by Congress, GATT would soon emerge as the basic institutional forum in which rules and regulations would be established to both foster and preserve multilateral world trade.

It was against this background that America's adjustment to freer trade was being deliberated in Congress. As we noted, traditional protectionist forces remained highly skeptical of the new economic order wrought by Roosevelt and war, and hence they sought a number of safeguards for their constituent interests. Chief among these were escape clause and peril point provisions, both of which by 1951 were written into the RTAA extension law.[11] They provided administrative means for the United States to abrogate contractual trade agreements if it could be shown that American industry was being unduly damaged by imports. Further indication of the uncertain public sentiment toward international trade liberalization is the fact that the restricted 1951 act was passed by a Democratic-controlled Congress; many Southern members, urged on by rising industrial constituents who had begun to relocate in their districts after the war, were now able to provide the Republican protectionists with the decisive margins in the closely contested congressional votes on the issue.[12] It appeared that further reductions in trade barriers were not much desired by the nation at large in midcentury, regardless of the president's support for such measures.

It was with some reluctance, then, that Truman signed the amended two-year RTAA extension in June 1951, noting testily the "cumbersome and superfluous" new protectionist provisions that the act now contained.[13] Yet, in the remaining eighteen months of his term, there would be no sign of change in the continuing congressional retreat from the act's 1945 language—the latter generally regarded as "the high-water mark of liberal trade sentiment in the United States."[14] And by mid-1952, as one presidential aide noted, the administration's "whole trade policy [had] tended to fall apart"[15] in the face of such forces. But despite the ambivalence of others, Truman himself was still firmly committed to liberalization of trade. Accordingly, he hoped to leave Congress with an agenda for action when the RTAA would again be up for reconsideration in 1953 under a new administration (by then Truman had already announced his intentions not to seek reelection). He therefore arranged to have the Public Advisory Board for Mutual Security (successor agency to the Marshall Plan program) undertake a broad review of America's foreign economic policy "to focus support for liberal import policies."[16] The board's subsequent report called for a streamlining and rationalization of existing trade policies along with a reduction in current U.S. tariff levels. The document was delivered to the new Eisenhower administration in February 1953, obviously meant to influence the latter's attitude toward the trade issue.[17]

Before considering the new president's response, we should first review the steelmakers' views on postwar trade legislation. Initially, their position reinforced the general ambivalence and uncertainty of the Congress. Walter S. Tower, president of AISI, had noted in 1945 that because

> more than half of the world's steelmaking capacity [is] in this country, the opportunity [to export] if not the obligation, is obvious. To meet that opportunity successfully can be of vital value to the whole world. Whether it can be met is a matter of national policy toward industry. [Yet] what that policy is to be in regard to overseas commercial activity is still unrevealed.[18]

Led by U.S. Steel, the leading domestic exporter by far, the industry supported the 1945 extension of RTAA.[19] But as the opportunity costs of underwriting liberalized trade to combat the spread of communism began to take effect, commitment to the policy wavered. This emerging resistance was lead by the smaller producers, traditionally the most protectionist-oriented sector of the industry. As one result, the AISI began to submit complaining briefs before the Committee for Reciprocity Information, an executive branch unit responsible for soliciting public opinion on tariff matters prior to new tariff negotiations.[20] Essentially, the steelmakers called for a reduction of foreign tariff barriers on their export products, a reduction of U.S. tariffs on raw material imports necessary to steel manufacture, and special treatment for certain items (such as tin plate) that were not economically competitive with foreign sources.[21]

The rapid recovery of European steel production from postwar destruction along with steadily rising costs in U.S. mills soon forced even the larger domestic firms to join their smaller colleagues in a reconsideration of the effects of freer trade. The industry's global advantages, so apparent in 1945, were fast fading by the end of the decade; also, it required no great feat of memory for company leaders to recall the competitive sting of lower-cost foreign steel prior to the war. The industry might agree in public with the general policy of free trade, but increasingly some voices began to demand certain conditions, exceptions, and modifications to meet their own particular needs.[22] Yet the prospects of any such specialized treatment for steel during the Truman years were minimal at best. As we noted previously, steel's business–government relations in this period were characterized by constant suspicion and hostility; in conjunction with the administration's desire to implement multilateral trade policies in place of outmoded bilateral pacts, it simply was out of the question that any industry-specific program contrary to this objective would be considered. In addition, it is questionable that the industry would have accepted such special treatment if in any way it required the imposition of further governmental controls over its behavior.

In short, sector-specific industrial policy was not a realistic alternative in U.S. foreign trade policy from 1945 to 1952. As a result, the industry sought the promotion of adjustment policies rather than any outright rejection of freer trade (and a concomitant return to high-tariff protectionism). This is perhaps most representative of the industry position as it awaited the inauguration of the new Republican president in early 1953. Certainly this was the view of the larger firms in the industry.

II

The eight years of the Eisenhower administration would not, however, meet the steelmakers' expectations, at least on the issue of foreign trade. If anything, the new president rejected adjustments proposed by the industry, opting instead for a continuation of the Truman policy of reducing trade barriers in order to promote multilateral economic interdependence and thus hold the spread of communism abroad.[23] By the time Eisenhower left office in 1961, the American steel industry was in far worse condition than it had been in 1953, and one of the most significant causes of that deterioration was the growing volume of cheaper imported steel into domestic markets. The development of this outcome under a nominally friendly Republican administration provides compelling insight into the nature of the steel industry's governmental relations; indeed, it sheds light on business–government interaction in general throughout the Eisenhower era.

The new president was forced to declare his views on foreign economic policy almost immediately after taking office. This was because RTAA was due to expire in June 1953. In his February 1953 State of the Union address, Eisenhower declared that "Congress [should] take the Reciprocal Trade Agreements Act under immediate study and extend it by appropriate legislation."[24] Although this recommendation was in conformity with the foreign economic policy review which Truman had commissioned, it more importantly reflected Eisenhower's own beliefs. In one of his frequent private letters to a trusted friend, he noted that he had recently read President McKinley's famous Buffalo speech of 1901 that called for liberalized trade:

> What [McKinley] discerned 55 years ago has grown more true with every passing year, especially as we became more and more a creditor nation. Yet an astonishing number of people today believe that our welfare lies in higher tariffs, meaning greater isolation and a refusal to buy goods from others. They fail to see that no matter what we do in providing, through loans, for the urgent needs of other countries in investment capital, unless we simultaneously pursue a policy that permits them to make a living, we are doomed to eventual isolation and to the disappearance of our form of government.[25]

Many of Eisenhower's Republican cohorts, however, were not so enthusiastic about this course. After initial consultation with congressional GOP leaders, Eisenhower was surprised to learn that "a few even hoped we could restore the Smoot-Hawley Tariff Act, a move which I knew would be ruinous."[26] Thus, although his own sentiments were clearly in favor of liberalizing America's international trade position, those of the Republican-led Congress often appeared just the opposite.

Eisenhower's response to this divided reaction displayed certain characteristics that were to mark other issues in his presidency: Uncertain of his following, he refused (or failed) to exert the strong and decisive leadership necessary to sway either lawmakers or the public to his side.[27] The result was a trade policy that, in the words of one recent study of the subject, "can only be described as contradictory and illiberal."[28] What this meant for the nation, mired in the throes of the Cold War, was numerous missed opportunities to provide creative leadership for economic progress. What it meant for the steel industry was the continuing necessity to seek an independent resolution to its international trade problems—and to suffer their domestic ramifications without much public assistance.

The first test of Eisenhower's stance, we noted, came in his efforts to renew the 1953 RTAA. After due consideration of the bill's shaky support in Congress, the president agreed to a one-year compromise extension that would effectively freeze the existing terms while a comprehensive review of America's foreign trade policy was again undertaken.[29] The review was proposed as "a broad-gauge study into the question of what our foreign economic policy should be," conducted by a temporary Commission on Foreign Economic Policy composed of both congressional and executive branch appointees.[30] To provide at least some bias toward his own liberalized trade views, Eisenhower named Clarence B. Randall, president of Inland Steel, as commission chairman. Randall had been a consultant to the Marshall Plan as well as the official public spokesperson for the steel industry in its seizure clash with Truman in 1952.[31]

It is somewhat ironic that Eisenhower chose a representative of the steel industry to lead his campaign for freer trade, since many in that business were blatantly opposed to the program. Randall, however, was no mainstream steel protectionist. Indeed, he was a confirmed advocate of a liberalized regime of world trade. Inland Steel, it should be noted, was based in Chicago, where it served primarily midwestern markets relatively isolated from foreign steel (the Saint Lawrence Seaway, which opened up the area to large-scale imports, would not be completed until 1959). Inland had thus never been a major participant in world steel trade, and its management was unfamiliar with, and certainly unsympathetic to the unique cooperative arrangements that often governed industrial competition in Europe. In 1952 Randall's book *A Creed for Free Enterprise* was published, which quickly became a national best-seller; in it the steel executive presented a forceful argument for the necessity of unfettered industrial competition if America were to retain its preeminence among nations.[32] Eisenhower was no doubt strongly attracted to this sort of thinking.

The Randall Commission's schedule was ambitious, to say the least. The group was not operational until September 1953, and it worked toward a termination date of January 1954 (timed to provide input to the 1954 congressional debate on RTAA renewal). As part of the data-gathering process, some fifteen hundred organizations and individuals were invited to submit written statements for the commission's consideration.[33] In December 1953 AISI, one of the invitees, delivered its views.[34]

The industry statement on foreign economic policy proved to be a well-articulated overview of the position of its larger firms.[35] Beginning with a brief review of prior American trade policies, the institute focused on the inadequacy of existing institutional mechanisms in this area. Accordingly, AISI stated, "We believe it is possible to create new instruments which will permit us to proceed in the direction of developing a coordinated foreign economic policy which will have sufficient scope, flexibility and continuity of action to serve the best interests of the country."[36] The statement then considered the steel industry's particular stake in foreign trade:

> In the eight years of post-war reconstruction the major steel producers in Western Europe and Great Britain rehabilitated, modernized and enlarged their capacity, many of them with both financial aid and the best technical advice from the United States. All of them have production capacities above pre-war levels, and their new capacity exceeds their current domestic demands. This means that not only are European markets closed to exports of steel products from the United States, but that these *modernized competitors are now in a position, more than ever before, to compete on a quality and price basis with United States producers in the world market* [emphasis in original]. In addition the Schuman Plan is deliberately designed to set up a vast supranational industry composed of all the major producing countries in Western Europe to regulate marketing within this enlarged area and to function more efficiently in export markets. It

> remains to be seen, probably in the very near future, just what kind of competitor this potentially powerful group will become in the free export markets left in the world.
>
> And our new competition is not confined to Western Europe and Great Britain. Many foreign countries which had little or no steel production prior to the last war have now built their own mills and, whether or not they can operate efficiently, in every case the local market for all that they can produce is reserved exclusively to them by their own laws. Chile, Columbia, Brazil, India and Australia are but a few of the countries which are building or enlarging their steel-making capacity. Many of these new plants are owned by their governments.[37]

This section was concluded with a note that

> unfair methods of competition from abroad are the subject of particular concern to our industry, especially subsidized and "state controlled" exports, and the practice of dumping.[38]

The institute proceeded with a listing of four basic principles it felt should guide the commission in its task:

> (a) The world leadership position of the United States demands a foreign economic policy that is far broader than a mere schedule of tariffs on individual import items. . . .
>
> (b) Our new foreign economic policy must be flexible. . . .
>
> (c) National security must not be overlooked in the formulating of economic policy. . . .
>
> (d) Our policy making in the foreign economic field must be stabilized by creating some sound basis for continuity of action.[39]

Finally, the statement said:

> Some continuing and unbiased agency must be made responsible for a constant review of changing factors in the foreign economic scene and for delineating policies which will serve the best long-term interests of the whole country.
>
> THE MACHINERY DOES NOT EXIST IN GOVERNMENT TODAY TO INAUGURATE AND CARRY OUT EFFECTIVELY A FOREIGN ECONOMIC POLICY WHICH WOULD BEST MEET THESE FOUR REQUIREMENTS [emphasis in original].[40]

To overcome these problems and to meet its four basic requirements, AISI recommended that a permanent and independent Foreign Trade Commission be created by Congress. This new commission would absorb the U.S. Tariff Commission, the Export-Import Bank, and the Committee for Reciprocity Information, thus becoming an institutionalized superagency to formulate, coordinate, and implement national policy. The statement concluded:

> The time has passed when a piecemeal tariff policy, administered by different and changing agencies, frequently under opposing theories, can be expected to determine this country's foreign trade course. Until a single permanently operating agency such as we suggest is established we will be without the machinery essential to effect a rational solution to our foreign economic policy.[41]

Observed in the contemporary context of the industry's political economy, this proposal was a commendable alternative for objectively reforming the chaotic structure of American foreign economic policymaking. More important, perhaps, it represented a call for governmental assistance by an industry that acutely perceived its future to be in the hands of an uncertain—and probably menacing—international marketplace. Although the climate surrounding business–government relations in the Truman years hindered any such programs, the industry seemed now to believe that conditions had changed: Not only was the foreign environment of steel trade becoming more precarious and thus in need of some form of business–state cooperation, but more significantly the new Republican administration was deemed capable of implementing such a cooperative steel–state policy without jeopardizing basic business freedoms.

This was not the first time (it should be noted) that industry leaders had turned to the new Republican government in search of a more reasonable approach to their long-term political–economic needs. In early 1953 new appointees to top posts in the Commerce Department proposed the establishment of twenty industry councils under their sponsorship, designed to enhance the coordination of business–government relations within the respective industrial sectors. These would be chaired by private businessmen from industry, although membership would include Commerce Department officials. The chair, however, would set the council agenda, and industry funds would finance group activities. Proposed topics to be discussed by the iron and steel council included production, marketing, distribution, capacity, and defense mobilization, among others. Steel leaders expressed hope that "a bright new era of intelligent and worthwhile cooperation" would result from this venture.[42]

Although drawing its immediate inspiration from the oil industry's National Petroleum Council (a business–government group operating under sponsorship of the Department of Interior),[43] the concept was, in fact, an extension of the Industry Advisory Committee (IAC) idea, a system of business–government cooperation that dated back to World War I. As noted in Chapter 3, Truman's Commerce Department briefly resurrected this mechanism in allocating scarce commodities during postwar shortages in 1948–49. These committees were again reestablished in late 1950 under the guidance of the National Production Authority, a Commerce Department unit created to spur mobilization for the Korean War. By the time Eisenhower took office in 1953, over 550 IACs were in existence. The steel IACs (there were many from this industry, organized along product lines) had been instrumental in numerous war mobilization issues, such as scrap steel conservation and coordination of industry capacity-expansion plans.[44]

The Justice Department, however, did not approve of Commerce's new proposal for these twenty high-level industry councils. Citing antitrust objections, Justice attorneys found the steel council (the first group to be organized under the program) to be "not in accordance" with existing standards, hence it and all of the other industry councils were rejected.[45] Although it is difficult to determine just how significant this venture might have been to the formation of any national industrial policies for the steelmakers, its establishment obviously would have provided a useful forum for the serious discussion of issues of abiding interest between the two groups. Yet such was not to be the case. As had occurred in all previous attempts in this direction, far too many public officials (as well as some others) felt that the inherent adversarial nature of the nation's legal system should prevail over any alternatives were the democratic capitalist tradition to endure. Business–government cooperation would have to be pursued outside the structure of antitrust law were it to gain a foothold in America.

The AISI recommendations submitted to the Randall Commission also failed to win approval from public policymakers. Although there was perhaps a defensive posture in these industry proposals (stemming from the steelmakers' realization that the historic policy of protectionism was at last over and that something new would be formulated in its place), one must nevertheless give the statement's drafters credit for their determination to look ahead to a fundamentally altered world economic order—and to face up to the consequences this would have on their affairs. One might argue that the industry felt it would be easier to influence (or capture) a single independent agency rather than a constantly shifting Congress or equally elusive Executive Office, and that this is what fundamentally accounts for its recommendations for change. This well may have been the case. Yet, in spite of such speculation, AISI's statement still stands as a perceptive and reasoned proposal to adjust America's foreign economic policy in the light of changing world conditions. It was a clear rejection of the protective philosophy of high tariffs and also a tacit admission that the federal government would necessarily have to become a major actor in the industry's more internationalized future. At the least, this statement deserved serious consideration by the Randall Commission, just as the proposed Commerce Department forum deserved a hearing from the Justice Department. It would be to the detriment of both the industry and the national interest that neither apparently received much attention.

The political timing of the industry's proposals was unfortunate. In the early 1950s the Congress seemed in no mood to drastically alter the nature, let alone the institutions, of foreign economic policymaking, and the president thereby refrained from pushing the issue much further. Moreover, other important and influential sectors of the economy were moving in a direction quite opposite to liberalization of trade, seeking instead a return to protectionism.[46] When the Randall Commission submitted its report in early 1954, it ended up a rather timid piece given the mandate its chairman had originally set (and nowhere did it recommend anything closely resembling AISI proposals).[47] The reason for this outcome was due to the many compromises that the politically naive Randall had to make with protectionist-minded congressmen who served on the commission; unless the chairman's liberal goals were watered down, he could not get unanimous agreement to the report, and above all else Randall wanted a unified commission in order to make his case with Congress and the public.[48] Yet even this tamed-down document found a rough reception. Randall vigorously lobbied important Republicans on the proposals and Eisenhower gave his clear endorsement, but it was still not enough to surmount the skepticism of legislators who feared committing themselves on such a controversial topic during an election year.[49] As a result, all the administration obtained from the commission's efforts was another one-year extension of RTAA, with a congressional promise to give the issue top priority in the next legislative session.[50]

In 1955 the president finally was able to secure a three-year extension of the act. By that time the national economy had pulled out of the recession that plagued the two previous years, which no doubt had affected congressmen reluctant to approve trade liberalization during a period of rising unemployment and earnings declines. But the 1955 measure was not strongly endorsed by free-trade advocates; it continued the retreat from earlier principles, chiefly through a new amendment that would allow the president to impose quotas on imports that threatened the health of domestic industries vital to national security (a clause that Eisenhower had opposed).[51] In 1958 the act was given a four-year extension, but by then other issues had temporarily overtaken the tariff issue in national politics, thus muting the controversy with which it was usually surrounded. Relatively few changes were made in its language.[52]

Although some progress was being made toward the president's goal of trade liberalization, he was stymied on other fronts. He could not convince Congress to ratify the Organization for Trade Cooperation, an international body designed to fill the role of the earlier disparaged ITO in administering world trade.[53] At the same time Eisenhower also had trouble in winning the admission of Japan into GATT, a move the administration considered indispensable to shoring up the fight against communism in Asia.[54] Finally, his aid programs to underdeveloped nations—also designed to keep friendly or neutral nations from falling into Communist clutches—were under fire from many quarters.[55]

The checkered pattern of success in Eisenhower's foreign economic policy thus clearly reflected the uncertainty that this subject aroused in the nation as a whole. Relatively inexperienced in the exertion of international economic leadership and still struggling to free themselves from isolationist ideological bonds, the American people faced up to the sharply revised postwar world order with reluctance. The president himself was obviously aware of the problems he faced in transforming public opinion in this arena, just as he realized that existing institutional structures were incapable of meeting the new demands. As one means of improving his control of these events, he undertook a reorganization of the administrative apparatus dealing with trade. In late 1954 Eisenhower accepted the recommendations of Joseph Dodge, a key official in the postwar U.S. occupation of Japan, to establish a Council on Foreign Economic Policy (CFEP) within the White House.[56]

Dodge's prior analysis of trade administration had uncovered many problems in both policy development and coordination; in existing agencies charged with overseeing international economic issues, for example, he found "there was no general recognition of the need to relate foreign economic conditions and policies with the similar factors of the domestic economy."[57] Dodge headed the CFEP until July 1956 when Clarence Randall (by then a White House consultant on trade) was appointed to the post. In theory, this body might have served a function similar to that proposed by the AISI in its 1953 statement to the Randall Commission. However, the CFEP was never able to achieve so lofty a stature; as Randall noted on leaving his job in 1961, "At no time had administrative responsibility been vested in my office. . . . My function has been solely one of proposing and coordinating ideas and policies."[58]

Even this latter objective seemed elusive. Eisenhower's foreign economic policy continued to engender painful frustrations throughout his term, at least for the steel industry. Perhaps it is unfair to blame the president for this outcome, since he was forced to deal with a situation in which so few precedents existed. The United States, as we noted, had little prior experience as manager of the global economy until World War II came to an end. Congress, meanwhile, was generally reluctant to embrace the new international order in any substantive way, and the business community (encompassing both large and small firms) often seemed equally divided on a proper role for America to play in this regard. Yet on the other hand, Eisenhower consciously chose to shape his overall foreign policy around one central vision: the need to protect the nation (and by extension the free world) in its Cold War struggle with the Soviet Union.[59] If there was one driving force in the Eisenhower presidency, surely this was it.

The ramifications of such a narrow philosophy were to prove disastrous for steelmakers. It is somewhat paradoxical that in order to animate his vision, the president felt it necessary to strengthen the steel industries of both Western and neutral nations (as we shall see later) even though doing so would have serious competitive effects on American producers. Given the acknowledged significance of the steel industry to national security, one might well have believed this sector would receive more favorable treatment at home. But just as a statist containment of international communism shaped Eisenhower's foreign policy, its antithesis—the protection and expansion of free-enterprise capitalism—sparked much of his domestic agenda. Under this contradictory regime, government could not step in to aid home industry no matter how vital (or how international) the industry's activities. Market forces were presumed to define domestic outcomes. Yet at the same time, the benevolent hand of America's government was deemed justified in assisting foreign industry in such cooperative (and quasi-socialist) programs as the Schuman Plan. Eisenhower's administration remained strongly committed to this inconsistent ideology throughout his eight years in office. As such, the president must bear ultimate responsibility for the serious long-term dysfunctions he created for many industries, steel included.

Against such a policy framework, the steel industry had to seek its own objectives—under its own power—in the international arena. Unlike several other sectors that were able to obtain particularized attention from government (such as petroleum, textiles, and certain strategic minerals and commodities), steel was unable to convince Washington that its plight was so bad, or its position so important, that it, too, deserved special treatment.[60] Consequently, the only role left for the industry in the shaping of foreign economic policy consisted of filing briefs before congressional trade hearings and the routine provision of testimony at sessions designed to elicit domestic views on GATT negotiations.

Yet at the same time it must be acknowledged that steel's record output and profits during the mid-1950s created little public sympathy for its requests for specialized benefits to help it adjust to a changing international environment. And as we shall see, persistent controversies surrounding steel's labor relations, allegedly monopolistic pricing practices, and their combined effect on rising national inflation also tended to sour audiences on the producers' viewpoint. Any help from government on international trade problems, it seemed, would first have to await a serious deterioration of the industry's economic situation (and probably a major revision of industry attitudes toward cooperation with the state). In spite of the CFEP's hope that "it should be possible to avoid crises and emergency decisions by anticipatory staff work," the United States was not yet ready to undertake so rational an approach to the management of its world commercial relations.[61]

III

Denied any special assistance from government, the nation's steelmakers continued to watch with alarm the growth of foreign steel output throughout the 1950s. The trade press constantly reported on rising threats from abroad and steel executives themselves frequently alluded to this development in public comments. There was good reason for concern. In 1930, 1940, and 1950 the United States produced respectively 44%, 43%, and 47% of the world's steel; by 1960 the proportion dropped to 26% (see Table 6.1).[62] In short, the seeds of revolution in the international steel industry were sown during the 1950s as more and more nations turned to internal sourcing from increasingly more efficient mills. And what was even more disturbing to American producers was the fact that, in many instances, these upgraded for-

Table 6.1 American Crude Steel Output as Percentage of World Production

Year	*U.S. Output*
1900	36.8%
1910	44.3
1915	48.9
1920	59.0
1925	51.0
1930	43.5
1935	34.8
1940	42.9
1945	63.7
1950	46.6
1955	39.4
1960	26.0
1965	26.1
1970	20.1
1975	16.4
1980	14.2
1985	11.1

Sources: 1900–40, Duncan Burn, *The Steel Industry, 1939–1959* (Cambridge: Cambridge University Press, 1961), endpiece pullout chart; 1945–85. AISI, *Annual Statistical Report* (various years).

eign industries were being subsidized by both the United States as well as their own governments. Eisenhower's early foreign policy dictum of "trade, not aid" was giving way to a new standard of "trade and aid."[63]

A number of political and economic factors underlay the growth of world steel output following 1945. Probably most critical was the severe shortages experienced in the immediate postwar era when nations could not obtain sufficient quantities of the metal to meet pent up internal demands. This prompted massive capacity expansion, not only in the United States and Europe, but elsewhere as well. In the industrialized areas this was to be expected, since most of the war-torn nations had enjoyed vigorous steel industries prior to 1940. But a somewhat novel element was the push by many nonindustrialized nations to move into the modern economic age. Steel production could serve a dual purpose in this regard: Not only would new mills be able to fuel local economic growth (insofar as the product was a crucial input to any industrialization plans), but its local production as well would alleviate foreign exchange problems that would otherwise occur had steel to be imported. In some nations, it should be noted, steel manufacture was justified as a means to obtain hard currency through subsidized production for the world market rather than for purely local demand.[64]

The prospects for new industry entrants were enhanced even more by the deepening Cold War. As the United States and the Soviet Union engaged in ideological conflict the world over, they vied with one another to supply the developing nations with economic aid and assistance. This would presumably demonstrate the superiority of the giving nation's political and economic system in the eyes of the receiver.[65] The result for recipient countries was easy availability of credit for capital needs and lowered barriers to the transfer of technology. In addition, the rapid expansion in the 1950s of raw materials needed in steelmaking production insured that new steel mills would be able to function once they went into operation (a condition that had limited world steel growth in the past because most known deposits of steelmaking materials were controlled by existing producers).[66] As a result of these changes, steel production in the developing world would grow from 2.7 million tons in 1948 to 22.7 million tons in 1962.[67] Although still only a minor portion of global output, this trend was clearly indicative of the coming shape of the industry; by the 1970s developing nation output would trigger a major reorganization of international steel trade.[68] Throughout the 1950s, the American government provided financial aid to domestic steel's free world rivals. From 1940 to 1956, in fact, U.S. assistance to foreign steel industries totaled some $866,172,000 in direct grants and loans delivered in eighty-one separate transactions to producing facilities in twenty-four countries.[69] (More than 78% of these funds, it should be noted, were authorized after 1950; see Table 6.2.) From 1957 to 1960, nearly $600 million more would be given.[70] These sums do not include the many millions more in indirect aid that the United States provided, such as technical and managerial assistance, access to scarce raw materials, and governmental purchases of foreign steel.[71]

America's steel producers were unable to prevent federal aid and assistance to Third World nations; indeed, few tried. This was perhaps to be expected, since most in the industry perceived little if any risk stemming from the steel capacity buildup in this sector, whereas the presumed benefit—containment of communism—was deemed more important. On the other hand, industry leaders should have been far more concerned with developments in industrialized Europe. The major European producers had been the leading factor in world steel trade throughout the twentieth century, capturing some 65% of total global exports by 1955. The United States, meanwhile, supplied only 14% of this trade, about the same proportion it had controlled since 1913 (see Table 6.3).

Yet in spite of this history there was a surprising lack of political concern displayed by the major American producers over steel industry events in Europe during Eisenhower's first term in office.[72] The one major exception, of course, was the AISI statement to the Randall Commission; but this made little impact and seems to have been quickly forgotten. Most of the complaints over iron and steel trade and tariff policy emanated from the smaller firms in the industry (a situation that would not change until the 1960s).[73] The operationalization of the Schuman Plan, brought to life in 1950 and fully functioning by August 1952, did raise some domestic concerns, but they did not come from the nation's two major steel exporters—U.S. Steel and Bethlehem Steel.[74] Only Clarence Randall of Inland Steel spoke out publicly on

Table 6.2 U.S. Government and International Agency Financial Assistance to Foreign Steelmakers, 1947–60

Annual Amounts of Assistance

Year	*Amount (000 000)*	*Year*	*Amount (000 000)*	*Year*	*Amount (000 000)*
1947	$28.3	1952	$42.1	1957	$145.8
1948	7.2	1953	11.2	1958	120.6
1949	203.9	1954	111.6	1959	195.7
1950	45.7	1955	88.3	1960	123.7
1951	66.1	1956	221.8		

Nations Receiving Assistance

Nation	*Amount (000 000)*	*Nation*	*Amount (000 000)*
Argentina	$87.5	Mexico	$87.7
Australia	13.4	Netherlands	14.9
Austria	56.4	Pakistan	0.6
Belgium	20.4	Peru	12.5
Brazil	84.7	Phillipines	0.5
Canada	5.7	Portugal	0.9
Chile	94.3	Spain	28.6
France	82.9	Taiwan	0.7
Germany	10.0	Turkey	144.6
India	156.6	UK	27.2
Italy	110.3	Uruguay	2.5
Japan	203.8	Venezuela	3.0
Korea	2.2	Yugoslavia	14.3
Liberia	45.6	ECSC	100.0

TOTAL AID GIVEN: $1,411,800,000

Source: U.S. Senate, 90th Cong., 1st sess, Committee on Finance, Committee Print, *Steel Imports* (Washington, DC: GPO, 1967), pp. 299–304.

the ECSC, fearing the effects it might have on free enterprise: "I therefore share the fears," he wrote in 1951,

> of those who believe that the Schuman Plan will weaken Europe by laying the dead hand of socialism and bureaucracy across its basic industry. I wish fervently that somehow some other way might be found for advancing the political unity of Europe.[75]

The domestic industry thus had little to say for the public record regarding the emergence of this new international actor in their midst. As one result of their silence, the steelmakers would have even less say in the formation of

Table 6.3 World Steel Trade and American Steel Exports, 1947–60[a]

Year	*World Steel Trade*	*U.S. Steel Exports*	*U.S. Percentage of World Trade*
1947	11.0	5.9	53.6%
1948	11.4	3.9	34.2
1949	13.5	4.3	31.9
1950	15.9	2.6	16.4
1951	19.1	3.1	16.2
1952	19.1	4.0	20.9
1953	18.4	3.0	16.3
1954	20.1	2.8	13.9
1955	28.8	4.1	14.2
1956	30.4	4.3	14.1
1957	33.9	5.3	15.6
1958	32.2	2.8	8.7
1959	35.7	1.7	4.8
1960	43.2	3.0	6.9

[a]All figures 000 000 net tons.

Source: UN Economic Commission for Europe and U.S. Department of Commerce. (Reproduced in U.S. Congress, JEC, 88th Cong., 1st sess, *Hearings on Steel Prices, Unit Costs, Profits, and Foreign Competition* [Washington, DC: GPO, 1963], p. 735.)

American public policy toward the ECSC. William Diebold, Jr., notes in his authoritative study of the Schuman Plan:

> United States policy toward the Community . . . was not primarily concerned with the problem of coal or steel or even with the economic consequences of pooling their production and trade. American approval of the Schuman Plan was based principally on its obvious importance as a step toward the establishment of a durable, constructive relation between France and Germany.
>
> . . . High American officials . . . for the most part . . . took the view that the political importance of French-German agreement on heavy industry outweighed economic features that might seem undesirable from an American point of view.[76]

President Eisenhower was a firm supporter of the ECSC. "While in Europe," he wrote in discussing the Schuman Plan, "I watched with keen interest the efforts to work out the first steps toward European federation. My experience there convinced me that the uniting of Europe is a necessity for the peace and prosperity of Europeans and the world."[77] Economic unification became even more important to Eisenhower when proposals for a six-nation European Defense Community (EDC) began to founder.[78] Soon afterward, an official U.S. representative to ECSC was appointed, and—following a visit to Washington by ECSC leaders in 1953—the president stated that

U.S. aid to the group "would foster European integration in a tangible and useful way."[79] The federal government subsequently authorized a $100 million loan to ECSC based on "straight political justifications" according to the State Department rather than on any economic grounds.[80] The domestic American steel industry, speaking through the AISI, acknowledged the political motivation of the loan and stated it would not offer any opposition.[81]

Yet, even had American steel interests chosen to complain about U.S. federal aid to their international competitors, it is unlikely that administration officials would have paid much attention. There is perhaps some truth to a remark by Clarence Randall that the steel industry's involvement in the Schuman Plan was simply a historical accident and that it could easily have been replaced by "textiles, or wine, or agriculture."[82] The unification of Europe and the suppression of further Communist advance was clearly the overriding objective in the administration's strong backing of the ECSC, and there was little the steelmakers could do to alter this larger political goal.

The European Community formally came into being in late 1952, just as an economic slump hit the international steel industry. The original Schuman Plan announcement (in May 1950) had occurred in a period of surplus capacity for the continental producers, yet this was quickly eliminated through the sharp increase in speculative demand brought on in late 1950 by the Korean War. By 1952, though, much of the boom's steam had been vented, and in 1953 orders rapidly began to fall throughout world markets.[83] The onset of this downturn caused the European steelmakers some worry, for over the previous two years they had been able to discriminate in pricing their shipments, charging more for surging export sales than for home consumption. With a transition to a buyer's market in 1953, strong pressures to reduce export prices began to arise.[84] This prospect caused others some worry as well. The High Authority (HA), governing council of the ECSC, fully realized the historic propensity of continental steel producers to turn to cartels to protect themselves during market slumps. Accordingly, the HA warned that it would "resort to various measures which are open" to insure competition under legal terms of the treaty binding the six ECSC nations together.[85]

But the producers would not heed these threats. In March 1952 private steel trade associations in Belgium, Luxembourg, and France announced plans for the Brussels Entente, designed to stabilize export prices and reduce "cut-throat competition" among continental firms in their sales to outside markets.[86] German and Dutch producers soon joined in the agreements, as did Austrian firms at a later date. After modifications to the initial pact, a tightly structured export cartel emerged in September 1952, and by 1954 export quotas as well as price points were being assigned to the participating nations. All of this was completely outside of any ECSC auspices.[87]

The Brussels Entente did not seem to have much impact in America; neither U.S. nor British firms publicly joined in this agreement, and U.S. producers never publicly complained about the new cartel's behavior. During the rebound in industry demand following the recession of 1953–54, the higher prices set by the entente for export markets would have benefited American firms, to the extent that they participated in export activities. Moreover, while ECSC producers consistently accounted for 75% to 80% of the steel imported into the United States during the early 1950s, this never totaled much more than 1 million tons per year (or less than 2% of apparent annual consumption); the pattern of U.S. exports of steel to world markets showed a similar consistency over this period, both in volume and destination (see Tables 6.4 and 6.5). At least on the surface, the Europeans' new cartel seemed not to effect the United States.[88]

But opposition to the Brussels Entente was not completely silenced. Instead, it came from the official governments of the Untied States and other third-party nations who traded with the ECSC. Responding to these protests, the HA formed a special commission to investigate the situation. The commission found, however, that the ECSC treaty enjoined anticompetitive behavior only within intracommunity trade and thus that the HA had no jurisdiction over member states' steel firm agreements external to ECSC borders. Reflecting the general attitude of most continental trade officials to steel cartelization, the investigators also suggested that the entente's export prices were eminently equitable, and thus not damaging to international steel trade.[89]

Surviving this legal challenge from outside governments, the Brussels Entente still faced mounting political pressure from detractors who argued that prewar European cartels were an indirect cause of World War II. As a defender of free trade, GATT took the lead in attacking anticompetitive behavior in steel. Because GATT had recognized the ECSC as the official representative of the six member nations for iron, steel, and coal, Denmark and Switzerland, also GATT members, demanded that the ECSC explain how GATT free trade principles squared with the entente's protectionist aims. But a subsequent study of export prices of all major steel-trading nations proved inconclusive; GATT never formally censured the ECSC for the private actions of its members.[90]

These attempts to constrain the entente did, however, have indirect effects. The exhaustive public scrutiny by official investigative bodies probably hindered the entente participants from achieving all their goals. Moreover, the High Authority of the ECSC could not tolerate any blatant repudiation by member-state steel firms of the ECSC's implicit commitments to the international commercial community. Internal pressures were probably applied to ameliorate the export-pricing scheme, if not to terminate outright the entire agreement. Support for this surfaced in late 1957 and early 1958 when (albeit in a softening market) export prices fell considerably below the levels set by the cartel. By the mid-to-late 1950s, then, strong public forces were working to reduce or eliminate vestiges of cartel activities in European iron and steel trade.[91] Moreover, by the late 1950s U.S. producers had begun to complain much more vigorously about European trade practices than they had earlier.

Table 6.4 Steel Imports into the United States from Country of Origin, 1950–65[a]

Year	*ECSC*[b]		*UK*		*Canada*		*Japan*		*All Others*		*Total Tons*
1950	802	74.5%	64	5.9%	189	17.6%	18	1.7%	4	0.4%	1,077
1951	1,906	84.2	131	5.8	75	3.3	113	5.0	39	1.7	2,264
1952	860	70.9	37	3.1	88	7.3	181	14.9	47	3.9	1,213
1953	1,248	74.7	84	5.0	180	10.8	119	17.1	39	2.3	1,670
1954	670	85.6	40	5.1	25	3.2	24	3.1	25	3.2	784
1955	609	62.3	47	4.9	185	19.1	96	9.9	33	3.4	970
1956	1,063	79.7	60	4.5	59	4.4	47	3.5	104	7.8	1,333
1957	890	77.2	58	5.0	52	4.5	32	2.7	123	10.6	1,155
1958	1,203	70.5	86	5.0	48	2.8	250	14.6	120	7.1	1,707
1959	2,898	65.9	215	4.9	377	8.6	626	14.2	280	6.4	4,396
1960	2,097	62.4	212	6.3	211	6.3	601	17.9	238	7.1	3,359
1961	1,951	61.7	166	5.2	304	9.6	596	18.9	146	4.6	3,163
1962	2,086	50.9	250	6.1	367	9.0	1,071	26.1	326	7.9	4,100
1963	2,245	41.2	349	6.4	583	10.7	1,803	33.1	466	8.6	5,446
1964	2,585	40.1	285	4.4	692	10.8	2,446	38.0	432	6.7	6,440
1965	4,191	40.4	720	6.9	644	6.2	4,418	42.6	410	3.9	10,383

[a]For each country: first column is 000 tons; second column is percentage of total imports.

[b]ECSC. Figures prior to 1953 are sum of the individual countries.

Sources: AISI, *Annual Statistical Report* (various years); AISI, *Steel Imports—A National Concern* (July 1970), pp. 63–64.

Table 6.5 U.S. Exports of Iron and Steel, 1945–60[a]

Year	*North & Central America*	*South America*	*Europe*	*Australia, Oceania*	*Asia*	*Africa*	*Other*	*Total*[c]
1945	1.507	0.984	1.663	0.151	0.424	0.380	—	5.109
1946	1.394	1.333	1.652	0.075	0.387	0.189	—	5.031
1947	1.920	1.873	2.048	0.109	0.677	0.250	—	6.876
1948	1.507	1.141	1.120	0.080	0.607	0.242	—	4.697
1949	1.659	0.963	1.156	0.121	0.867	0.289	—	5.055
1950	1.326	0.440	0.582	0.105	0.528	0.113	—	3.095
1951	1.748	0.635	0.487	0.123	0.430	0.120	—	3.543
1952	1.993	0.632	1.227	0.127	0.348	0.132	—	4.459
1953	1.619	0.678	0.555	0.049	0.321	0.125	—	3.345
1954	1.130	0.748[b]	0.644	0.108	0.371	0.133	0.023	3.158
1955	1.205	0.914[b]	1.657	0.062	0.569	0.124	0.022	4.553
1956	2.009	1.114[b]	0.998	0.046	0.897	0.096	0.005	5.166
1957	2.462	1.327[b]	1.048	0.039	1.951	0.109	0.017	6.953
1958	1.567	0.667[b]	0.604	0.012	0.544	0.052	0.005	3.449
1959	0.920	0.462	0.316	0.007	0.364	0.020	0.014	2.103
1960	0.981	0.521	1.127	0.102	0.664	0.078	—	3.473

[a]All figures 000,000 net tons.

[b]Between 1954 and 1958, exports to Central America are included with totals for South America (not North America).

[c]Rows may not equal "Total" due to rounding.

Source: AISI, *Annual Statistical Report* (1945–1960).

IV

Although European steel firms may have entered into collusive marketing agreements during the 1950s, this did not result in any relaxation of their robust expansion and modernization programs undertaken after cessation of the war. By the end of the 1940s most of the affected nations had rebuilt their steel industries. Continued high demand, however, brought even further growth.[92] From 1950 to 1955 ECSC steel ingot production grew by 66% while total world output expanded by 43%. In comparison, U.S. production was up only 21%, though it, of course, was operating from a much larger base (see Table 4.3). The increased ECSC output was accomplished through a rise in its steel capacity to 62.5 million tons in 1956, representing an increase of more than 33% form levels prevailing as recently as January 1, 1952. This remarkable program was financed with over $2.6 billion in new investment, much of it obtained on highly favorable terms from a variety of both public and private sources[93] (see Table 6.6). More than sheer volume is of note here, however; the reshaped industry that emerged was characterized by both greater efficiency and more economically rationalized operations than had existed before. A wave of large-scale mergers swept over continental producers in the early-to-mid-1950s, resulting in much higher levels of industrial

Table 6.6 Investment in Steel Facilities, 1952–60[a]

Year	*ECSC*[b]	*UK*	*Japan*	*U.S.*
1952	$545	$168[c]		$1,298
1953	542	165	1951–55:	988
1954	454	144	$355	609
1955	524	163		714
1956	570	210		1,311
1957	708	265	1956–60:	1,723
1958	644	295	$1,739	1,137
1959	590	277		934
1960	785	409		1,505

[a]All figures $000, 000 in American currency.

[b]European Coal and Steel Community.

[c]Based on conversion rate of £1 = $2.80.

Sources: For ECSC: ECSC, High Authority, *Investment in the Community Coalmining and Iron & Steel Industries, 1961* (Brussels: author, July 1961), p. 18, Table 15. For UK: Duncan Burn, *The Steel Industry, 1939–1959* (Cambridge: Cambridge University Press, 1961), p. 453 (for 1952–1953); [UK] Iron and Steel Board, *Iron and Steel Annual Statistics for the United Kingdom, 1960* (London: author, 1961), p. 99, Table 91 (for 1953–1960). For Japan: Ken-ichi Iida, *History of Steel in Japan* (Tokyo: Nippon Steel Co., 1973), p. 32. For the United States: American Iron and Steel Institute, AISI, *Annual Statistical Report* (various years).

concentration.[94] This seemed surprising in the face of ECSC policies nominally opposed to concentration and in favor of greater competition. Yet to a degree the very specter of the community's presence brought on the merger movement: Producers, unsure of the future within ECSC constraints, chose to unite in defense and thus protect their positions against uncertainties that the changing political environment might hold.[95]

Other nations were also expanding and rationalizing their steel industries during the 1950s. One of these was Great Britain. Following the return to power of the Conservative government in 1952, the recently nationalized British steel industry was ordered denationalized under terms of a new Iron and Steel Bill.[96] Great Britain continued to reject membership in the ECSC after denationalization, but it correctly viewed the new organization as a powerful threat to both traditional British export markets as well as domestic supply.[97] As a result, a national steel development plan that had been initiated in 1950 by the Labour government was allowed to proceed. The program called for a capacity expansion to 20 million metric tons by 1957, a significant boost over the 15.5 million tons of crude steel that the industry had turned out in 1949. In 1955, however, the goal was further revised upward, projecting a 25-million-ton-capacity level by 1958.[98] Thus, even though reprivatization had occurred, the British steel industry continued to rely heavily on governmental coordination and planning in its principal activities (such as pricing, which had been controlled by the government since 1932 and remained so after 1952). The privately owned steelmakers thus approved and even encouraged state oversight of their industry as an efficient, if not indispensible, means to achieve both the nation's ambitious expansion plans as well as to fend off foreign intruders.[99]

One other nation also deserves attention for its steel planning. This is Japan. As we noted in Chapter 3, the Japanese initiated a program of state-led growth following surrender. The boom in steel demand brought on by the Korean War, though of relatively short duration, greatly added to these efforts by allowing the nation to reach prewar output levels far earlier than had been expected.[100] Japan's First Modernization Program, covering the period 1951–55, was intended to expand capacity by modernizing and rationalizing existing mills. The program was successful: Ironmaking capacity grew by 214% and steelmaking capacity was up 93% (the latter to 10.4 million tons in 1955). Yet, when the program was completed, it was obvious that demand would remain strong (capacity utilization in 1955 was over 97%). Consequently, a Second Modernization Program was launched in 1955, to run through 1960. Rather than merely rationalizing output, this new offensive would seek to greatly expand capacity through the construction of a series of huge integrated steelmaking complexes employing the latest and most efficient technology.[101] When completed, raw steelmaking capacity would increase by 297% from its 1955 level, up to 28.2 million tons. This would allow Japan to nearly triple its share of world production over what it had been in 1950.[102] American involvement in this record of achievement was conspicuous, especially in the choice of technology used by the Japanese (as noted in Chapter 9).

The entire world was beginning to expand steel output to meet burgeoning demand. From 1950 through 1960 the compounded annual growth rate for all producers was 5.8%; if U.S. output is excluded, then the rate increases to 8.8%.[103] Much of this activity was either coordinated or financed through public agencies. Indeed, after the war formal state participation in steelmaking was more the rule than the exception for most producing countries. Given the significance of steel to an industrialized (or industrializing) economy as well as the historic reliance of many nations on state economic planning, it is not surprising to observe this development.[104]

In spite of the impressive record of growth registered by foreign steelmakers, opinion among U.S. producers remained strongly in favor of free enterprise rather than state planning. This applied to most economic endeavors. Businesspeople, it was argued, should undertake their activites according to the rules of free competition. The proper role of government was to stay out of the marketplace to as great an extent as possible (though vigorous public efforts in the prevention of collusion or other anticompetitive behavior were encouraged). The administration was a firm believer in these principles even if it realized the difficulties that often complicated their implementation. Notwithstanding such complications, Eisenhower still chose to allow business the freedom to pursue decisionmaking independent of public controls. Economic planning by government was never an important part of this agenda.

For the steel industry this outcome is not surprising. The industry was enjoying record production and profits by 1955, hardly conducive to building a public consensus for specialized economic assistance and trade protection. Moreover, the many public confrontations that steelmakers had had with the government over the past decade also reduced the level of sympathy that might otherwise have accompanied such requests. The door to governmental assistance with their trade problems being closed, the steelmakers now had to turn to other means to obtain market stability and insurance against financial deterioration. These involved activities within the domestic environment of the early Eisenhower presidency, the subject of Chapter 7.

7

A New Strategy for Competition: Eisenhower and the Steelmakers, 1953–56

The Eisenhower administration did not provide the aid and assistance to international problems that large steelmakers had hoped for. On the other hand, the administration did initially fulfill industry expectations in the domestic arena. Unlike the previous Democratic regime, there was relatively little overt presidential intervention in steel industry decisions regarding prices, labor relations, capacity expansion, and distribution during the years 1953–56—though Congress, it must be acknowledged, remained an irritant as far as producers were concerned. "When we came into office," the president wrote his brother in late 1954, "there were Federal controls exercised over prices, wages, rents, as well as over the allocation and use of raw materials. The first thing this Administration did was to set about the elimination of those controls."[1]

It was in the context of a noninterventionist public policy that the steel firms formulated their domestic economic strategy for the 1950s. That strategy was essentially founded on returning the industry to sustained financial strength, something it hadn't enjoyed with any consistency since the 1920s. Prices would be raised as necessary to obtain a favorable return on investment, dividends would be stable, and plant and equipment would be prudently modernized. Input-factor costs, on the other hand, would be held in check as much as possible, while any increases would be promptly recovered through higher prices if productivity improvements were not forthcoming.

Large-scale expansion, however, was still considered too risky in view of market uncertainties and existing capacity levels. Walter S. Tower, president of the American Iron and Steel Institute (AISI), stated at the group's 1952 annual meeting that a projected capacity level of 120 million tons by the end of that year (something the Truman administration had advocated) was too high: "There is growing doubt whether all of that capacity will be needed when completed, and there is fear that tough problems are ahead if it cannot be used profitably."[2] The fear, of course, was that overcapacity would lead to price cutting and subsequent financial red ink, just as it had in the 1930s. This was the same concern that industry leaders had voiced throughout Truman's frequent urgings for steel expansion. "There isn't any question about it," stated Benjamin F. Fairless, U.S. Steel Corporation chairman, in May 1953, "we are going to find ourselves with too much capacity in a very short time. What are we going to do about it? Are we going to brag that we lost less money than a competitor?"[3]

A key ingredient in the steelmakers' plan for economic recovery under Eisenhower was the avoidance of any rapid capacity increase that might result in idle furnaces and low mill utilization rates. Domestic overcapacity might not have been a problem in terms of foreign competition if an understanding with such producers was in place, nevertheless it could still prevent a return to financial stability owing to intraindustry competitiveness stemming from soft domestic markets. Steel leaders therefore insisted that any major capacity expansion would take place only if tax laws were changed to allow for an accelerated recovery of investment, thus mitigating the negative financial impact and reducing the risks.[4]

By the end of Eisenhower's term in office, however, the industry's financial position was in serious disarray and its domestic strategy in shambles. Expansion—so carefully avoided in talk—was enormous. Forty million tons of ingot capacity were added from 1953 to 1960, representing an increase of 37% and costing more than $8.9 billion in new investment (compared to new plant and equipment outlays of only $5 billion from 1945 to 1952; see Tables 2.1, 2.3, and 7.1). This rapid growth would have severe consequences for the industry's future, not only in terms of the burden required to finance it, but also because of the nature of the expansion program that was undertaken.

I

The addition of new steelmaking capacity during the 1950s must be seen in terms of two overriding and related constraints: First, expansion was initially undertaken only with reluctance, and, second, as a result output was increased not by the erection of entirely new integrated mills, but rather by the rounding out of existing faciltiies. This pattern of growth was fundamentally different from the expansion taking place overseas. In the latter case, whole industries were being rebuilt according to a calculus of efficiency that would often insure the lowest possible cost of operations utilizing the latest technology. New mill sites were chosen not for historical reasons, but rather for their proximity to efficient bulk-handling facilities to ease the heavy transportation costs involved in steelmaking; this usually meant the selection of deep-water coastal locations for new mills.[5]

In the United States, however, U.S. Steel's 1.8 million-ton Fairless Works

Table 7.1 American Steel Industry Financial Data, 1953–60[a]

Year	*Revenue*	*Before-Tax Earnings*	*After-Tax Earnings*	*Dividends*	*Depreciation*	*Retained Earnings*	*New[b] Investment*	*Long-Term Debt*	*Stockholders Equity*	*Debt/ Equity Ratio*
1953	$13,156	$1,732	$ 735	$324	$614	$411	$ 988	$1,327	$ 6,781	19.6%
1954	10,593	1,230	637	343	670	294	609	1,486	7,140	20.8
1955	14,049	2,204	1,099	436	737	662	714	1,547	7,920	19.5
1956	15,272	2,159	1,113	508	748	606	1,311	1,568	8,665	18.1
1957	15,592	2,213	1,132	566	766	566	1,723	1,802	9,466	19.0
1958	12,551	1,523	788	540	673	247	1,137	2,145	9,898	21.7
1959	14,233	1,637	831	553	665	277	934	2,303	10,249	22.5
1960	14,221	1,583	811	564	692	240	1,505	2,488	10,545	23.6

[a]All figures $000,000.

[b]New investment in plant and equipment.

Source: AISI, *Annual Statistical Report* (various years).

on the East Coast was the one exception to a program of capacity expansion by means of the rounding out alternative. Indeed, capacity growth was often only an undesired side effect to the process of modernization. Rounding out is achieved by the addition of piecemeal facilities to a mill already in existence; often this can occur only at the expense of efficiency because new operations must be shoehorned into already cluttered grounds, thus sharply increasing product handling and other related costs. Mills that gained capacity as a result of rounding-out programs were obviously less efficient than comparably sized mills designed and built from the ground up as an integrated unit (i.e., so-called "greenfield" mills).

The question that must be raised is why American producers chose the rounding out rather than the greenfield alternative. Several reasons appear significant: first, industry forecasters initially miscalculated long-term demand; second, when expansion finally was undertaken, a premium was placed on the speed with which a new plant could be put into operation; and third, the cost of rounding out was considerably lower than that required to build a wholly new integrated mill that provided the same amount of capacity addition. All three of these factors are closely interrelated. The first two will be considered below, whereas the issue of cost will be deferred to later in this chapter (where it can be better examined in conjunction with industry pricing problems and governmental intervention).

We have already pointed to the initial reason for rounding out: Industry leaders remained convinced that steelmaking was a mature industry in which per capita demand would not increase much over existing levels; therefore, it seemed unreasonable to sink large sums of money into expensive new mills for which there might be little long-term demand. This position was supported by the same reasoning that animated industry arguments with Louis H. Bean in the late 1940s (discussed in earlier chapters). It was not until the mid-1950s-or-so that company attitudes began to change, but the reasons for this are difficult to explain in any systematic fashion. In one careful analysis of the question, economist Henry W. Broude states:

> All of [my evidence] . . . suggests that change in perspective, revision of expectations regarding demand, and some more deep-seated and perhaps far-reaching revision of outlook with regard to growth of the economy *did* manifest itself in selective expansion within specific firms.[6]

Perhaps the collective historical consciousness of the industry regarding the overcapacity problem of the 1930s finally began to evaporate in the face of continued high demand. As we noted before, this recollection of history was a powerful factor in the constrained record of growth exhibited by the industry from 1947 through the early 1950s. But while it was not until the mid-1950s and beyond that any consistent beliefs about higher demand schedules began to predominate with industry planners, by then it was too late to make much difference.[7]

The pattern of expansion can be seen in Table 7.2. Four periods are recognizable: (I) 1945–49, (II) 1950–54, (III) 1955–56, (IV) 1957–60. From 1945

Table 7.2 Investment and Growth in Steel Capacity, 1945–60

Phase		*Investment ($000 000)*	*Net Capacity*[b]	*Net Capacity Addition*	*Phase Total*[b]
I	1945	$ 115	95.5	1.6	
	1946	365	91.9	(3.6)[a]	
	1947	554	91.2	(0.7)[a]	
	1948	642	94.2	3.0	
	1949	483	96.1	1.9	2.2
II	1950	505	100.0	3.9	
	1951	1,051	104.6	4.6	
	1952	1,298	108.6	4.0	
	1953	988	117.6	9.0	
	1954	609	124.3	6.7	28.2
III	1955	714	125.8	1.5	
	1956	1,311	128.4	2.6	4.1
IV	1957	1,723	133.5	5.1	
	1958	1,137	140.7	7.2	
	1959	934	147.6	6.9	
	1960	1,505	148.6	1.0	20.2

[a]Worn-out war-time capacity was retired in these years.

[b]Net tons 000 000.

Source: AISI, *Annual Statistical Report* (various years).

to 1949 (as we saw earlier) overall capacity growth was low while the industry fought Truman's calls to expand. But beginning in late 1950 and lasting through early 1954, expansion was vigorous: Capacity grew by 29.3% in the five-year interval. The primary stimulus, however, was not a change in industry attitudes, rather, it was the Korean War boom and the temporary availability of accelerated depreciation for defense-related investments. The latter was administered through certificates of necessity issued by the Office of Defense Mobilization (ODM), which estimated military demand in light of existing national capacity and then provided fast writeoff authority as well as subsidized funding to firms that would enlarge to meet projected shortfalls. The special legislation covering this program was in effect from mid-1950 through December 1953 for steel furnace construction and up to December 1956 for certain other types of steelmaking facilities deemed in short supply.[8] (The extensive capacity additions shown on Table 7.2 for the years 1953 and 1954 reflect the completion of projects begun earlier under ODM certification.)

The nature of this defense-led expansion, however, was strongly tilted toward balancing existing plants (i.e., rounding out). This usually occurred through the addition of new finishing facilities to match existing ingot furnace facilities already in place or through the addition of new furnaces to supply existing finishing mills.* Of the first $1.8 billion in ODM certificates granted

*A finishing mill converts the slabs, billets, or blooms that come from the furnace into the various shapes that go into final products, such as hot strips, plates, structurals, and rails.

to steelmakers, only the Fairless Works involved a wholly new integrated facility (it received $326 million in certification).[9] Some observers criticized this rounding-out method of expansion, charging that the steel producers seemed to be pursuing purely commercial goals at the expense of the nation's military objectives.[10] But there was no doubt that steel output could be brought to market faster through such piecemeal expansion rather than by the longer procedure of first designing and then building a completely new steelmaking complex on virgin ground. Under intense pressure to turn out sufficient metal for national security purposes, the industry chose the most expeditious route available during the 1950–54 boom years.

With termination of the certificate of necessity program, a third phase of industry expansion got underway in which investment slowed considerably. This new phase, however, was in keeping with the industry's domestic strategy under the Eisenhower administration, in which investment was to be moderated. Responding to criticism that his firm did not spend enough on improvements and innovations in the early 1950s, a Bethlehem Steel Corporation vice president explained:

> The technical improvements we wanted to install had not been fully proved to our satisfaction. . . . We don't move until we are sure we can get the best possible yield out of a new technique or facility. Then we move as fast as we need to and no faster. Unlike many who can't wait, we don't have to add to capital unless we are sure it will pay out.[11]

This reply was fairly representative of most American steelmakers at that time. Thus, not only was their capacity expanded by rounding out rather than through construction of wholly new mills, but the technology utilized in the process remained conventional. The open-hearth furnace, the original version of which was pioneered in the late nineteenth century, accounted for 88.4% of U.S. ingot capacity in 1946. By 1955, it still had 87.8% of the total.[12] Only one small domestic producer chose to install a revolutionary new basic-oxygen-furnace (BOF) technology that had been developed in Europe in the late 1940s. This occurred in 1954 when tiny McLouth Steel blew in a small 540,000-ton BOF mill at its Detroit plant.[13] The second American BOF facility was not built until 1957, when Jones & Laughlin opened a 675,000-ton installation. It remained until the 1960s before the leaders of the domestic industry turned to BOF in volume.[14] Yet, the economic advantages of BOF relative to open-hearth technology seemed striking: BOF installation costs were on the order of 50% lower and operating costs were $3.00 to $12.00 per ton less than open hearth operating costs. A "heat" of steel using BOF technology could be made in forty minutes, whereas a minimum of six hours was required by the fastest open-hearth furnaces in operation.[15] Given these advantages, some steel industry critics have charged that the American producers sowed the seeds of their future decline by passing up the BOF technology when it first became available and by foolishly opting instead for the outmoded open-hearth method.[16]

Numerous analyses of the diffusion and adoption of the BOF process have been made, of which most are critical of domestic producers.[17] The general thrust of this literature is that the major firms enjoyed monopoly pricing powers in the market and thus had no incentive to innovate or reduce costs; only after 1959, when foreign competitors had built up efficient new capacity and exported volume tonnage into the United States at steep discounts, did the domestic producers feel compelled to undertake their own cost-reducing innovation programs. But by this time, according to the charges, the financial stability of the firms had been seriously eroded, and they could never fully recover their once predominant position in the world market. "Does it not follow," asks perhaps the most influential of the critical commentaries, "that [the large steelmakers' problems] are in part the result of self-inflicted injury?"[18]

We have already noted, however, that the steelmakers had substantial and compelling incentives to reduce costs given the improved state of foreign competition. This was true throughout the decade, especially for the largest firms (U.S. Steel and Bethlehem) who, due to their exposure in coastal markets, had the most to lose from import invasions. The monopoly-power argument may be a necessary cause of the producers' inertia toward innovation, but it is not sufficient (and doubts may be raised as to the former). On the other hand, there is no question that the industry was slow to adopt BOF; prior to 1960 only two small installations were built in the United States. But what seems to best account for this phenomenon is that the internal administrative routines that prevailed at the major producers were simply not well attuned to quick changes of any kind.[19] It was far easier for managers to deny the efficacy of BOF than to recommend it: The process had never been tested on large-scale plants comparable to those prevailing in America; there were no ready sources of bulk oxygen available; the process emitted unacceptable amounts of pollutants into the atmosphere; significant progress in open-hearth oxygenation procedures was being made; total BOF costs were uncertain; the rush to expand, especially during the Korean emergency, favored utilization of known and reliable methods; and so on. The combination of these factors considered against typical decisionmaking patterns found in most large firms (and characterized generally as "technological momentum" by our leading historian of technology) appears to provide a sufficient explanation of why BOF was not adopted earlier by major American steelmakers.[20]

Such a conclusion does not, of course, excuse the firms' failure to innovate (although it does remove the image of duplicity surrounding industry behavior implied in the monopoly argument). There is obviously a relationship of some sort between large firms with fixed resources like U.S. Steel and their passivity toward new ideas.[21] Yet to suggest that antimonopoly remedies are the only answer—as many industry critics tended to do—reveals a strikingly bare arsenal of public policies to meet such an important national problem.

We shall return to the issues surrounding innovation in Chapter 9. Meanwhile, suffice it to say that a new phase of expansion got underway in late 1957. Like previous growth, this too was based on a rounding out of existing mills rather than any greenfield developments or utilization of new technologies. Indeed, no such fully integrated mill project would take place in America until the early 1960s when Bethlehem Steel began construction of a 2-million-ton–complex at Burns Harbor, Indiana, to service the Midwestern market (the last greenfield mill to be built in the United States).[22] The post-1957 growth spurt would carry on into the early 1960s (with the exception of 1958–59 when recession and a long labor strike chilled industry investment programs). Moreover, expansion occurred even though the demand cycle for steel had turned down in 1958 and remained there for over five years. Following that recession, capacity grew only slowly, edging up but 7% over the entire decade of the 1960s compared to a total increase of 48% during the 1950s.[23]

In the face of long-held skepticism by industry leaders toward growth, one must naturally ask why so much expansion occurred during the Eisenhower years. A number of reasons present themselves. As noted before, industry attitudes were changing by the 1950s; Arthur B. Homer, president of Bethlehem Steel, stated in 1952:

> Our studies showed that since 1900 there is a long-term, close correlation between steel production and GNP corrected for inflation. If we accept the Gross National Product Index (and, since 1919, the Federal Reserve Board Index of Industrial Production) as being reflective of the country's rate of growth and increased requirements—the indications are that steel capacity is not now excessive. . . .
>
> Over a period of years, then, we see no reason why the overall trend in steel production shouldn't bear about the same relationship it has historically to GNP and the FRB production index. If that holds, as it should, our expansion has worked out in a way that should be welcomed generally.[24]

Even U.S. Steel's Benjamin F. Fairless, a veteran of the no-growth philosophy of the 1940s, could state by 1957:

> Since steel is the basic ingredient in practically every field of advance, we in the steel industry have a special responsibility that is vast and vital. We must supply the increasing flow of steel needed to build thousands of miles of new highways, thousands of new schools, and other resources required by a swiftly expanding population. . . .
>
> The industry can meet these needs only by going forward steadily with the expansion of its capacity to produce. . . . And we know that, over the long range, if the goal should be to keep output abreast of population, even at today's per capita use, capacity must be added at a rate of at least three million tons a year.[25]

These remarks from the industry's two largest producers express a remarkable change in attitude from the no-growth sentiment that characterized company positions before Congress and the public in the late 1940s. One major reason for the change, no doubt, was the high industry capacity-utilization rates scored up through 1957. In the eleven-year span from 1947 to 1957, the rate averaged 89.2%. This performance record would obviously tend to disabuse those industry skeptics who predicted imminent return to the depressed output levels of the 1930s. Another reason for expansion, however, could be found in the maintenance needs to replace worn and outmoded plant. Much of the replacement process necessarily involved the installation of larger facilities than had previously existed owing to considerations of economic scale that had accompanied new process technologies innovated over the years.[26] There were also powerful intraindustry competitive pressures behind the expansion moves: Once one firm increased capacity, its product and market rivals felt compelled to do the same in order to maintain competitive equilibrium.[27] This same logic also could be applied to the rising industrial rivalry between the United States and the Soviet Union during the Cold War years of the 1950s. The trade press was constantly alert to the narrowing steel ingot gap between Russia and the West and growing production in all of the Iron Curtain countries tended to raise national security concerns in America. One result was a push by military interests to increase domestic steel capacity as a hedge against potential Soviet threats.[28]

Finally, there was also political motivation inherent in steel expansion. Although the Eisenhower administration generally adopted a policy of noninterference with the steelmakers, other branches of government (such as Defense) were not so reluctant to voice their concerns with industry behavior. Congress was also not an uninterested bystander on this topic. With output running at near-capacity levels throughout much of the 1950s (and with imports generally unavailable in large amounts owing to supply-and-demand conditions abroad), some customers found they could not obtain sufficient supplies of steel to remain in business. These firms naturally complained, and they found a receptive audience in the members of the congressional small business committees, much as they had in the 1940s. In conjunction with other objections about supply problems in steel, talk of governmental controls over output consequently resurfaced in the mid-1950s.[29] Although no overt pressure was ever applied on this issue, it seems reasonable to assume that the steelmakers had no desire to give Congress any excuses to reconsider controls legislation. Accordingly, perhaps one of the best means to avoid such a move would be through increased capacity to meet market demands.

In sum, there were many reasons behind the industry's impressive record of growth in the 1950s, a development generally unanticipated by most company leaders at the outset of Eisenhower's presidency and, in fact, relatively opposed by them at the time. But while expansion did eventually occur, the particular shape it took in America never emulated the industry rationalization that often could be found abroad. In the United States capacity additions were achieved through rounding-out procedures rather than by the erection of more efficient new greenfield sites. Moreover, existing technology was utilized, not the new BOF variant. We have already enumerated the industry rationale behind these choices. Our principal focus was on the earlier miscalculations of demand by industry planners and the subsequent need to catchup to real demand as quickly as possible. Another important factor, however, also warrants attention: the high cost of new plant investment and the necessary relation this had to steel industry pricing policies. At existing price levels, steel leaders claimed, they could not afford to invest in new mills.

II

The American steel industry attracted considerable attention in the 1950s from a wide variety of constituencies. Although many issues competed for recognition by these groups, the one that raised the highest level of public interest was pricing. President Eisenhower may have chosen to forego his predecessor's penchant for public castigation of price hikes in the industry, but this did not deter others from taking up the slack. Indeed, at times it must have appeared to industry officials that public dissatisfaction with steel company behavior so common in the late 1940s had never ended. Yet the concern of the critics was justified insofar as the steel sector was an important part of the national economy whose behavior had repercussions beyond its immediate participants. A common problem, however, affected most of those who chose to comment: They restricted their critique to the singular phenomenon of prices. A far more rewarding approach lies in consideration of steel prices in terms of their relationship to overall industry competitive strategy. When this is done, perhaps a better balanced picture of firm behavior emerges than that contained in the political conclusions of so many of the critics who focused only on rising price levels.

Steel-pricing policies, it will be recalled, were central to the 1952 seizure case. Following an end to the fifty-four day strike that the seizure failed to halt, the industry was granted authority from federal price controllers to raise base steel prices by $5.20 per ton (equal to a 7.9% increase in the index of basic steel prices). The new union contract that ended the walkout provided for an approximate wage hike of twenty-five cents per hour scheduled over two years, the first year of which would raise steelworkers' straight-time hourly earnings by 8.8%. In addition, however, the settlement contained a wage reopener clause to be effective before June 30, 1953.[30] Soon after Eisenhower took office, he lifted the emergency wage-and-price controls stemming from the Korean War and returned autonomy in these areas to management and labor.[31] Accordingly, on June 12, 1953, the industry granted the United Steelworkers (USWA) an approximate raise of ten cents per hour under terms of the reopener clause (up 5.1%). On June 16 U.S. Steel announced new prices effective the next day; they would increase the average quote for finished steel by $4.30 per ton (up 4.9%), an amount quickly matched by the rest of the producers.[32]

This surprisingly peaceful conclusion to the industry's seventh wage-price round since the end of World War II triggered a new phase in steel-pricing policies. Economist Gardiner C. Means, a critic of the industry, has described it as follows:

> The rise of steel prices since 1953 presents a different problem from that of the rise in the war inflations. . . . The bulk of the rise before 1953 can be attributed to the excess demand associated with the monetary expansions of the two wars. But from 1953 on there was no monetary inflation and no general excess of demand. As far as wholesale prices are concerned, the war inflations had come to a halt and the price readjustments following them were in large measure complete by 1953. For two and a half years both the wholesale and the consumer price indexes were remarkably stable, though with considerable shifting of relative prices within this stability. Then in the middle of 1955 the wholesale price index started up, followed by consumer prices in 1956. Steel prices shared or even spearheaded this rise.[33]

In the 1930s Means had coined the term *administered pricing* to represent a situation in which oligopolistic firms were able to set prices on criteria other than traditional supply-and-demand forces. The policy of the steel industry in the 1950s, Means argued, was clearly a case of administered prices.

From 1953 through 1959 the industry performed what amounted to an annual ritual of negotiated wage increases quickly followed by price hikes to compensate the producers for their higher labor costs. The index of finished steel prices rose much faster than wholesale prices in general over this period; according to one analysis, steel price hikes alone were responsible for 40% of the inflation in wholesale prices of all finished goods in the United States betweeen 1947 and 1958.[34] The implications of this pattern were to prove disastrous to industry strategists who had counted on a policy of noninterference in their private affairs by the government. With inflation emerging as a major political issue by the mid-1950s, steel-pricing behavior drew critical congressional attention. Charges of monopoly practice were again often heard, and official inquiries into industry structure were once more undertaken.[35] Though steel leaders worked diligently to improve communications with the public, they could not match the highly publicized setting of a congressional hearing designed to bolster a politician's standing with his constituents. Moreover, the plethora of industry data that fell into the public domain as a result of the many governmental inquiries provided evidence for numerous academic studies. Most of these tended to bolster the charges of a monopolistic industry pricing its products without regard for the larger public interest.[36] Indeed, some even went so far as to conclude that price policies in the 1950s were the chief culprit behind the subsequent economic decline in steel; in the words of one analysis:

> The persistent price escalation of steel prices during the 1950s was the primary cause of the industry's lackluster performance during the 1960s—resulting in the erosion of dometic markets by substitute

> materials and imports, the loss of export markets . . . , and the decline in return on investment.[37]

Given these sweeping charges by leading critics of the steel industry, it is important to understand why the producers acted as they did in regard to prices. At the outset, however, one must note that there can be little doubt that the firms did price according to oligopolistic models of behavior. The United States Steel Corporation (up to 1958, at least) usually signaled the price for each of the major product lines in the industry; remaining firms then fell quickly into agreement. Discipline was relatively tight throughout the 1950s, with little detectable price shading even during the soft markets of 1953–54 and 1958–59.[38] And (as we have seen) steel prices did march inexorably upward throughout the decade, regardless of demand conditions, price movements in other goods, or anything else (see Table 2.4).[39]

To understand this behavior one must turn to an analysis of the large steelmakers' perceived competitive position, not only in the United States (where the firms competed with substitute products as well as with one another), but also in the world market.[40] The price strategy of the industry throughout the 1950s reflected a persistent belief that it must modernize if it were to remain a leading factor in the future configuration of the global steel trade. With the major steel-producing nations of the world rapidly expanding productive capacity, it was well understood by domestic producers that only the most efficient firms would dominate in the years ahead. The restraints put on steel prices by the Truman administration, however, inhibited the firms from generating sufficient capital to finance new investment in plant and equipment; not only were internally derived funds inadequate, but the industry's weakened financial position that resulted also crippled its standing in the external capital markets, thus limiting both debt and equity routes to the raising of needed investment funds[41] (see Tables 7.3 and 7.4).

Moreover, the artificially low price of steel that governmental restraints had mandated tended to stimulate excess demand for the good—which, in turn, could not be met through existing capacity, thus bringing further pressures to expand and adding to the financial burden of the producers.[42] Given this situation, it was concluded by industry leaders that a strategy of price increases had to be pursued to improve internally generated investment funds. It was also argued that the government should alter depreciation rules to allow for accelerated amortization of steel investment and that replacement cost accounting procedures should be substituted for existing historical cost methods.[43] In addition, of course, the firms also believed that a program of cost reduction should be pursued wherever possible; but if costs could not be contained, the plan called for them to be passed along in the form of higher prices.

The dominating factor that motivated this industry strategy was a belief that prior governmental policies had hindered efforts to raise investment capital for modernization. Through public jawboning over prices and a perceived Democratic party bias toward labor in wage disputes, industry leaders felt

Table 7.3 Iron and Steel Industry Income and Investment Compared to Total Manufacturing Income and Investment, 1947–60

	(A) Total Manufacturing Investment[a] (000,000)	(B) Iron and Steel Investment[a] (000,000)	(B)/(A)	(C)/(D)	(C) Iron and Steel Income[b] (000,000)	(D) Total Manufacturing Income[b] (000,000)
1947	$ 8,703	$ 638	7.3%	na	na	na
1948	9,134	772	8.5	6.6%	$4,377	$ 66,777
1949	7,149	596	8.3	6.3	3,951	62,702
1950	7,491	599	8.0	6.9	5,155	74,371
1951	10,852	1,198	11.0	7.5	6,614	88,495
1952	11,632	1,511	13.0	6.1	5,545	90,172
1953	11,908	1,210	10.2	6.9	6,762	97,953
1954	11,038	754	6.8	6.1	5,547	91,057
1955	11,439	863	7.5	7.2	7,510	104,490
1956	14,954	1,268	8.5	7.0	7,652	109,268
1957	15,959	1,722	10.8	7.3	8,266	112,476
1958	11,433	1,192	10.4	6.7	6,987	103,817
1959	12,067	1,036	8.6	6.6	7,908	119,929
1960	14,480	1,597	11.0	6.6	8,075	121,987

[a]Investment refers to plant and equipment expenditures.

[b]Income refers to that portion of national income originating in the iron and steel industry and that portion originating in total manufacturing.

Source: U.S. Department of Commerce and Securities and Exchange Commission. (Reproduced in U.S. Congress, JEC, 88th Congr., 1st sess., Hearings, *Steel Prices, Unit Costs, Profits, and Foreign Competition* [Washington, DC: GPO, 1963], pp. 177, 189).

Table 7.4 Stock Issues and External Long-Term Financing, Primary Iron and Steel and all Nonfinancial Corporations, 1951–60

	Primary Iron and Steel			All Nonfinancial Corporations		
Year	External Long-term Sources[a]	Stocks[a]	Stocks as Percentage of External Long-term Sources	External Long-term Sources[a]	Stocks[a]	Stocks as Percentage of External Long-term Sources
1951	$432	$ 67	15.5%	$ 7,747	$2,700	34.9%
1952	445	(43)	(9.7)	9,397	2,987	31.8
1953	2	69	3,450.0	7,551	2,266	30.0
1954	277	(8)	(2.9)	6,341	2,065	32.6
1955	263	222	84.4	8,625	2,696	31.3
1956	254	230	90.6	11,061	3,147	28.5
1957	382	257	67.3	11,913	3,457	29.0
1958	497	122	24.5	10,854	3,564	32.8
1959	285	63	22.1	9,541	3,706	38.8
1960	316	147	46.5	9,775	3,014	30.8

[a]All figures $000, 000.

Source: U.S. Congress, JEC, 88th Congr., 1st sess., *Hearings on Steel Prices, Unit Costs, Profits, and Foreign Competition* (Washington, DC: GPO, 1963), p. 281.

prices had been kept artificially low and costs artificially high. So now under the new Republican regime, it was argued, the industry had to catch up for lost time incurred under Truman. The threat of foreign competition was a critical variable in the industry's strategic decisionmaking at this time, just as it had been throughout the twentieth-century history of American steel.[44] This threat appears to have been far more powerful in motivating industry improvements than any desires for monopoly profits that might accrue to shareholders or managers (as some charged).[45] Dividend policy in the 1950s appears to have been principally designed to enhance the attractiveness of firm shares to potential investors, thus easing barriers to capital formation. In turn, price policy was essentially designed to strengthen the firms' ability to compete in the long-run against traditional offshore rivals (i.e., by generating sufficient cash flows to allow for investment in plant modernization).

Various public pronouncements made by steel leaders (although obviously self-serving at times) are consistent in their emphasis on the need to raise capital to fund modernization. "Steel's greatest problem today," characteristically stated Benjamin F. Fairless in early 1956, "is to get the money required to carry the expansion that the country expects."[46] A similar theme runs through industry testimony both before and after this date.[47] One example, provided by Bethlehem Steel, well illustrates the problem:

> To expand to meet the nation's needs, including a safe margin for security purposes, is one of the problems the industry is facing. . . .
>
> [But] with today's facility costs inflated in relation to the average cost of existing plant, the industry is running out of opportunities to add additional capacity at costs which would justify the investment at present earning levels. Some of us are in better shape than others in this regard, but all of us who are going to expand will be faced sooner or later with the necessity of adding new fully integrated facilities—from raw materials to finished products—at costs ranging to $300/400 per ton of annual ingot capacity. . . .
>
> One of our officials put the matter this way. . . . Given a 90 per cent [utilization] operation, with the current price structure and a representative mix of rolled steel products, a million tons of ingot capacity would produce billings of some 85 million dollars. Take a profit margin on sales such as Bethlehem had last year—8.5 cents net income per dollar of total revenues—and your net profit per ton is around $7 per net ton of ingot capacity, which isn't too good a result on an investment of $100 per ton [rounding out costs] and three times as bad if the investment is $300 [greenfield capacity costs]. Since we may assume that new facilities will be more efficient, the $7 figure could be raised somewhat, but the story would remain essentially the same.[48]

According to industry spokespersons, the answer to this dilemma could be provided only by higher prices, otherwise expansion could not proceed.[49] Some help, however, might be derived from changes in depreciation schedules, so the industry lobbied on their behalf. The certificate-of-necessity program that had expired in 1953, it was suggested, should be reinstated. Ironically, the person most instrumental in killing this proposal was Secretary of the Treasury George M. Humphrey—a businessman with strong ties to the industry who, on leaving public office in 1957, would assume the chairmanship of the National Steel Corporation. Humphrey's position on the subject was that any extensions of accelerated depreciation represented unwise governmental meddling in the economy, hence, they should be opposed.[50] Eisenhower agreed philosophically with his persuasive secretary. Lacking a coordinated policy for steel, the president was never much committed to pursuing any new paths that might stabilize the producers' financial health. The steel industry, although high on nearly everyone's priority list of key domestic sectors, was clearly expected to earn its own way into prosperity according to traditional economic rules for success.

The pressure of inadequate finances also played a major role in the industry's choice of the less efficient rounding-out method to meet expansion needs. Rounding-out capacity could be obtained for approximately $80 to $100 per ton in the 1950s; greenfield capacity, however, required construction costs of $300 to $400 per ton under standard technology.[51] From 1953 to 1960 after-tax profits per ton of steel shipments averaged $12.08 for the industry; assuming an investment cost of $350 per ton and a 90% utilization rate, returns would thus average only 3.1% after taxes for greenfield output. Yet, with an investment of $90 per ton and again a 90% utilization rate, rounding out would yield 12.1% (see Table 7.5 for details of the calculation). Given

Table 7.5 Steel Industry Return on Investment, 1953–60

Year	*Net Profit (000, 000)*	*Net Shipments (000 tons)*
1953	$ 735	80,152
1954	637	63,153
1955	1,099	84,717
1956	1,113	83,251
1957	1,132	79,895
1958	788	59,914
1959	831	69,377
1960	811	71,149

Total after-tax profits: $7,146,000,000 in 1953–60.
Total net shipments: 591,608,000 tons in 1953–60.
Therefore, profits ÷ shipments = $12.08 per ton.

New mill construction cost = $350 per ton of capacity. (But assuming a 90% capacity utilization, then $350 ÷ 0.90 = an adjusted construction cost of $389 per ton of capacity.) Therefore, $12.08 ÷ $389 = 3.1% return on investment, *ceteris paribus,* between 1953–60.

Rounding out construction cost = $90 per ton of capacity. (But assuming a 90% capacity utilization, then $90 ÷ 0.90 = an adjusted construction cost of $100 per ton of capacity.) Therefore, $12.08 ÷ $100 = 12.1% return on investment, *ceteris paribus,* for 1953–60.

Source: Data used in calculations are from the AISI, *Annual Statistical Report* (various years).

these returns, the productivity benefits deriving from a greenfield mill would have to be substantial to overcome the relative financial attractiveness of rounding out. Company decisionmakers obviously did not believe that greenfield mills could provide these, and, as a result, modernization and expansion were limited to the less efficient rounding-out alternative when they occurred.

The firms, of course, tried to put the matter in a positive perspective. As one example, Bethlehem Steel in 1946 had an ingot capacity rating of 12.9 million tons in all its plants, derived from a total of 143 steelmaking furnaces. By 1957 the number of furnaces had grown to only 147, but capacity jumped to 20.5 million tons (as 13 new furnaces were installed, 9 retired, and 123 redesigned and rebuilt). "In simple terms," wrote the company, "this illustrates the nature of Bethlehem's growth. It has resulted largely from improvements accomplished through good management at a notably low cost per ton."[52] To demonstrate its efficiency the company presented the data in Table 7.6 showing the average cost per ton of capacity expansion (both rounding out and greenfield) for the top eight steel producers between 1945 and 1956.

The average cost for U.S. Steel was not notably low, yet it was skewed upward by the impact of building the industry's only greenfield mill.[54] Thus adoption of the rounding-out approach and the avoidance of new technology may have resulted in lower investment costs and also may have put additional capacity into operation faster, but these advantages were not achieved without serious drawbacks. One analysis noted:

> Capacity [by rounding out] is increased by adding to existing facilities—to the point where most mills are now so hopelessly cluttered that any attempt at efficient operation in the charging-room floor is hopeless. Rounding-out is popular because it costs only about $100 a ton of capacity, but is obviously no long-term solution to the production of steel. Eventually, in those steel plants, something has to give.[55]

What had to give, of course, was efficiency. The short-term benefits of rounding out would in the longer term prove unwarranted. But as far as the industry was concerned, the producers had no choice in the matter: External con-

Table 7.6 Per-Ton Steel Expansion Costs, 1945–56

Firm	*Average Expansion Cost (per ton)*
Armco Steel Company	$138.64
Bethlehem Steel	150.26
Inland Steel	166.48
Youngstown Sheet & Tube	174.83
National Steel	249.67
Republic Steel	270.42
U.S. Steel	278.88
Jones & Laughlin	341.66

Source: Bethlehem Steel Corporation.[53]

straints, they believed—primarily governmental policies that favored lower prices—had forced them to select this alternative. Within a relatively few years' time the miscalculation would be apparent to all.

The opposed basic industry view—that higher prices were necessary and justified in order to finance expansion—was not without its own flaws, of course. For example, if the competitive strategy of the producers was formulated in relation to perceived threats from European steel, then a policy of raising prices to better meet cheaper foreign competition is obviously illogical. Yet this is precisely what the industry argued. Speaking in 1962, U.S. Steel Chairman Roger M. Blough (successor to Benjamin Fairless) made an interesting comment that had equal relevance for industry price strategy in the 1950s:

> While the [recent] price rise might have appeared to intensify our competitive difficulties with cheaper foreign steel, that steel is usually priced in relation to ours anyway, and in the long run, the increase would have improved our competitive strength. By using the added profits produced by the price increase to help obtain the most modern and efficient tools of production, we could hope *eventually* to narrow the gap between American and foreign steel prices. [Emphasis added.][56]

The critical miscalculation here is that for Blough—and indeed for the entire American industry—foreign competition seemed to consist only of European producers. If this had been the case, then his comments regarding joint movements in prices would be generally correct (in terms of pricing, domestic and continental producers displayed remarkable coincidence when it came to their behavior in U.S. markets). But to the dismay of both the American and European steel industries, another nation would soon rise up to challenge the historic hegemony of the West. This, of course, was Japan, whose steel industry had no previous involvement with international pricing agreements nor any shared philosophy on export pricing. The price strategy of the American steel industry, so carefully formulated during the 1950s to raise capital in order to address the threat of future European competition, essentially ignored Japan as a potential rival. This would prove to be a fatal mistake, as we shall see.

The domestic steel industry's price strategy failed in the long run insofar as it did not sufficiently address all foreign competitors. But it also failed over a shorter time span. This was due, interestingly enough, to mounting criticism of steel prices from the government, a development that industry leaders found surprising given the ideological posture of the political party then in power. It is thus ironic that governmental pressures influenced the steelmakers into adoption of their ill-fated strategy and that governmental pressures also narrowed its chances for success. To complete the paradox, this failure occurred under the general auspices of a nominally friendly Republican administration.

The public pressures that brought about the shorter-term failure were applied on three main issues: prices, labor wage rates, and industry profits. The adhesive that bound them together, however, was inflation. As noted earlier, the rise and attempted containment of inflation played a central role in President Eisenhower's domestic economic program throughout his term in office. Inflation was initially taken up as an issue from which political advantage might be seized. Administration economic counselors wanted to demonstrate that inflation could be managed by Republicans without resort to wage–price controls, as had been the case during much of the Truman years.[57] Unfortunately, the approach did not work as planned, and the inflation issue turned into a political lightning rod for attacks by the opposition party. Throughout Eisenhower's eight years in the White House (but especially after 1955), he presided over a climb in nearly all prices save farm products and processed foods. This occurred despite three recessions in his presidency, and it led to fears that unless mandatory controls were imposed, persistent inflation would be a necessary if unwelcome side effect to a full-employment and growth-oriented economy.[58]

The inflationary surge of the Eisenhower years did not get underway until 1956, although it had been building up for some time before that. The president often voiced his personal concern with the problem, noting that "we watch it closely every day."[59] Yet behind this remark was a lack of consensus among White House advisors over the real nature, and dangers, of inflation. Arthur Burns, Eisenhower's first Council of Economic Advisors (CEA) chairman, argued that a degree of inflation was unavoidable in the economy, hence extraordinary efforts to eradicate it would be useless. His successor, however—Raymond J. Saulnier, who served as CEA chair from 1956 to 1961—took a decidedly more aggressive posture, declaring that inflation was the principal economic problem facing the nation. This latter viewpoint dominated much of the president's domestic program in his second term. Long-run price stability, Saulnier counseled, should be the objective of federal policy, and the government should undertake no action that might detract from such a course.[60]

The 1950s rise of inflation as a national problem sparked a serious academic debate about its causes. Previous economic theory generally considered inflation to be a function of macroeconomic policy, thus subject to control through appropriate fiscal and monetary measures (the specific choice of tools, however, being a matter of further debate within the discipline). But the "new" inflation of the post-1955 period seemed impervious to traditional treatment. In consequence, several economists turned to an examination of the institutional structure of the economy for insight into inflation's roots. Yet opinion here was no less divided, though the split was more often along political rather than theoretical economic grounds. Liberal economists found the cause of inflation to be administered prices as practiced by concentrated, monopolistic industries. Firms in this environment, it was argued, jacked up prices purely for the sake of profit improvement. Accordingly, wage-and-price controls (along with stricter antitrust enforcement) were prescribed as the necessary remedy. Conservative economists, however, often found the stimulus to rising prices not in price-pull inflation but rather in wage-push movements. In this case powerful unions were indicted as the culprit and a policy of governmental intervention into wage negotiation procedures was recommended.[61] Under Saulnier's lead, traditionalists in the administration reluctantly turned to this latter approach after 1956.

Congress, however, tended to favor the administered-price interpretation. In a highly publicized series of hearings conducted from 1957 through 1961 by Senator Estes Kefauver (D-Tennessee), a number of leading industries were thoroughly examined to determine the nature of their price policies.[62] Congress, it should be noted, grew progressively more Democratic from the eighty-fourth through the eighty-sixth seating (1955–60); only in the first two years of his first term did Eisenhower enjoy Republican majorities in either house.

It was against this background of rising politicization of the inflation issue that the steelmakers' price strategy was launched. The industry, we noted, was determined to raise prices as a means to improve cash flows necessary for capital improvements. The inauguration of the Eisenhower administration was believed to signal a more favorable political climate in which to accomplish this objective. Accordingly, prices rose steadily from June 1953 up until mid-1956 with relatively little public opposition from the White House.[63] As can be observed in Table 2.4, the index of basic steel prices rose 18.6% from 1953 to 1956, whereas the nation's wholesale price index (WPI) was up only 3.8% in the same period. In conjunction with accelerated depreciation benefits (five-year Korean War certificates of necessity were still in effect during these years), industry profits were also up handsomely, especially after the recession year of 1954; after-tax profits rose by 51% from 1953 to 1956. In light of these developments, the mood of the industry began to improve. By late 1955 *Iron Age* could state:

> Basic to this new feeling of confidence is that the recession year 1954, during which industry-wide ingot operations fell as low as 61 per cent, was nevertheless a good year for steel profits. It's axiomatic in any industry that if profit margins will stay up when operations go well under capacity, the future is secure.[64]

Thus, by the end of 1955—a record year for both output and profits—the steelmakers' strategy for restoring financial strength under a friendly administration seemed to be working well.

Such optimism, however, would soon turn illusory. This is because public opposition began to mount regarding the perceived effect that steel prices were having on the national economy as a whole. Inflation (we saw) began to climb upward in early 1956. With elections to be held in November of that year, the president did not wish the problem to escalate into a campaign issue favorable to the Democrats. As early as mid-1955, in fact, Eisenhower had privately expressed his displeasure over steel prices, stating after a new round of hikes that he "was pretty disgusted with business men [who] . . . can't wait more than a few hours to raise the price of steel 7 dollars a pound [*sic*]" and he "didn't know when he would get over it."[65] Throughout his second term the president's disenchantment with the activities of Big Steel leaders would grow.

But Eisenhower was, of course, not the only public representative to register disgust with the industry. Congress also became involved in this matter. In November 1955 the Joint Economic Committee (JEC) announced that it would begin examining the facts behind the prior July price hike alluded to by Eisenhower. The committee was particularly concerned with the question of why the steelmakers hadn't absorbed the higher wage costs they granted the steelworkers' union in contract renegotiations preceding the hike.[66] Public hearings on the wage–price spiral in steel would be held soon, it was promised.

The proposed congressional inquiry represents a renewal of public concern with the industry policy of granting relatively high wage increases and then immediately recouping them through price hikes. When Eisenhower was first elected to office in late 1952, both management and the union were still recovering from the bitter seizure and strike of the previous summer. In that episode the producers were adamant in their intention to grant raises only on the condition of accompanying price boosts because they feared the international economic repercussions of rising costs and shrinking margins. The union, for its part, seemed equally determined to improve the economic status of its members. Yet the unexpected death of Philip Murray, head of the United Steelworkers of America (USWA), just before Eisenhower's election raised hopes that a less strained labor–management relationship would soon develop. David J. McDonald, veteran secretary-treasurer of the union who was named to succeed Murray, seemed less committed to the adversarial posture of the past and more likely to seek labor gains through quiet discussion rather than hostile confrontation.

Steel management officials were pleased with this change in USWA leadership, and they appeared willing to help the generally conservative and less cerebral McDonald strengthen his somewhat tenuous hold over the union. Accordingly, in the 1953 wage reopener negotiations a settlement was struck well in advance of deadlines, a generous raise was provided the workers, and both sides of the bargaining congratulated each other on their "statesmanlike conduct" and "honorable compromise."[67] In late 1953 and early 1954 McDonald and Benjamin Fairless of U.S. Steel made a joint tour across the nation to that firm's mills, generating favorable publicity about the new mood in steel's labor relations.[68] Even President Eisenhower liked the affable McDonald; following a meeting between the two in April 1954, he wrote in his diary that "my own opinion of [him] is very high. I . . . have found him to be completely trustworthy and a man of great judgment and common sense."[69]

In June 1954 a new contract between the industry and the USWA was again renegotiated peacefully; even though recession had set in and capacity-utilization rates were down to 70%, labor received another raise. *Iron Age* commented, "The fact that generous terms of the settlement helped McDonald politically within his own union and as a leader in the national labor movement was a welcome by-product to some industry policy makers and an influence on industry thinking" about the final contract terms.[70] Fairless had personally been involved in the 1954 contract talks, an unprecedented step for a U.S. Steel chairman. The traditional pattern of steel industry–labor negotiations, it might be noted, called for the union to settle first with U.S. Steel and then apply the terms of the new contract to the rest of the companies. This system had long been agreed to by both management and labor representatives.[71]

During this period McDonald maintained an intense personal rivalry with Walter Reuther, head of both the United Automobile Workers of America (UAWA) and successor to Philip Murray as president of the Congress of Industrial Organizations (CIO). This rivalry carried over into contract negotiations of their respective unions. McDonald often appeared determined to win better terms for the steelworkers than Reuther could obtain for the UAWA, thus bolstering his claim to national labor leadership in the eyes of suspicious rank-and-file USWA members.[72] The 1954 settlement "put his steelworkers away out in front of the autoworkers" (who had yet to negotiate their contract; see Table 7.7 for comparisons).[73] As in the previous year, this wage raise, too, was quickly followed by higher prices (though for the record,

Table 7.7 Gross Average Hourly Earnings of Production Workers in Selected Industries, 1949–59[a]

Year[b]	*(A) Basic Steel*	*(B) Automobiles*[c]	*(C) Coal Mining*[d]	*(D) All Manufacturing*	*(A)/(B)*	*(A)/(D)*
1949	$1.656	$1.702	$1.947	$1.405	97.3%	117.9%
1950	1.675	1.715	1.933	1.418	97.7	118.1
1951	1.89	1.84	2.04	1.55	102.7	121.9
1952	1.91	1.98	2.24	1.64	96.5	116.5
1953	2.15	2.10	2.48	1.74	102.4	123.4
1954	2.18	2.19	2.48	1.80	99.5	121.1
1955	2.27	2.25	2.48	1.84	100.9	123.4
1956	2.47	2.28	2.70	1.93	108.3	128.0
1957	2.66	2.43	2.95	2.05	109.5	129.8
1958	2.76	2.48	3.04	2.11	111.3	130.8
1959	3.04	2.66	3.16	2.19	114.3	138.8

[a]Gross average hourly earnings include incentive earnings, premium pay for overtime, holiday work, shift differentials, and effects of changes in skill levels, but they exclude payroll costs of vacations, pensions, insurance and other fringe benefits.

[b]As of January 1.

[c]Motor vehicles and equipment workers.

[d]Bituminous coal mining.

Source: Bureau of Labor Statistics. (Reproduced in U.S. Congress, JEC, 88th Cong., 1st sess., Hearings, *Steel Prices, Unit Costs, Profits, and Foreign Competition* [Washington, DC: GPO, 1963], p. 35.)

the president of U.S. Steel stated "competitive conditions . . . require . . . a lower price adjustment than would otherwise be justified by all considerations.")[74]

The 1955 wage renegotiations proved less harmonious, which would be a precursor of events to come. For one matter, the UAWA and Ford Motor Company announced a very generous new agreement, including substantial improvements in nonwage benefits, only one day prior to the opening of the steel talks. For another, Benjamin Fairless had retired as chairman of U.S. Steel a month earlier, to be succeeded by Roger M. Blough. The latter, a Yale-trained lawyer who had worked his way up through the legal hierarchy of the corporation, did not appear to have the image of warm relations with hourly employees as did Fairless (a former steelworker himself).[75] An additional factor, of course, was the record profits earned by the industry in 1955, as after-tax net topped $1 billion for the first time. As a result of these matters, the 1955 settlement was signed only after difficult bargaining that was not completed until the day after the old contract in force had expired; some four hundred thousand workers, in fact, were idled for part of that day. The union was unable to win any changes in its benefit package, but it did obtain a steep wage increase (almost double what the autoworkers had received).[76]

It was the price hike of $7.50 per ton announced by U.S. Steel only one day after the settlement that had prompted Eisenhower's private outburst (noted earlier) against industry leaders. Perhaps it was also this incident that caused him to begin a temporary reconsideration of his policy of nonintervention in industry affairs. "This government," the president stated in June 1955 (before the steel contract was settled), "has gone on the theory that the Executive department . . . will not project itself into the details of private negotiations between employer and employee."[77] This position, of course, had been in effect since his inauguration. It had held up in mid-1954 despite the warning from Arthur Burns that the absence of a public White House statement on the need for peaceful wage settlements could lead to an imminent steel strike; Burns, however, was overruled by the secretaries of labor and the treasury, both economic conservatives who counseled silence.[78] Throughout Eisenhower's first term, the government would remain an interested bystander, but nothing more. Labor and management would be left to resolve their problems alone regardless of how significant the ramifications might be to national economic objectives.

By 1956, however, the climate surrounding steel negotiations began to change. Not only was prior union–management harmony eroding, but the White House decided as well that public intervention was justified to break the inflationary wage–price spiral that past settlements always seemed to invoke. No doubt, this new policy was influenced by election-year political considerations.[79] Workers throughout the nation had won sharply increased benefits in 1955, a year in which the national economy made a strong recovery from the recession of the previous two years; union leaders appeared determined to do as well in 1956.[80]

In March U.S. Steel gave an indication of matters to come when it charged that recent inflationary trends had been caused by ambitious labor leaders "who, with power to bring about industry-wide strikes, seek always to outdo each other in elevating employment costs in their respective industries."[81] David J. McDonald, of course, rejected this partisan accusation, but his public rebuttal was moderated by a private meeting held with the president in late March; Eisenhower requested the exercise of "statesmanship" in the upcoming negotiations in order to minimize inflationary tendencies that "an unrealistic wage settlement" might aggravate.[82] The president also probably did not wish to have a major steel strike mar his reelection campaign.

Much was also at stake for McDonald at this time. He seemed intent on winning significant contractual gains in 1956 to shore up his still-divided following among union rank-and-file members. He therefore submitted a long list of demands to industry bargainers, yet these, in turn, elicited equally ambitious counterdemands (and the major integrated firms, it might be noted, this time bargained together as a single unit rather than permitting U.S. Steel to strike the initial deal and then having the rest fall in line). The USWA primarily desired more money and premium pay for weekend work, though it also wanted the installation of supplemental unemployment benefits (SUB) similar to agreements reached with the can-manufacturing industry the previous fall. For its part management seemed resigned to another costly settlement in consequence of another record-earnings year, so instead, it pushed for long-term labor stability. A five-year contract was demanded, without annual wage reopeners but with fixed and predetermined wage boosts scheduled over the life of the agreement.[83]

Despite its earlier intentions, the administration played an active role in behind-the-scenes arrangements to secure a peaceful settlement to the steel negotiations. Secretaries James Mitchell of Labor and George Humphrey of the Treasury were the principal go-betweens, but a very interested president was kept informed of all developments as they occurred. Yet the administration still seemed somewhat unsure of its part: It did not care to suggest any specific wage-and-price ceilings nor any specific compromise terms, but it did desire to facilitate matters up to these limits; a strike, the White House felt, should not be necessary.[84] Unfortunately, the administration could not have it both ways. Negotiations broke down and, without any forceful public intervention to keep the talks moving, the workers walked off their jobs on July 1.

The administration continued its middle-of-the-road efforts during the strike. The neutral offices of the Federal Mediation and Conciliation Service were offered to the disputants, while Mitchell and Humphrey constantly urged compromise. But Taft-Hartley legislation, one available means to end the impasse, was never seriously considered.[85] What eventually resulted was a settlement that strongly favored labor—an outcome to which Eisenhower was problably quite opposed. The limited intervention that the White House pursued no doubt had a much stronger influence on management than it did on the union; industry leaders were generally closer to, and more disposed to, the persuasions of a Republican president (especially in an election year).

Labor leaders, on the other hand, took the soft administration intervention for just that: No specific terms were proposed, so nothing specific had to be pledged. The outcome of a strike, meanwhile, did not appear to have as urgent an effect on Democratic-election chances as it did on the Republicans. Thus, when the White House finally demanded action to end the impasse, it was management that was more likely to back down the most.[86]

Amid these considerations the strike came to an end on July 27. This was only one day after President Gamal Abdul Nasser had provoked a new crisis in the Middle East by seizing the Suez Canal. Although there is no evidence that this event directly affected the steel negotiations, the immediate national security tensions it raised no doubt hastened the administration's determination to end the walkout in the militarily important steel sector. The industry compromised on a three-year contract running through mid-1959, with the preset annual pay increases it had originally proposed. In addition, a SUB clause and a twice-a-year cost-of-living escalator were also included (the latter apparently as somewhat of an afterthought without much pressure from the union). When the total package was added up, it provided about forty-five cents per hour in pay hikes over the life of the contract, resulting in annual wage-cost raises of 8% to 9% for the three-year term. This was considered a great victory for McDonald, as it boosted the steelworkers significantly past the UAWA and behind only the United Mine Workers of America (UMWA) in national pay scales; compared to average hourly earnings in all manufacturing industries, steelworkers' pay was now 28% higher (see Table 7.7).[87] A press statement drafted by Secretary of Labor Mitchell commended the settlement as " a major milestone in the history of industrial relations in this country [agreed on] without the intervention of Government."[88] It would, indeed, prove to be a milestone in the steel industry's recent history but hardly one that could provoke optimism for those concerned with the long-run industrial competitiveness of the nation.

The producers, of course, upped their prices soon after the new contract was signed, but not before the White House forced some moderation. This latter development sparked rumors about political deals in this election year.[89] But notwithstanding the settlement terms or the subsequent price hike, the question that must be explored is why the industry interrupted its prior ritual of wage-and-price increases by taking a long strike. One reason may have been the new industrywide bargaining procedures: With many players now directly involved, decisionmaking became more complicated and agreement more difficult for management. Another compelling reason was that the steelmakers—principally U.S. Steel—realized that the spiral could not go on indefinitely without causing serious damage to their competitiveness in the world market. This implies that the industry's earlier strategy of slowly improving its comparative advantage position vis-à-vis the Europeans was now under pressure to proceed more quickly—the reason being that foreign steelmakers were improving and expanding faster than had been anticipated.[90] Total steel imports may have amounted to only 1.8% of domestic U.S. shipments in 1956, but their volume was up 38% over 1955 levels because of strike-hedge buying by American customers, a pattern that would soon be repeated.

It is therefore possible to argue that one reason behind management's emboldened labor posture in 1956 was the firms' conviction that imported steel threats would increase in the near future—far earlier than had been predicted only a few years before. If so, then production costs would have to be better controlled and mill productivity more rapidly upgraded. Yet, in view of growing political pressures to restrict price increases, there might be serious limitations to the generation of investment capital through internal sources, the basis of the large steelmakers' previous financial strategy for modernization. This left only two choices: restrict labor costs and reallocate the savings to investment or turn to the government for assistance (through improved depreciation schedules or other such means). Of course another alternative would be to seek a détente with labor and enter into some shared-benefit program, but the potential for such a move must have seemed remote in the context of contemporary labor relations as well as the industry's historical reluctance to share any of its power with outside constituencies. The implications of the industry's dilemma must have been disturbing to Big Steel leaders before the 1956 strike was settled; following announcement of the terms, the dilemma could only have become worse.

The 1956 steel strike thus stands as an important turning point in the industry's strategic response to a changing domestic and international environment. The pass-through price strategy adopted at the outset of the Eisenhower presidency would now have to be revised, that much seemed certain. But the direction of strategic change is not so clear. Should the industry press ahead with modernization plans and hope the task could be completed before European competition was at full strength? If the latter approach were chosen, the question of a sufficiency of expansion funds as well as the kind of expansion to pursue would have to be answered—a problem that remained despite the record earnings that occurred between 1955 and 1957. Could the government be induced to aid the industry, much as foreign rivals were turning to their governments for state benefits? Or should the industry undertake a political campaign to seek restrictions on imports? Yet perhaps more basic for large American steelmakers was the question of whether or not the government should even be considered as a legitimate partner in any endeavor or under any circumstances.

But there could be no doubt that industry officials were facing decisions that would fundamentally affect the future shape of American steel production. The choice to sustain a month-long labor strike in order to hold down wage increases was one part of the strategy reformulation process. But as we shall see in Chapter 8, this move—like others that would be tried—ultimately proved insufficient to the task. Far more would have to be done before the industry could be restored to global supremacy.

8

The Beginning of the End: Eisenhower and Steel, 1956–60

The years following 1956 would become progressively more painful for America's large steelmakers. Their strategy for international stability withered amid the growing strength of offshore producers, while their domestic strategy of raising prices to fund modernization collapsed over governmental objections to inflationary wage–price escalations. Both options, it will be recalled, had been chosen only after alternative proposals involving various forms of governmental assistance to the industry proved fruitless. In neither the domestic nor international arenas was the federal government prepared to offer substantive programs for the resolution of industrial problems; likewise, most steel leaders were disinclined to modify their traditional attitudes of independence by seeking any cooperative accommodations with Washington. Organized workers, for their part, remained committed to a traditional posture of adversarialism versus the companies. And although nearly all concerned remained in agreement that a healthy steel industry was necessary to national well-being, especially for security needs, no one appeared to have any creative new ideas for rectifying the ideological gridlock that continued to impede improvements in the relations of steel with its institutional partners.

I

The announcement by U.S. Steel in early August 1956 that it would raise prices an average of $8.50 per ton (up 7.4%) to offset the new labor contract was greeted with anger in many quarters.[1] The strike that preceded this hike, it seemed, did nothing to halt steel's wage–price spiral that was increasingly viewed as a principal cause of the rising inflationary trend in the American economy. Yet it was Congress, not the White House, that now took the lead in demanding an explanation for industry behavior. The administration, following its convincing reelection victory in 1956, temporarily retreated from concern with steel's internal affairs.

This decision would perhaps be regretted. By abandoning the campaign to limit wage–price hikes in steel, the administration inadvertently signaled industry officials that they were free to renew their earlier strategy of a rapid pass-through of labor costs to customers. When this began to occur, the Democratic Congress stepped up its condemnation of the large steel firms, bringing increased national attention to bear on their behavior. Much of the congressional criticism, however, appears to have been motivated by partisan political ends rather than by any sincere interest in seeing the national economic situation improved. As a result, the industry was cast into the role of a political symbol, to be extolled or abused not in accordance with long-term national interests but rather only in terms of the partisan political value that might accrue from such attention. The Eisenhower administration must bear the ultimate responsibility for allowing this situation to develop, as it alone possessed the power necessary to initiate a more balanced analysis of industrial needs and public policies.

This was particularly true in the case of steel's international competitive position. After the election of 1956, the administration chose not to undertake an interventionist course, rather it resorted to its previous commitment to reduce the level of governmental involvement in the economy and thus allow private firms greater freedom of economic decisionmaking (a choice with which steel industry leaders enthusiastically agreed). Ironically, the economic policy that would soon bring grief to the steel industry was thus formulated not only by a friendly Republican White House and supported by company officials, but also was spearheaded by one of the industry's own leaders—Secretary of the Treasury George M. Humphrey, one of the most influential members of the Eisenhower cabinet until his resignation in July 1957 to assume the chairmanship of National Steel, one of the larger integrated producers.[2]

The Joint Economic Committee (JEC) of Congress (as was noted in Chapter 7) announced an investigation into steel prices in late 1955. Although committee staffers were directed to gather data, formalized hearings were delayed by the mid-1956 strike.[3] Meanwhile, other congressmen undertook their own inquiries; small business committees in both houses, for example, were concerned with the effect that large-firm behavior might have on the small entrepreneur. The chief issue for steelmakers in this context was distribution, much as it had been in the late 1940s when the same congressional committees subjected the industry to scathing examination.[4]

But the most penetrating of the congressional inquiries into steel would be that launched by the Democrat-controlled Senate antitrust and monopoly subcommittee. From 1955 through the early 1960s this unit, initially under the leadership of Senator Harley Kilgore (D-West Virginia) but after 1956 by Senator Estes Kefauver (D-Tennessee), subjected the industry to continual questioning and denunciation by its majority party members. What the subcommittee's program often amounted to was ideological opposition to Big Business. Senator Joseph O'Mahoney (D-Wyoming), chairman of the late-1930s Temporary National Economic Committee (TNEC) investigations and a leader in the Truman era antitrust faction (where he served as chair of the JEC) remained an influential subcommittee member in the 1950s. Both John Blair and Paul Rand Dixon, the subcommittee's chief economic advisors, were committed to late New Deal attitudes toward business concentration and monopoly.[5] This attitude, of course, tended to closely mirror marginal pricing theories that animated most conventional economic analysis of steel industry behavior—in sharp contrast to market realities that shaped the competitive practices of the domestic producers' international rivals.

But it was Kefauver himself who would soon prove to be the most tenacious combatant in the group's investigation of steel (just as he was in concurrent inquiries into other allegedly monopolistic industries, such as automobile production and drug manufacturing).[6] Throughout the 1950s, it should be noted, Kefauver was a leader of his party, and in the 1952, 1956, and 1960 Democratic nominating conventions he was a serious contender for presidential candidacy. The frequent publicity that the senator received through his subcommittee (generally known as the Kefauver Committee) was a major source of his national reputation as a defender of traditional populist values; indeed, it provided a prominent platform from which to seek public recognition for his beliefs.[7]

The subcommittee's concern with the competitive behavior of the steel industry surfaced early in the Eisenhower years. In fact, the congressmen even agreed with the administration on one particular steel industry issue. In mid-1954 Bethlehem Steel announced that it had proposed a merger with Youngstown Sheet & Tube, fifth-largest firm in the industry. Bethlehem (second-largest) had long been interested in access to Youngstown's markets in the Midwest. This interest was heightened in 1952 when U.S. Steel brought its new Fairless Works on line at a site near Philadelphia, thus challenging Bethlehem's traditional dominance in eastern markets. An appropriate response, Bethlehem felt, would be to enter the U.S. Steel-dominated midwestern market, an area in which the number-two producer had no steelmaking facilities.[8] Union with Youngstown, it was claimed, would not only strengthen the latter firm's ability to compete in the region, but it would also provide for a better overall industry competitive balance. The combined ingot capacity of the proposed new Bethlehem–Youngstown entity would account for 19.6% of the industry total, thus making it a more formidable challenger to U.S. Steel's dominant 30.9% market share.[9]

Bethlehem executives said the chief reason behind their plan was the high cost of capacity expansion. It was simply too expensive for the firm to undertake construction of its own new greenfield mill in the Midwest, while Youngstown was too small to finance much expansion independently. Therefore, it was reasoned, the demonstrated need for more steel capacity would be best met by merger (since Bethlehem would financially assist Youngstown's expansion); also, this move would enhance industry competition by providing a stronger challenge to U.S. Steel.[10]

Eisenhower's Department of Justice did not agree with this analysis, since it felt the steel industry was already too highly concentrated. In 1950 the antitrust statutes had been strengthened by the Celler-Kefauver bill that amended Section 7 of the Clayton Act, but the new restrictions had yet to be tested in court. The Bethlehem–Youngstown case, believed Attorney General Herbert Brownell, would be a fine test of the new law. President Eisenhower firmly backed Brownell on the matter, indeed even encouraged the action.[11] The Justice Department subsequently announced its formal opposition to the merger. The Senate Antitrust and Monopoly Subcommittee held its own hearings on the proposal and then joined the Department in opposition.[12] In December 1958 a federal court ultimately agreed with Justice, denying approval to the plan.[13] Shortly after the court's decision, Bethlehem announced plans to build a new greenfield mill at Burns Harbor, Indiana.[14] This modern facility, eventually completed in 1970, provided the national economy with many benefits. Had Bethlehem been allowed to acquire Youngstown's assets earlier, reasonable doubts about whether or not this new mill would have been constructed can be raised.

Yet what is of perhaps more significance is that antitrust lawyers were effectively shaping national industrial policies in steel to a far larger degree than were any other public officials. Moreover, they were doing so according to a calculus of decisionmaking that perhaps reflected less interest in the economic particulars of high fixed-cost steel production, the long-term outlook of the industry, and recent changes in the international competitive environment than in their own highly particularized set of legal constraints as applied to industry-concentration levels. Rather than guiding economic rationalization of the American steel industry or even promoting it, the Justice Department chose instead simply to apply the conventional wisdom that characterized much of the Eisenhower administration's attitude toward economic policy throughout the 1950s.

The Senate's antitrust subcommittee would soon pursue the steelmakers' alleged anticompetitive behavior on its own. Under Kefauver's leadership a broad investigation into administered pricing was launched in 1957. Because of its ability to set prices based on profit goals rather than any underlying market-demand factors, said the subcommittee, concentrated big business was most responsible for the nation's rising inflation rate. Kefauver stated:

> No better example of this behavior can be found than that exhibited by the steel industry itself. Because of this reason, and also because of its importance as the nation's basic materials industry, the subcommittee has chosen steel as its first specific field of examination.[15]

The probe began on August 6, 1957. This date followed in the immediate wake of yet another round of wage–price hikes in steel. Under terms of the

1956 United Steelworkers of America (USWA) contract (it will be recalled), the producers were obligated to give workers a predetermined raise on July 1, 1957. Effective that same day, U.S. Steel announced price boosts averaging $6.00 per ton (bringing the increase over 1956 prices to 11.4%). This amount, however, was considerably lower than had been predicted by industry analysts.[16] The administration had earlier voiced concern about the inflationary dangers of higher prices, but it is more likely that the impending congressional investigation was responsible for the producers' moderation. Smaller firms, especially National Steel, had been loudly calling for even higher quotes in order to generate the investment funds necessary to plant modernization and expansion; increases of up to $10.00 per ton were mentioned.[17] But U.S. Steel, traditional price leader in the industry, opted for the lower levels and remaining firms dutifully went along. It would be difficult, however, to attribute this action to any considerations for the public interest (even though this was one reason company executives publicly gave for their behavior).[18]

The price move won the firm no favors with the Kefauver Committee once public hearings got underway. The U.S. Steel officials spent six days before the lawmakers, answering in detail questions concerning investment, costs, prices, profits, competition, inflation, and corporate power.[19] At times their testimony appeared quite unconvincing, especially when the subject of pricing was discussed. This seemed most apparent in several exchanges between Roger M. Blough, chairman of U.S. Steel, and Senator Kefauver:

> MR. BLOUGH: There are all kinds of theories about pricing. . . . What I am trying to indicate to you is that, if we offer to sell steel to a customer at the same price as a competitor offers to sell to the customer, this is very definitely a competitive price.
>
> SENATOR KEFAUVER: Wait a minute; if you offer to sell at the same—
>
> MR. BLOUGH: If we offer to sell steel at the same price as a competitor at the customer's door, that is very definitely a competitive price with our customer. Now, it isn't a different price and, therefore, it isn't a noncompetitive price, but it is a competitive price.
>
> SENATOR KEFAUVER: You mean it gives him the right to decide whether he will buy his steel from you or Bethlehem Steel? That is the only difference; isn't it? That is the only competition? If you offer to sell a customer at the same price as Bethlehem, the only point of competition is whether they buy from you or whether they buy from Bethlehem, isn't it, or some other steel company?[20]

More than anything else, perhaps, this debate indicates how the nature of steel competition as viewed by industry officials (who saw price as only one of many factors considered by buyers) continued to diverge widely from the conventional opinions of most others in the nation, Congress included.

Testimony from the USWA as well as other producers was also taken by the subcommittee. Dr. Otis Brubaker, research director for the union, condemned the recent price increase in steel, stating (though with little proof):

> United States Steel Corp., which initiated the 1957 steel price increase (as it has normally initiated the increases in prior years), could have put into effect a price cut of $6 a ton, instead of a price increase of $6 a ton, absorbed the cost of the wage increase that occurred in July, and could still have earned greater net profits after taxes in 1957 than were ever earned in the history of the corporation.[21]

By the end of the hearings in November 1957 it was difficult to determine how much damage the industry had sustained to its public standing. Although more observant followers of the tedious and complex arguments may have found some enlightenment regarding steel's problems, it is probably safe to assume that the general public—insofar as it followed the news at all—remained unconvinced that the industry was competitive. When Kefauver released the subcommittee's report on steel the following February, he castigated the 1957 price hike as one that "substantially exceeded" cost increases (total employment costs were up 8.9% that year). Accordingly, he recommended that both the Justice Department and the Federal Trade Commission (FTC) undertake antitrust investigations into the pricing practices of the steel industry.[22] He did not suggest that the industry and the government attempt to work together toward any long-term mutual solution to national industrial goals; yet, for that matter, neither did U.S. Steel's Blough nor any other industry leaders.

One of the accusations leveled against steelmakers in the hearings was that their prices tended to rise regardless of market conditions. This was true in 1957, but it became even more pronounced a year later. By 1958 the economy was in recession. Steel's capacity-utilization rate for the entire year was only 60.6%, and at one time or another four hundred thousand workers were laid off from the mills because of declining orders.[23] In spite of this, both prices and wages went up again that year. Industry profits in 1957 had set a record even though shipments in the last two quarters of the year were weak; in 1958, the firms still managed to show "a better-than-average return on sales despite the worst steel slump of the postwar era."[24]

It was perhaps inevitable that the 1958 steel-pricing round would raise intense interest among those concerned with the nation's political economy. The industry's 1956 labor agreement called for another wage boost in 1958 as well as cost-of-living escalators (the latter growing in value due to rising inflation). David McDonald, USWA president, was in no mood to recommend any moderation in these contractual obligations, since in early 1957 he had narrowly survived an election challenge to his office from a virtually unknown local union official. He therefore felt he had to deliver to the rank-and-file if he were to maintain his position; accordingly, the full wage boost was demanded despite the recession and the widespread layoffs.[25] At the same time Kefauver and his congressional supporters began a well-publicized campaign to force the companies to absorb the entire wage boost that would come due in July.[26] Of particular interest was Kefauver's demand that the White House intervene in some way to help obtain this objective. The senator telegraphed the president:

> I again urge you to use the full powers of your office in order to prevent this disastrous occurrence [i.e., price hikes]. I am confident that if you would bring together the leaders of the steel industry and the United Steel Workers of America, a realistic hold-the-line price–wage program could be developed.[27]

Eisenhower, however, would not accept this recommendation. His administration remained committed to a policy of nonintervention, trusting instead to the goodwill of the involved parties to settle their differences in the national interest. This approach was taken even though Eisenhower had earlier adopted a program of limited exhortation to help moderate the nation's inflationary trend.[28] Perhaps the president felt that any public acquiescence in the Kefauver-initiated plan would be interpreted as a lack of leadership in the White House (an issue to which he was sensitive). Administration pressures to hold down price hikes would certainly not endear the president to steelmakers as they felt his office was largely responsible for the onerous terms of its present labor contract. Thus as the situation in the industry worsened, business, government, and labor leaders remained largely intransigent in their positions.

Still some progress toward improvement could be found. Even if there was no public demand by the president for industry absorption of the pending wage increase, there can be little doubt that U.S. Steel was being pressured in private to act with caution. Absent any secret White House prodding (and there is no evidence that this occurred), the firm's leaders were facing powerful public scrutiny of their every move.[29] In response to this spotlight, U.S. Steel inaugurated an historic change in industry-pricing traditions: It refused to take the lead in announcing changes. It would remain until July 29, one month following the wage hike, before Armco Steel announced a price boost of $4.50 per ton in certain cold-rolled product lines. Republic Steel then advanced prices in many other lines on August 4 (the two firms were, respectively, third- and fourth-largest industry entrants). Soon afterward all the remaining competitors in the industry joined in with the new quotes, U.S. Steel included.[30]

It was obvious at this point that U.S. Steel's long-established price-leadership role within the industry was in jeopardy. Public opinion and governmental pressure had become overwhelming burdens for the company to shoulder; many were blaming U.S. Steel for the entire nation's recent inflationary problems. The administration's laissez-faire approach to economic management, however, meant that it could offer no substantive aid to the company's dilemma. Kefauver had clearly seized the initiative with well-publicized statements, and he quickly followed these up with new subcommittee hearings in August 1957.[31] As the recession intensified over the following months amid continuing nonintervention by the administration, the stage was being set for yet another dramatic confrontation between labor and management in steel. The current USWA contract would expire in July 1959, and nothing on the horizon seemed to promise any harmony for the negotiations.

II

In many respects 1959 marks a symbolic turning point for the American steel industry. In that year the producers suffered through one of the longest strikes in U.S. industrial history, indicative of a failed domestic policy that had been developing for several years. Indeed, one could easily apply this judgment to the entire twentieth-century history of the industry. At the same time the volume of steel imported into America in 1959 exceeded exports for the first full year in the century. This reflects the domestic producers' deteriorating international position that had been apparent since the late 1940s. A striking fact, however, joins these two trends together: On both counts a fair degree of responsibility for the industry's problems can be ascribed to governmental policies—either by commission or omission. The firms themselves cannot, of course, be exculpated from guilt, yet one must surely acknowledge the contributory role of governmental policy that led to steel's decline in this period.

The roots of the 1959 strike should be clear by now. The three-year 1956 labor agreement would expire at the end of June, and the industry—especially U.S. Steel—was determined to arrest spiraling wages that had caused so much acute embarrassment over the previous years. The union, under David McDonald's shaky leadership, seemed just as intent on protecting workers against mounting inflation as well as the continuing economic slump. And finally, the Eisenhower administration would not seriously budge from a steadfast policy that refused to recognize specific sectoral needs in the economy. This combination of forces would intersect in such a way to prove debilitating for all three parties involved, not only in 1959, but well into the nation's future.

The months preceding termination of the contract were politically active. In April Kefauver's antitrust subcommittee held hearings on a bill introduced by Senator O'Mahoney to require certain large firms to notify the government prior to any price increases (and to wait thirty days after filing before implementing the raise, thus allowing time for public hearings). Testifying for the industry, Roger Blough flatly rejected the plan as tantamount to "socialism," while McDonald mildly endorsed it (though with some reservations).[32] O'Mahoney's proposal bore similarities to a plan considered by the JEC as far back as 1950, when Congress earlier investigated the steel industry over pricing practices.[33] History again seemed to repeat itself when Senator Albert Gore (D-Tennessee) introduced a measure to impose a utilitylike regulatory structure over the steel industry—another idea borrowed from the industry's 1950 congressional confrontation.[34] Neither of these proposals received serious consideration, as they clearly derived from political partisanship given the surrounding atmosphere of debate. (This was unfortunate, for the utility proposal in particular might well have produced a means to overcome the industry's serious financial burdens involved in funding new çapacity.)[35]

It would remain for the administration itself, however, to play the most important public role in steel's impending labor negotiations. Despite the White House's aversion to intervention, persistent inflation demanded that leadership be exercised on this issue. At first the president believed he could help matters by offering objective third-party analysis of data that otherwise might be manipulated by the involved interests. Accordingly, in January 1959 he established a Cabinet Committee on Price Stability for Economic Growth; its main purpose was to conduct studies that would "help prevent price increases."[36] But the committee was given no authority beyond its relatively innocuous research mandate, and little came of its efforts. At no time did the administration seriously consider tougher measures such as wage–price guideposts or controls to hold down inflationary increases.

In fact, by March of 1959 the president was reiterating his belief that the government should not interfere in labor talks between private parties. But this was hedged somewhat when he also stated that any final agreement in the steel industry contract should be concluded "in such a way that the price is not compelled to go up."[37] This ambivalence, or indeed even contradiction, in the White House position would again seriously impede the bargaining process, much as it had in 1956. Eisenhower seemed trapped between the poles of theory and pragmatism in his dealings with steel. At one end, the ideology of economic conservatism preached a simple hands-off attitude; at the other, political–economic reality demanded that some high level third-party leadership be interjected if the nation's long-run interests were to be even recognized by the principal disputants in the matter.

Steel production, of course, remained of critical importance to the national security as the president well realized. The Berlin crisis and new shelling in the Formosa Strait off China only reinforced the depth of Cold War tensions surrounding America in early 1959.[38] It was vital that the nation's steel output not be interrupted owing to relatively minor concerns. But by failing to acknowledge that his administration had any desire, let alone responsibility, to provide active leadership in this impasse, Eisenhower thereby eliminated himself from effective consideration by the principal parties. At most, he would undertake only *ad hoc* intervention in the steel dispute—a public exhortation here, a secret meeting there, designed to remind labor and management of the larger national issues involved.[39] But beyond this, nothing more would be done. Given the size and power of the two principals (and the significance of the stakes involved for both), Eisenhower's approach, of course, would not work. This would become painfully apparent in succeeding months.

A detailed recounting of the subsequent lengthy bargaining process need not be provided here, as it has been amply described before.[40] But some points are illuminating for the overall purposes of our analysis. The industry (as we noted) was determined to obtain an agreement that would not require any further price hikes. In April, well before the June 30 contract deadline, the companies announced their position: a one-year contract at existing pay levels and elimination of the cost-of-living escalator clause.[41] The surrounding economic situation seemed at first to support these objectives. The 1958 recession had put many steelworkers temporarily out of work, and there was no improvement in the underlying structure of the downturn by 1959. In such a depressed market, management would normally be assumed to have the upper hand over labor when it came to granting pay raises.

But to a large degree this bargaining advantage was mitigated by the widespread fears of a steel strike in the minds of industry customers. As a result, users bought record amounts of steel in the first two quarters of 1959 in order not to be caught short if the mills were shut down (an outcome most considered inevitable).[42] The financial impact was embarrassing to the producers. U.S. Steel, for example, set a first-half earnings record and produced and shipped more steel between April and June than in any previous quarter in its history.[43] For the entire year, in fact, industry output and earnings levels would be above the prior year's totals.[44] The union leadership was quick to exploit these financial reports as indicative of the firms' ability to grant substantial pay hikes in addition to improved fringe benefits.[45]

But the industry would not be easily deterred in its determination to suppress wages. Negotiations were conducted in an atmosphere decidedly more hostile than previous sessions. McDonald (as we noted) was under intense pressure to manifest leadership to his constituents, a condition presumably satisfied only by the winning of a greatly improved new contract. Management, for its part, presented the union with a tough new bargainer: R. Conrad Cooper, an executive vice president of U.S. Steel, who replaced the retired John A. Stephens as chief industry negotiator. Stephens had presided over all of the previous labor-bargaining sessions since 1945, and he seemed to enjoy genuinely warm and friendly relations with McDonald. Cooper, on the other hand, had no such ties with his adversaries.[46] A further indication of the seriousness with which the parties approached the talks was the impressive public relations campaigns that were launched; both sides understood the importance of general public opinion in this contest, and each was determined to win popular support.[47]

The magnitude of the public's stake failed to prevent what most informed observers believed unavoidable: the fifth industrywide steel strike since 1945. By the end of June little progress had been made in the formal talks. On June 25 McDonald wrote to the president that "the steel negotiations are deadlocked"; he requested Eisenhower to appoint "an impartial fact-finding board" to help resolve the dispute. On June 27 the president rejected the idea, noting that the Taft-Hartley Act called for such boards only in emergencies, which he said this was not, and that "free voluntary collective bargaining" without governmental involvement remained the best solution.[48]

McDonald's suggestion to the president for a board of inquiry was probably only a ploy for time, since there was little sentiment on anyone's part for meeting the USWA's inflated demands. Nevertheless, McDonald formally reacted to the president's reply by postponing the strike date to July 15.[49] Moreover, in subsequent bargaining he offered to reduce his demands to an approximate twelve cents per hour increase (equal to a 3.4% hike in total hourly employment costs), all of which was to come in the form of supplemental benefits, none in basic wages. He also agreed to a one-year contract.[50] But the industry, sensing the weakness of McDonald's position, remained adamant in its call for a complete freeze on labor costs. The only acceptable condition for granting even a small wage increase, Cooper said, was that the union agree to modify a number of work rules contained in the existing basic agreement that governed contract negotiations. These concerned such items as local working conditions, incentives, employee scheduling, and seniority—all factors that industry officials felt would add to steel mill efficiency if they could be better controlled by management. Cooper's logic was that, with the promanagement Eisenhower in the White House and with recently retired U.S. Steel labor negotiator John A. Stephens now serving as an administration aide, the time was ripe for a company victory.[51] The union would have none of this, however, as it believed that any retreat in work rules would compromise the basic integrity of the USWA. McDonald blustered that his members would not "be turned into a company union."[52]

With no softening of the industry's stance, the expected strike was called. Some 511,000 union members walked out on July 15. They would remain out for 116 days until November 7 when the United States Supreme Court ordered them back to work under the 80-day cooling off provisions of the Taft-Hartley Act. This amounted to the longest industrial strike in American history up to that point in time.[53]

In the interim the White House continued to demonstrate its ambivalence about the proper governmental role in this matter. Eisenhower monitored the situation daily; Labor Secretary James P. Mitchell served as the administration's principal spokesperson. The industry, however, did not wish for any closer White House participation; it perhaps feared a repeat of the 1956 outcome when Eisenhower's involvement resulted in overly generous labor gains.[54] Thus a personal letter from the president sent in mid-August to George Humphrey, now serving as chairman of National Steel, seems almost pathetic in tone:

> What I am trying to say below is not to be interpreted as any effort to inject the government into the current strike situation in the steel industry. I merely have put together in my own mind a "formula" that makes sense to me. Admittedly I have no intimate acquaintanceship with the steel business, but I do necessarily study daily a whole list of reports and I do have a good many indications of differing convictions and opinions which have induced me to set forth the bare bones of what would seem to be fair to everybody.[55]

The president then outlined some specifics for agreement, but asked his long-time friend not to answer. This behavior was in marked contrast, of course, to his predecessor's high-profile role in previous steel–labor negotiations, just as it would differ from the activism of his two Democratic successors to office.

Meanwhile, management and labor negotiators failed to progress.[56] It was not until two months into the strike that Eisenhower publicly demanded action. At the end of September he invited the two sides to meet with him, separately, at the White House. At the same time his staff began to seriously consider the use of a Taft-Hartley injunction to break the deadlock.[57] It might be noted that the White House's renewed interest in the steel strike coincided with the departure of Soviet Premier Nikita Khrushchev from the United States following a celebrated two-week visit in late Spetember; it was speculated that Eisenhower did not wish to have the antilabor Taft-Hartley sanctions in place during the Russian's visit. When resumed talks following the White House conference ended in less than a week, without settlement, the president initiated injunction procedures by appointment of a Board of Inquiry. The board's hastily prepared report was submitted by October 19, but its conclusion was not optimistic:

> As we submit this report, the parties have failed to reach an agreement and we see no prospects for an early cessation of the strike. The Board cannot point to any single issue of any consequence whatsoever upon which the parties are in agreement.[58]

On the same day, Eisenhower instructed Attorney General William P. Rogers to file for a strike-ending Taft-Hartley order (its first application ever in the steel industry).[59] The president continued to meet with the principals involved, at one time even displaying sympathy for McDonald (whom he felt was an "actor" but who still had a "political problem" with his membership that worked to limit any "rational" settlement to the strike).[60] Nevertheless, the quick disposal of several legal challenges to the injunction paved the way for a court order temporarily ending the strike; by November 7 the workers were back in the mills. The timing was propitious for many steel customers as prestrike inventories had been heavily depleted by then; there can be little doubt that had the strike continued much longer it would have produced serious ripple effects throughout the economy.[61]

The terms of Taft-Hartley stipulate that management submit a final offer to labor by the sixtieth day of the eighty-day cooling-off period; if the workers reject this, they are then free to strike again. The time pressures involved in this procedure had an effect on the industry, for the strike was ultimately settled on January 4, 1960, only days prior to the deadline. Other pressures, however, were also involved. In early December Eisenhower departed on a world tour, leaving the steel situation in the hands of Mitchell and Vice President Richard M. Nixon (the latter expected to be his party's presidential nominee in the upcoming 1960 elections). Nixon presided over several secret meetings with the disputants in late December, which eventually led to a settlement.[62] But similar to the 1956 situation, this settlement would also end in terms unfavorable to the industry.

The turning point for the union's victory was management insistence on changes in work rules. McDonald's demand for a large increase in wages and benefits was generally unsupported by all parties, initially including even rank-and-file workers of the USWA; not only would existing economic conditions in the steel industry dictate against such a settlement, but the prevailing national concern with rising prices would also provide powerful incentive for a noninflationary agreement. Given this situation, Cooper and his aides made a serious blunder when they thrust work-rules modifications into the negotiations. As McDonald was to state later, this "handed us an issue. I couldn't have written the script better myself."[63]

In November and December, following the court's ruling on Taft-Hartley, McDonald was able to portray the industry demand for work-rules changes as an attempt to destroy the USWA. The union's skillful manipulation of this issue soon gained McDonald the solid backing of his members, and by late December management faced the clear prospect of a renewed strike once the injunction ended. Under prodding by Nixon and several key industry customers, a settlement was then reached in time to avoid the final offer vote (which, according to all observers, would have been won by the union). The final settlement called for a thirty-month contract to run through June 1962. The terms were in labor's favor insofar as wages were to be raised (though not immediately), benefits improved, the cost-of-living escalator extended, and the work-rules issue shunted off to a joint management–labor committee for further study. Management's original bargaining position was completely ignored.[64]

Eisenhower sought to portray the agreement as a victory for voluntary bargaining free of governmental involvement.[65] Roger M. Blough of U.S. Steel was less evasive, noting:

> The union's refusal of [our earlier] offer created a serious deadlock. The union not only refused to bargain lower but after negotiating settlements in other industries, withdrew its previous offer and raised its demands very substantially.
>
> So it was at this point—and under these circumstances, that administration officials in Washington sought to bring the parties together and eventually recommended a settlement which both parties ultimately accepted.[66]

That settlement of course granted labor more than most had expected it could win. But at the same time one must also acknowledge that union gains were far less than obtained in the 1956 contract;[67] total employment costs per hour rose only 0.6% in 1960, whereas they averaged nearly 9% in the previous three years. For government inflation fighters, this was good news. Another benefit to the public interest came the day after the settlement when U.S. Steel informed the administration that no price increase would be forthcoming (a proviso, according to some, stemming from a Nixon–Mitchell deal with the principals that brought an end to the strike).[68]

Deal or not, steel prices were not raised in 1960. Much of this was due to economic conditions: The industry fell back into the deep slump that had begun in 1958.[69] Another reason for price moderation was congressional pressure. There was talk of new legal restraints on steel during the strike, just as there had been prior to the walkout. In January Senator Kefauver stated that his subcommittee would watch prices very closely and take whatever action was appropriate if they again started to climb.[70] Not until 1962, in fact, did industry prices go up, providing one of the longest periods of steel price stability in the postwar era. By then, however, it was too late. The damage from the earlier wage–price spiral could not be undone, and producers, workers, and the nation as a whole would have to pay the costs of their previous intractability. To more than a minor degree, these participants are still paying that price today.

9

An Industry in Decline: The Growth of Steel Imports into America

The conclusion of the 1959 strike brought with it an end to the industry's strategy for financial rejuvenation. That is, producers would no longer be able to pass on all costs to customers with relative impugnity. Much of the impetus behind this change was growing public concern with the inflationary impact of steel-pricing decisions. Not all might agree with this analysis, however; one economist has written:

> What happened in steel [in 1959] had nothing to do with inflation; it was a purely sectoral phenomenon of joint labor–management monopoly approaching the point of maximum return. Once having attained it, the ceiling was reached, and no further increases were possible. Because the two parties were in a posture of both co-operation and conflict in sharing the fruits, communication between them was necessary but difficult, and broke down in 1959.[1]

According to this view, changing economic conditions forced steelmakers to accept a more competitive posture than had hitherto existed. Substitute materials, such as aluminum and prestressed concrete, were eroding traditional steel markets, and foreign competition was growing. As a result, higher costs could not be passed on blithely to consumers.

The present study would argue that economic analysis alone is insufficient to explain either steel's strategy in the 1950s or the demise of that approach; institutional factors also played a leading role in these events, thus they must be given equal consideration. Yet no one should doubt that rising imports were not of real concern to the industry. From the outset of the 1959 negotiations—indeed, even earlier—steelmakers stated unequivocally that changed conditions in world trade were a primary reason for their demands for improved efficiency at the mills and noninflationary wage settlements.[2] The union, however, rejected this argument as a hoax, nothing more than a bargaining ploy and propaganda.[3] When the strike was finally settled, however, it should be noted that there was no public mention that international factors had anything to do with the agreement.

Nevertheless, there is clear evidence that steel leaders were more than aware of their deteriorating comparative international position. We have emphasized throughout this study that such concerns were always in the forefront of the industry's strategic planning; in fact, the decision to raise prices to generate plant modernization funds—a basic motivation behind the industry's domestic strategy after the mid-1950s—could only have been devised in the context of heightened global competition. The breakdown of international market stability during the same period lends further support to this contention. To fully understand the implications of the 1959 strike and its settlement, one must therefore address the issue of imports and their impact on domestic producers.

I

We have previously described the early Eisenhower administration's role in providing assistance to foreign steel industries. In conjunction with their own governments' efforts on behalf of rationalization and modernization, foreign firms created a profoundly transformed international steel sector by the late 1950s. One result was a decline in the proportion of world steel produced by the United States (see Table 6.1), though this was perhaps expected and even inevitable given the reconstruction that followed World War II.[4] But an unanticipated consequence was a steep rise in the level of steel imports into the United States (see Table 6.4). These trends would only accelerate further after 1960. While of course many factors contributed to this development, one must not exclude governmental policies as a primary source.

Eisenhower's second term saw no relaxation of his staunch anticommunism. A principal tool in the implementation of this policy (we noted earlier) was the provision of economic assistance to both America's Western Allies and emerging nations in the nonindustrialized world. This was intended to help them build sufficient economic strength to ward off any encroachments of socialism (and also, some would claim, to tie their economies to American capitalist institutions).[5] But for whatever ideological reasons, by the beginning of Eisenhower's second term there had developed an even greater emphasis on economic aid as an anticommunist weapon. In part this was due to congressional pressures that mitigated against purely military aid, which the administration had sometimes pursued in the past.[6]

In response, the White House in late 1956 appointed a blue-ribbon citizens' advisory panel to reevaluate the administration's entire foreign aid and security programs. Benjamin F. Fairless, retired head of U.S. Steel and then serving as chairman of the American Iron and Steel Institute (AISI), was named the panel's leader.[7] The Fairless Commission issued its report to the president in March 1957. Although adopting a more conservative approach toward mutual security and aid than Congress or the president seemed to favor, nevertheless the group came out squarely in favor of continued foreign trade. The Commission wrote:

> We should, in our own interests, seek the expansion of this mutually profitable and beneficial trade. We should proceed with the program of gradual reciprocal tariff reduction, and liberalization of customs procedures. At the same time, however, we must use the full power of bargaining and persuasion to seek reduction of foreign quantitative restrictions on our trade. The elimination of these restrictions must accompany tariff revision. . . .
>
> We also believe that to the extent consistent with strategic considerations, the Government, in purchasing supplies needed abroad, should encourage the international development of productive efficiency by buying in the most advantageous markets.[8]

The Eisenhower administration's continuing commitment to foreign aid and trade was thus given a strong personal endorsement by the domestic steel industry's leading (albeit retired) executive. As one result, the administration extended its program of direct aid to foreign steelmakers, generally considered to be the linchpin of their economies. In the four years between 1957 and 1960, the federal government and various international agencies in which the United States played the dominant role provided $585.8 million in such assistance; this accounted for over 40% of all U.S. aid given foreign steelmakers between 1947 and 1960.[9] European producers, however, received only 12.1% of the total in Eisenhower's second term, far less than their share over the earlier postwar period (when they had 46.1%). The major recipient now was Japan, which obtained over $176 million in aid to her steel industry between 1957 and 1960.[10] This accounted for approximately 10% of total Japanese steel investment in the period, which coincided with Japan's Second Modernization Program for the industry. Though appearing to be a relatively small proportion, the amount and terms of the American aid were critical to Japan's overall steel-development program; it signaled the soundness of the investment to other lenders and thus opened up larger funding.[11]

The United States provided more than just direct financial assistance to Japan's steel companies. Technical know-how was liberally offered, and numerous trade missions made their way between the two countries.[12] More important, the United States also insured that the Japanese steel industry's scrap metal needs would be met. This is an event that deserves illumination because its consequences would be of significance to domestic producers.

Scrap is a major input to the open-hearth process of steelmaking; it accounts for up to 50% of the furnace charge that produces raw steel. Scrap can be obtained from two principal sources: home scrap, a by-product of the blast-furnace operations of a steel mill (i.e., the making of pig iron), or purchased scrap from outside sources, such as dealers who collect, sort, and sell it to the steelmakers. Following World War II scrap was in high demand and relatively scarce throughout the world, which was one reason behind the Europeans' decision to initially reconstruct their steel industries with Bessemer-process plants.[13] (The open-hearth process, developed in the 1880s, provides a superior-grade steel for many purposes in comparison to the older Bessemer process, but the latter requires only negligible amounts of scrap.) Japan, however, chose a different strategy: Not only did that nation install open-hearth converters in its early postwar reconstruction program, it also invested relatively little in blast-furnace facilities. Japanese firms spent instead for downstream rolling and finishing operations.[14] Scrap needs were met by dealers or the U.S. government, which controlled local sources of residual military scrap left over from the war.

Scrap prices tend to fluctuate with global demand. As the postwar steel boom gripped the world, scrap prices naturally rose; during the Korean War they reached record heights. The export of scrap from the United States (the world's leading supplier) had been controlled by the government since 1949, but in October 1953 restrictions were lifted. Not only was Eisenhower philosophically opposed to economic controls, but the ending of the Korean War boom tended to mitigate such needs for national security purposes. American steelmakers, of course, were also affected by scrap prices; the huge industry-capacity expansion that began in the United States after 1950 was almost totally committed to scrap-consuming open-hearth furnaces.[15] With scrap prices rising, domestic steelmaking costs necessarily went up as well. As we have noted, these higher costs were generally passed on to consumers in the form of higher prices.

But we also noted the growing political concern with steel pricing. As a result producers began to monitor their costs more closely. By 1955 this led to some industry calls for renewed export controls over scrap; not only were domestic scrap stockpiles allegedly being depleted, but the heavy demand from abroad (it was claimed) was forcing up scrap prices and thus contributing to America's problems with inflation.[16] It was only through the diligent efforts of free-trader Clarence B. Randall that AISI decided not to undertake a major lobbying and national public relations campaign in favor of new scrap controls.[17]

The problem, however, remained. Smaller producers and steelmakers without much pig iron capacity continued to demand governmental action to hold down scrap prices. By 1955–56 these prices again reached record heights in the face of a renewed worldwide boom in steel production. Europe, in its expansion programs initiated after 1950, had begun to add more open-hearth capacity, which of course only added to scrap demand. This prompted Treasury Secretary George M. Humphrey to write Commerce Secretary Sinclair

Weeks (who was strongly opposed to any scrap controls, but whose department had previously administered them):

> I don't think it is good for them [i.e., Europeans] or for us to have them so dependent on us. They ought to build up their own steelmaking capacity to supply their own finished requirements and thus relieve the pressure of demand for our scrap. I think it is part of wisdom for us to put the pressure on them to accomplish this both for their good and ours. The way to accomplish it is to begin promptly to reduce the amount of scrap they can count on from here. Otherwise, if the present situation continues, they will be draining our scrap away in lieu of building steelmaking capacity of their own. If an emergency really arises, we will be short of scrap that will be badly needed here and will have to shut them off entirely, which will substantially reduce their own steel production when they need it most.[18]

It is interesting to note Humphrey's concern for his steel industry colleagues in this scrap-controls issue because he was usually strongly committed against any governmental meddling into market-denominated outcomes. But the rest of the administration, in spite of Humphrey's concern, would not bend. When Senator Prescott Bush (R-Connecticut), the target of procontrols lobbying by small-steel producers in his state, demanded an explanation of the scrap export situation, he received the following reply from trade officials in the administration:

> There are technical considerations, but probably the most difficult is in the area of fundamental policy. As you well know, the Administration favors trade liberalization, free enterprise, a free market, and minimizing and avoiding Federal controls wherever possible. Eliminating controls was one of the first actions of this Administration. It would be contrary to fundamental beliefs to use governmental control devices to satisfy the demands for modification of each domestic economic problem.
>
> To do so, without clear justification and unquestioned necessity, would only lead to multiple similar demands in other instances based on the precedent. I know you believe in the principle that the best form of adjustment of unbalanced situations is in the action of the free market forces.
>
> It is apparent there are problems. But it is not yet clear if these are short-run or long-run problems. That there may be a long-run problem is recognized. It has been discussed with foreign governmental authorities and officials of the European Coal and Steel Community where action has already been taken to modify their demands on our scrap resources.
>
> In my opinion, the Government should not be forced into hasty action by the usual pressures to meet a special problem, and should

> not impose the kind of control advocated until it is unquestionably clear that the problem is of serious proportions and of such long range nature as to make it unavoidable.[19]

This reply could easily serve as a proxy for the administration's entire response to the steel industry during the 1950s. In general there would be no sectoral policies for special problems, no public intervention to achieve national economic objectives, no industrial policy to shape and implement long-term American goals regarding the nation's basic economic infrastructure. The natural forces of the market, it was presumed, was the best corrective to any disturbances that might occur.

The AISI, still reflecting Randall's influence, also endorsed the administration's position on scrap and refused to go along with its members who sought controls. In late 1956, following a thorough study of the problem, the institute concluded:

> The following considerations should be kept in mind in establishing a policy for both the government and the industry to follow:
>
> 1. *U.S. foreign economic policy regarding metallics* [i.e., scrap]. It is obviously not in the long-run interest of the United States Government or the American Iron and Steel Industry for the U.S.A. to . . . neglect the interests and aspirations of friendly foreign countries. . . . Steel is required to build strong economies throughout the Free World. Therefore, the availability of adequate ferrous metallics [i.e., scrap] in one form or another to meet the requirements of friendly foreign countries should be a matter of national concern.
> 2. *Avoidance of economic controls.* To the maximum extent practicable, the American Iron and Steel Industry should be committed to the avoidance of economic controls and the maintenance of free markets, subject to the laws of supply and demand, competitive prices, comparative costs, etc. This policy should apply in particular to the international movement in metallics. . . . Therefore, only as a last resort, when the nation might be subjected to substantial injury, should the industry advocate a government embargo or other limitation on the export or import of metallics.[20]

Thus just as the administration would not pursue sectoral industrial policies, so too would leaders of the steel industry seek to avoid any governmental intervention that might abrogate the free market and its presumed benefits for economic prosperity.

II

The scrap problem as addressed in these documents was mainly concerned with Europe. Efforts to limit Japan's drawdown of scrap, however, were somewhat more successful. Unfortunately for domestic steelmakers, they had an unanticipated effect on that nation's choice of steelmaking technology—a choice that in the long run would do far more damage to America's steel industry than mere short-term price fluctuations in scrap. By late 1956 Japan had begun to replace Europe as the primary export market for U.S. scrap; Japan's growing heavy industrial sectors (e.g., shipbuilding) needed steel in large volumes; this, in turn, required increased quantities of scrap given the open-hearth technology traditionally utilized in Japanese mills. In prior years most of this had been supplied by West Coast American dealers, but with growing demand it was now procured in Gulf Coast and East Coast U.S. market as well. As a result, U.S. scrap prices hit their all-time high in the fall of 1956.

Political pressures for action to limit Japanese scrap purchasing mounted. In response, the Commerce Department in February 1957 temporarily suspended export licenses for certain grades of scrap after a congressional study seemed to indicate that domestic shortages were imminent.[21] A voluntary quota system was suggested as one solution, whereby the Japanese would agree to restrict their level of scrap imports from the United States. But while this was eventually agreed upon, free-trade proponents in the administration were unhappy with its implications. The Council on Foreign Economic Policy (CFEP), now headed by Clarence B. Randall, took the lead in seeking to find another solution less injurious of free trade.[22] A CFEP report on a trade meeting held in March 1957 provides an indication of what it desired:

> Randall is red hot to help Japan build blast furnaces and develop sources of iron and coal.
>
> Prevailing idea of seeking voluntary restraint on part of scrap importing countries is abhorrent to him and others but best we can do in the circumstances, which involve great pressures on the White House.
>
> But in doing so we are treating symptoms and not the disease.
>
> Disease is that Japan (and he wanted to limit this discussion to Japan) is not developing furnace capacity or sources of metallics and will probably not do so as long as they can look to U.S. for scrap.
>
> Japan has made mistakes of putting what capital is available in the finishing end of the steel business instead of building from the ground up. . . .
>
> Question: Can we help them expand capacity?
>
> Answer: Looks like there is no great ambition in Japan to do this because they are "not sure of market" and Japanese sources of capital and the industry itself may be dragging their feet (dragon feet, Hah) especially so long as they can rely on U.S. for scrap.[23]

Japan, however, would soon reverse this reported position and move toward the direction desired by Randall. At a governmental conference in Tokyo in June 1957, called to discuss the needs of the Japanese steel industry, the following report was filed by a Commerce Department official in attendance:

> In general, all persons contacted gave the impression of being most anxious to present a clear picture of the problems confronting the steel industry in Japan. They stated that for sometime they had suspected that the supply of scrap from the United States to Japan might be curtailed, either by USA regulation or price. Consequently, they have been working on plans for producing steel, issuing a minimum of scrap. In general, this involved the construction of additional blast furnaces (for pig iron) and the installation of top blown oxygen converters [i.e., basic oxygen furnace (BOF)]. It is their plan, in the future, to use top blown oxygen converters. This method of steel production will not require more than 10% of the total scrap requirements for this type of operation when compared to the conventional open hearth operation. . . .
>
> Four of these top blown oxygen converters will be install [*sic*] in August or September 1957. They will have a total daily capacity of 4,500 tons of steel. In 1958, four more . . . will be added. . . .
>
> [The Japanese] further stated that it was their hope that by installing more oxygen converters and not rebuilding old and antiquated open hearth furnaces that they might eventually get to the point where they would be able to operate on the scrap that could be obtained in Japan and the small amount that might be available in nearby countries. They realize that if this is achieved it will be necessary for them to greatly increase their blast furnace capacity in the future. They stated that this move was entirely motivated by the realization that scrap would be in short supply and high in price in the future. From the above, *it could be assumed that Japan is planning its future steel production in accordance with the desires of the USA,* meaning they are planning in the future to greatly reduce the steel scrap required from the United States of America. [Emphasis added.][24]

Japanese steelmakers, working in close concert with their government policymakers, made good on these plans. In October 1959 the U.S. Embassy in Tokyo wrote to Randall, "The industry has gone far in putting into effect the good suggestions you made in the spring of 1957 that Japan move significantly toward fuller establishment of a metallic base [i.e., scrap self-sufficiency] for its steel industry."[25] The policy of urging a reduction of U.S. scrap exports to Japan continued through the end of the Eisenhower administration. By 1962, 30% of Japan's steel production was in BOF converters, compared to only 6.5% in 1958.[26] In retrospect one can conclude that the impetus for this change in Japan's steelmaking technology owed much to U.S. foreign economic policy, just as it did to Japanese foresight and wisdom. Such an observation stands in sharp contrast to the prevailing sentiment of contemporary critics of the steel industry, who viewed early Japanese adoption of BOF—and lagging U.S. adoption—as indicative of both monopolistic conditions and managerial inadequacies in the American market when compared to the wiser foreigners.[27]

There were, of course, larger political considerations behind the administration's desire to see the Japanese steelmakers do well. The most important was the maintenance of close ties between Japan and the West. As one White House briefing to CFEP staffers noted, "A friendly Japan is vital to the interests of the United States and . . . in this connection it would be highly undesirable for Japan to become dependent on Communist China for its raw materials for its vast basic industry."[28] This was an especially important concern to American policymakers who dealt with the Japanese steel industry. One Commerce Department official working on these matters reported:

> Red China has an adequate supply of both iron ore and coking coal which is available in Manchuria very close to Japan and this could be obtained for use in Japan if a trade agreement could be entered into between USSR and Japan. Therefore, I consider it most important that the USA cooperate to the fullest extent with Japan in working out an economical supply of raw materials for their steel industry which would come from the free nations.[29]

Another governmental official confirmed this, noting:

> Japan is [presently] being approached by Russia on the subject of negotiating trade agreements between the two countries. It is my understanding that Russia has suggested to Japan that if trade between the two countries is consummated that Japan could obtain large quantities of both coking coal and iron ore [both steelmaking inputs] from Manchuria or possibly Hainan Island in Red China.[30]

Japan, of course, was highly dependent on foreign sources not only for scrap, but for almost all other raw materials needed in the production of steel. President Eisenhower was keenly aware of this vulnerability, and he gave it serious consideration in terms of the complex questions of global political economy that the issue raised for America. Responding to a reporter's question about Japanese trade with Communist China (in April 1957), he stated:

> Now, I would like for you to consider the Japanese problem just a minute. . . .
>
> In Japan there are 90 million people that we want to be our friends. . . . They are inventive, they are industrious, they are good workers. Now, we don't believe that there is any prospect of keeping Japan a viable economy merely by giving them some cash each year.
>
> More and more our own industries come to us, come to the Government and insist on either higher tariffs or stiffer quotas, to stop Japanese goods flowing in here; and then we say to Japan, "Now, you mustn't trade with any of the great area right next to you which has been your traditional trading area."

> Now, what is Japan going to do? . . . How are they going to keep . . . going? I do not say that the sole answer is in any one of the three directions I have just briefly mentioned: in aid or in us taking more of their stuff or in them trading with the neighboring areas. But I do say if we are going to keep Japan our friends, on our side of the Iron Curtain, we can't look at it just in any one of the separated roadways and say, "You mustn't do that and you mustn't do that." Finally, you just block them, and they have no place to go except into the arms of somebody where we don't want them to go.
>
> So we must approach these things with intelligence and with a regard for our own future long-range welfare, as well as some immediate direction or some immediate advantage that we think we see.[31]

In the calculus of decisionmaking pursued by his administration, this meant encouraging the industrial buildup of Japan so that it might stand on its own feet as a capitalist bulwark against potential Communist expansion in Asia. Yet, regardless of what the president chose to do for Japan and Japanese steel, he did not sufficiently consider the ramifications that his foreign policies would have on his own nation's steel industry. It was perhaps easy to dismiss any such external threats to domestic steel in the context of the mid-1950s when both output and profits were high. But a closer examination of the industry's dwindling comparative advantage vis-à-vis its foreign rivals—a fact fully realized by the larger steelmakers—might had led federal policymakers to reconsider the economic implications of their political goals. The suspicion remains that a more systematic appreciation of the long-term needs of the industry would have forced Washington to temper its enthusiasm for foreign reconstruction or perhaps to adopt policies that would help the industry in its adjustment to the emerging new world order in steel. Neither action was taken during Eisenhower's presidency (or, for that matter, his predecessor's). The task of containing perceived Communist expansion was given a higher priority, despite its effect on future American economic development.

III

By the late 1950s the postwar reconstruction of foreign steel mills began to have a tangible effect on domestic markets. Tariff barriers continued to fall in America in accord with the administration's trade liberalization policies. Both the 1955 and 1958 extensions of the Reciprocal Trade Agreements Act (RTAA) gave the president increased authority to reduce duties on metals, which he did.[32]

As one result, imports of steel products, such as wire and rod, increased in the late 1950s. Both of these commodity-type items were relatively simple to manufacture and inexpensive in price, and both enjoyed strong demand throughout the United States; as such, they were ideal candidates for importers.[33] Up through 1959 West Coast markets had been the fastest-growing target for foreign penetration, since prices in that area were traditionally higher (and thus more attractive for importers) and access was easy for cheap water-borne freight from abroad. Between 1953 and 1957 steel imports rose by 88% on the West Coast, going up 20% in the latter year alone.[34] Following the opening of the Saint Lawrence Seaway in 1959, midwestern markets also became vulnerable to foreign competition. By 1964 more imports would funnel through the Great Lakes than any other region in the country, thus depriving domestic steel producers of their traditional dominance of this largest of American markets.[35] Yet regardless of the port of entry, the European Coal and Steel Community (ECSC) continued to supply the bulk of America's imported steel throughout the 1950s; over 70% of all imports between 1953 and 1959 originated in the six community nations (see Table 6.4). But Japan's share was rapidly growing. Over this seven-year span, it doubled.[36] Moreover, Japan also began to ship higher value-added steel, such as plate products. Indeed, by 1958 the federal government—attracted by lower prices—was purchasing plate steel from the Japanese.[37]

The mounting threat of foreign steel was, of course, a serious concern to industry leaders. We have already noted how their vast expansion and modernization programs undertaken since 1950 were directly related to improvements in steelmaking abroad.[38] The smaller producers were initially most vulnerable to imports, however, so they were the first to address the threat in other than economic terms. The chairman of small Atlantic Steel Company, testifying before the House Ways and Means Committee in 1958, stated, "Unless this trend of imports is sharply turned by Congress, it's only a matter of time before every American producer, regardless of where located, will suffer." Atlantic produced less than 150,000 tons of steel in 1958, only 37% of its rated capacity.[39]

The major firms, however, were unsure of the wisdom of this political response, even though they clearly agreed that imports were a problem.[40] One large firm, Armco, stated, "Our domestic industries are certainly entitled to protection against unfair foreign competition," but it would not go beyond this general platitude by requesting anything more specific.[41] Roger Blough perhaps best captured the sentiment of this group when he noted in 1958:

> Already in steel, there is a straw in the wind. It may be only a straw, but I give it for what it may be worth. Exports of steel from the United States in 1958 will fall off roughly 40 percent from 1957. But what is more significant, in this year when our industry operated far below its capacity, our steel imports are expected to be 10 to 15 percent above those of a year ago.
>
> And may I add that while recognizing the value of such expedients as tariffs and quotas, increased tariff protection is not the ultimate or basic answer because free nations must depend upon trade. Trade is mutually beneficial. Moreover, if experience teaches us anything, then certainly we have learned that the power to erect tariff walls is not one in which America enjoys a monopoly.[42]

Industry leaders would adhere to this view for the remainder of the Eisenhower era. In the months preceding the 1959 labor strike, company officials often pointed to the disparities between domestic and foreign wage rates in steel as the primary reason for the growth of imports (a contention rejected, of course, by the union). U.S. Steel said:

> It was evident to the Steel Companies at the outset of the negotiations that a further inflationary wage rise, resulting in a further widening of the international wage gap, could only add to the difficulties that American producers and American steelworkers were already experiencing at the hands of their foreign competitors.
>
> It was also clear to them that while tariffs and other artificial trade barriers might provide temporary relief, such devices could not afford any basic or fundamental solution to the problem—and Mr. Blough pointed this out distinctly in all of his discussions of foreign competition and the international wage gap.[43]

By 1959 steelworker hourly earnings were nearly 40% above the all-manufacturing average in the United States; this obviously was one reason why domestic prices were higher than those abroad.

Given their generally favorable views on free trade, the large steelmakers therefore rejected a political strategy of lobbying for import restrictions as a means to combat the rise in foreign competition. Instead, other strategic options would be pursued. We have already discussed the most favored ones: an attempt to halt rising labor costs, and modernization of existing plant and equipment to insure technological parity with offshore mills. Capital expenditures were high for the industry in the late 1950s and would remain so into the early 1960s (steel's capital output ratio exceeded the all-manufacturing average by 125% in 1960, see Table 9.1). Between 1956 and 1966, investment averaged $1.36 billion per year for the industry as a whole, compared to only $720.9 million in the 1946–56 interval; moreover, this investment was carried out despite the deep industry recession of 1958–62.[44]

There were other prongs to the industry's economic strategy beside investment in new technology and a cap on labor costs. One involved proposals to recapture export markets increasingly being lost to foreign mills. Both industrial and governmental leaders became sensitized to the export issue during the late 1950s because of rising problems in the nation's balance of payments.[45] As deficits mounted and gold reserves dwindled, an alarmed administration sought to assist domestic industries in their export efforts.[46] In early 1960 the Department of Commerce met with steel officials to map plans to boost exports. Unfortunately the discussions focused only on an easing of export licensing requirements, which was a somewhat peripheral issue to the real problem.[47] Also, most industry representatives seemed relatively unconcerned with the issue. This lack of attention may have been due to the indifference with which the majority of steel firms approached exports; only U.S. Steel and Bethlehem had previously given this market much emphasis.[48] On

Table 9.1 Capital Output Ratios: U.S. Steel Industry and All Manufacturing Firms, 1947–60[a]

	(A) Steel Industry	*(B) All Manufacturing*	*Percentage by Which (A) Exceeds (B)*
1947	1.26	0.95	33%
1948	1.20	0.97	24
1949	1.44	1.07	35
1950	1.24	0.96	29
1951	1.09	0.91	20
1952	1.46	0.95	54
1953	1.26	0.90	40
1954	1.74	1.01	72
1955	1.43	0.93	54
1956	1.49	0.98	52
1957	1.67	1.02	64
1958	2.25	1.13	99
1959	2.23	1.02	119
1960	2.32	1.03	125

[a]Ratios in columns (A) and (B) are based on constant 1958 dollars. The capital output ratio is defined as dollars of undepreciated plant and equipment per dollar of value added. Value added equals the net billing value of products shipped and other services less purchased materials, supplies, freight, other services, and so on.

Source: AISI, *Annual Statistical Report* (various years), and U.S. Department of Commerce. (Reproduced in U.S. Congress, Senate Finance Committee, 90th Cong., 1st sess., *Steel Imports,* Committee Print [Washington, DC: GPO, 1967], p. 175.)

the other hand, uncertainties surrounding the limits of the Webb-Pomerene Act may also have contributed to domestic reluctance to enter export markets in any systematic manner. Unlike their foreign rivals, American firms could not easily discriminate between prices charged locally and abroad. At any rate, no sustained program of business–government cooperation emerged at this time to improve the export situation even though public officials often stated their intentions of doing something helpful.[49]

One reason for the low level of exports, of course, was the relatively high price of American steel. Since 1951 U.S. export prices of steel had constantly risen, dipping lower only in 1958 when competition began to erode American positions in certain world markets. Export prices of foreign steelmakers by contrast were observed to be more flexible during this period. However, because U.S. mills generally did not produce for foreign markets—instead taking export sales only as incremental gains when they could be accommodated—it was considered unwise to price products in terms of such marginal business; the domestic market, after all, dwarfed the size of potential offshore sales. In addition, fair trade laws prohibited American producers from practicing as much price discrimination between export and home market sales as their foreign rivals could, while the existence of foreign cartels gave these rivals certain advantages in the international market unavailable to local producers. The result was that American export steel prices generally moved in correspondence with domestic prices—that is, usually up throughout the 1950s. Under such circumstances, it was perhaps to be expected that only those export markets subsidized by the federal government's Agency for International Development (AID) would continue to buy American steel (since the program mandated the purchase of American-made goods by AID recipients).[50]

The concern for exports among steel officials was visibly heightened by the impact of the 1959 strike and the resultant surge in imports. "The world steel map," wrote *Iron Age* at the end of that year, "is going through an era of marked change. Possibly not since the rise of the U.S. as a manufacturing nation has there been a comparable period of shifting of the balance of power in steel production."[51] The real turning point in America's international standing, in fact, was the year before (1958), when the worldwide steel recession curtailed exports but brought in unexpectedly higher imports owing to steeper prices that prevailed in the United States.[52] But most officials viewed both the 1958 and 1959 figures as only temporary phenomena. An executive of the American Steel Warehouse Association noted in mid-1959:

> Our export market is rapidly drying up because of labor rate differentials. . . . We will have to learn to live with a steadily increasing tonnage of foreign steel but I doubt if the percentage of this tonnage in relation to domestic production will show more than a moderate increase during the next ten years.[53]

For a time this view seemed prescient: By mid-1960 the tonnage volume of U.S. steel exports once again surpassed imports (for the first time since late 1958).[54] Steel's foreign trade in dollars also recovered from the negative balance of $158 million in 1959; for all of 1960 it was a positive $152 million (see Tables 9.2 and 9.3). But the recovery proved only transitory; by the end of 1960 the total annual tonnage of imports topped exports for the second year in a row. Moreover, by 1962 the balance of payments in steel trade also slipped back into the deficit column, where it remained throughout the decade (reaching a deficit of over $1 billion by 1968).[55]

An expansion of exports was thus not a viable strategy for improving the steelmakers' import problems during the late 1950s. Another approach, however, did find some favor with domestic producers. This was direct investment abroad. In late 1960 two American specialty (i.e., alloy) steel firms announced plans to buy into or build plants in Europe.[56] The steel industry had long been active in foreign ownership, but this was limited to operations involved in sourcing raw materials.[57] The changed environment of world steel after 1960, however, indicated that basic manufacturing might also be a worthwhile offshore endeavor for U.S. producers. Roger Blough of U.S. Steel stated in late 1960, "It is absolutely necessary to remain competitive. If that means some companies have to operate abroad, it is natural they do."[58] Blough's company did, in fact, undertake some manufacturing operations in Italy, Spain, and Central America soon afterward.[59] Other domestic steelmakers expressed interest in doing the same.[60]

Table 9.2 American Foreign Trade in Steel, 1945–65

Year	Imports (000 net tons)	Exports (000 net tons)	Trade Tonnage Balance (000 net tons)	Imports as Percentage of Domestic Consumption[a]	Exports as Percentage of Net Shipments
1945	54	4,354	4,300	0.1%	7.6%
1946	23	4,375	4,352	0.1	9.0
1947	32	5,919	5,887	0.1	9.4
1948	148	3,950	3,802	0.2	6.0
1949	291	4,344	4,053	0.5	7.5
1950	1,014	2,639	1,625	1.4	3.7
1951	2,177	3,137	960	2.8	4.0
1952	1,201	4,005	2,804	1.8	5.9
1953	1,703	2,991	1,288	2.2	3.7
1954	771	2,792	2,021	1.3	4.4
1955	973	4,061	3,088	1.2	4.8
1956	1,341	4,348	3,007	1.7	5.2
1957	1,155	5,348	4,193	1.5	6.7
1958	1,707	2,823	1,116	2.9	4.7
1959	4,396	1,677	(2,719)	6.1	2.4
1960	3,359	2,977	(382)	4.7	4.2
1961	3,163	1,990	(1,173)	4.7	3.0
1962	4,100	2,013	(2,087)	5.6	2.9
1963	5,446	2,224	(3,222)	6.9	2.9
1964	6,440	3,442	(2,998)	7.3	4.1
1965	10,383	2,496	(7,887)	10.3	2.7

[a] $\% = \frac{\text{imports}}{\text{shipments} - \text{exports} + \text{imports}}$. Shipments based on net shipments of carbon, alloy, and stainless steel products.

Source: AISI, *Annual Statistical Report, 1965,* p. 8.

But these thrusts into multinationalism were relatively minor compared to the total scope of operations of the firms involved. Unlike many other sectors of American industry, the steelmakers did not embrace a global production strategy with much enthusiasm.[61] This is particularly noteworthy in that diversification, either into new products or new geographical markets, has often been regarded as a natural strategy for firms competing in mature or declining industries.[62] Little documentary evidence is available to indicate why this option was not pursued more aggressively by steel. One likely reason, however, was the heavy financial burden involved: Capital intensive steel firms would have had to invest hundreds of millions of dollars into diversification moves to make much of a dent in their overall structure and sums of such magnitude were simply not available. As noted by AISI's Benjamin F. Fairless in 1960:

> Even if we were of a mind to do so, such a move [i.e., geographical diversification] is hardly feasible for the American iron and steel industry. We can't very well scrap our existing plants, representing an investment of many billions of dollars, and spend more billions to build new plants overseas.[63]

Table 9.3 Dollar Balance of U.S. Foreign Trade in Steel, 1950–65

Year	*Iron and Steel*[a] *Exports (000,000)*	*Iron and Steel*[a] *Imports (000,000)*	*Iron and Steel Trade Balance (000,000)*	*Total U.S.*[b] *Merchandise Trade Balance (000,000)*
1950	$466	$87	$379	$1,122
1951	602	280	322	3,067
1952	609	189	420	2,611
1953	484	225	259	1,437
1954	464	103	361	2,576
1955	639	130	509	2,897
1956	762	212	550	4,753
1957	993	212	781	6,271
1958	563	231	332	3,462
1959	365	515	(150)	1,148
1960	601	449	152	4,906
1961	423	382	41	5,588
1962	424	484	(60)	6,260
1963	470	633	(163)	5,241
1964	622	749	(127)	6,831
1965	507	1,177	(670)	4,942

[a]Excludes ore, pig iron, ferroalloys, and scrap.

[b]This represents the U.S. balance of international merchandise payments only; the balance of other goods and services normally included in balance-of-payment data is excluded.

Source: U.S. Department of Commerce.

Although economic theory suggests that financial markets will make available investment funds for worthwhile projects to any firm at any time, the clear reality was that steel's financial decisionmakers were highly reluctant to leverage their balance sheets any further in deference to existing obligations.

Moreover, it remains problematic whether foreign governments would have been sympathetic to large-scale American investment in their nations in so strategic an industry. But most importantly, the producers seem to have concluded that their best hopes for combating imports lay not in multinational operations abroad but rather in upgrading home mills to improve competition with foreign sources. One may criticize this decision in hindsight, but, no doubt, it had powerful support at the time such decisions were being considered. Not only was such a choice in conformance with the traditional parameters of steel industry thinking, but any move to invest abroad would probably have raised strenuous objections from the United Steelworkers of America (USWA) and would also likely have brought complaints from the politicians who sided with the USWA.

Yet, as the industry pursued its domestic revitalization strategy, it perhaps realized the limitations of such a strictly economic approach to competitive renewal. Accordingly, steelmakers began at least to reconsider more politically oriented steps after 1959. Movement in this direction, however, would be taken only slowly and cautiously. Several reasons account for this. For one, the predominantly Republican steelmakers would face consistently Democratic public policymakers throughout most of the 1960s. For another, the industry remained divided in its attitude toward international trade policy and somewhat inept in the management of its political goals. We have already noted the general reluctance with which leaders of the larger steel firms embraced tariff barriers. The prewar ideology of high protectionism was essentially muted from 1945 onward as steel officials—led by Clarence Randall—joined the ranks of those business interests who called for freer trade as America's best long-run strategy for continued economic prosperity.

It was only the smaller firms who rallied for tougher import restrictions during the 1950s. This was perhaps natural insofar as they usually bore the brunt of such trade in that period.[64] Near the end of the Eisenhower era these firms managed to focus some political attention on their problems. In 1958 four small steel companies filed a complaint with the U.S. Tariff Commission that charged steel wire products from abroad were injuring the domestic industry. The four sought either higher tariffs or the imposition of quotas on foreign wire coming into the country.[65] This action was filed under provisions of the antidumping laws and marked the beginning of a long campaign by the steelmakers to have the government both streamline and actually implement existing trade sanctions. Over time the issue would be joined by the larger firms; indeed, it would eventually develop into the industry's major political tactic in its fight against imports.[66] By the early 1960s both U.S. Steel and Bethlehem would join in the steel wire case, calling it "a showdown of enormous proportions" and a "major test of the anti-dumping route to protection against imports."[67]

Although unsuccessful in this particular case,[68] the industry would continue to resort to the antidumping statutes for import relief in the future. More than anything else, however, this strategy seemed only to illuminate the continued absence of any substantive industrial policies on the part of the government and the steelmakers. Forced to define their relationship through the intricacies and ambiguities of international trade law rather than some more mutually conciliatory mechanisms or institutions, it is perhaps no wonder that the level of cooperation between the American actors paled in comparison to their foreign counterparts.

The turn to the law thus marked a new phase in the industry's response to expanding levels of imports. This line of defense would be tied to demands for import quotas by the late 1960s; as such it represents a deepening politicization of the once economically oriented strategy of the steelmakers. Indeed, by the mid-to-late 1960s the revised strategy would be overwhelmingly political in nature; this was perhaps indicative of an abandonment of any hopes for matching the competitive advantages of the industry's international rivals—especially those in Japan—despite higher levels of domestic investment.[69]

But the real roots of this political approach extend back to experiences suffered in the 1950s when the industry tried—and failed—to win public support for its strategy of domestic economic improvement: Neither the government nor the general public could be convinced of the severity of the steelmakers' problems. Unwilling to compromise in its historical reluctance to work more closely or cooperatively with public policymakers or organized labor, American steelmakers therefore had to bear the burden of matching foreign competitors without the benefits that such business–government cooperation might have created. In the end it would prove a burden far too heavy for them to shoulder successfully. The ultimate result, played out unceasingly over the next twenty-five years, was the decline and fall of the American steel industry from global competitiveness.

10

History, Ideology, and Steel

On January 1, 1962, Benjamin F. Fairless, past chairman of both U.S. Steel and the AISI, died. His passing perhaps marks a symbolic end to the old order of managerial prerogative in the steel industry with which he had been so closely associated since the 1930s. It would remain, however, until the mid-1960s before steelmakers began to seriously reevaluate their relations with the external environment, principally labor and government. This change was due to a continuing deterioration of American steel's performance in the international market, as well as a highly publicized 1962 confrontation between the Kennedy administration and the large steel firms over prices (a contest the firms lost).[1]

These events seem finally to have convinced management that its institutional efforts in dealing with external constituencies needed fundamental repair. Moves were therefore undertaken to restructure relations with the United Steelworkers of America (USWA), and to build closer bureaucratic ties to the government.[2] Although such initiatives would not yield immediate results, they at least signaled a beginning. By 1969 the federal government was persuaded to impose import restrictions over foreign steel, and in 1973 a no-strike agreement was reached with the union.[3] Progress in both these areas, though at times erratic, has continued since. Unfortunately for the steelmakers these changes came too late to offer much help: By the mid-1980s the domestic industry was in obvious decline and relegated permanently to a far less important role than before in global competition.

The twentieth-century history of business–government–labor relations in steel would thus seem to hold few positive lessons for the formation of an enlightened policy of sectoral cooperation that yields economic dividends for America. Indeed, the most characteristic impression that emerges from this history is one of failure—market failures, managerial failures, and public policy failures. Neither steel managers, union leaders, nor several administrations much distinguished themselves by their actions in this arena. Instead, they provide a striking example of how embedded ideologies of suspicion and distrust could contribute to the decline of a major and vital sector of the national economy.

Our investigation of these events has revolved around the central concept of institutional divisiveness. We have endeavored to show how the absence of any ameliorating public programs for steel, combined with the steadfast intransigence of corporate and labor leaders, played an important role in the industry's eventual decline. By contrast, we noted how several foreign nations proved far more successful in establishing collaborative relations with their own steel producers; one result was improved international performance by these offshore competitors. By 1959—less than fifteen years after war's end—these foreign producers began to attain international dominance in steel despite what previously had appeared as formidable comparative advantage for American firms.

Prior scholarship, we noted, has generally placed the primary burden of blame for these outcomes on company management. Due to errors in expansion planning, neglect of technological innovation, and perhaps the arrogance of corporate power in relation to price and labor policies, critics find the industry responsible for its own problems. However, we have taken some exception to this rather narrow conventional analysis. While not denying that managerial inefficiencies did exist, we nevertheless found serious shortcomings in the foresight of labor leadership as well as in various public policies that affected steelmakers in the postwar era. The federal government's continuing failure to appreciate the special circumstances surrounding the manufacture of tonnage carbon steel, including the nature of international competition that faced this sector, contributed significantly to the subsequent diminishment of industry performance. Moreover, the government's propensity to financially assist offshore producers despite the negative effects this might have on long-term domestic interests only worsened the situation.

The reader may have inferred that the purpose of this study was to insinuate that implementation of some form of industrial policy for American steel could have somehow altered these outcomes.[4] This was not, however, my intent; indeed, it is quite the opposite. What I have instead tried to demonstrate is that the American system of institutional relations as it pertained to steel was, in fact, never amenable to any such cooperatively defined regime of decisionmaking as is assumed under industrial policy. As we have seen, from 1901 onward there was no question of the significance of the steel industry to national economic welfare; this was clearly and consistently recognized by public officials, company leaders, and spokespersons for other concerned constituencies. Yet what distinguishes the steel case from almost every other major sector of the American economy is that despite this acknowledged importance and despite frequent clashes between the concerned parties, nothing of substance ever developed to improve matters. Why, one must ask, was this?

A number of factors would appear crucial to an answer. Historically, the modern American steel industry emerged in a single step as the largest and most salient industrial sector in the nation. This occurred in a period of heightened political tension over the proper role of the large corporation in national life. Because of initial differences between industry and governmental officials regarding the nature of competition in steel, the producers (particularly U.S. Steel) became a convenient symbol for corporate power and other such pejorative assumptions surrounding the transformed political economy of modern America. Steel industry behavior toward the rights of labor—behavior which sought, as did nearly all other activities undertaken by the firms, to stabilize its competitive environment—did nothing to mitigate the negative impressions held by critics. Indeed, it only deepened the perception. Over time, and because of continuing friction regarding these competitive and labor policies, the historic divisions between the involved institutions widened. By midcentury these differences had become so embedded in the respective ideologies of both the steelmakers and the state that at times it often seemed few if any issues could bridge the participants together in agreement.

At the end of World War II the American government began to pursue new internationalist objectives that would further isolate the position of the steelmakers within overall national priorities. The dominant theme of postwar American foreign policy has been the containment of a perceived expansionist Soviet state. One tool toward the implementation of this policy was the provision of economic assistance to our allies, either real or potential. Unfortunately for American steelmakers, the outcome of this new policy often meant a reduction in their global competitiveness, for in numerous cases U.S. economic assistance abroad translated into benefits for foreign steel industries, usually the backbone of their own national economies. As a consequence, the calls by domestic steelmakers for aid to themselves to help meet this growing foreign threat went largely unheeded. A major reason for this was the historical burden of distrust that surrounded steel–government relations in America; the adversarialism that often characterizes business–government relations throughout the nation seems particularly virulent in the steel case. Not only were there traditional antistate attitudes held by steel leaders,[5] but as well there was a strong sense of antisteel sentiment on the part of public policymakers. The end result of this adversarial posture became only too apparent after 1959 when imports of foreign steel began to take off and the financial performance of domestic steelmakers went into stagnation.

Is this sufficient reason to place the blame for steel's decline on public policymakers? To answer affirmatively would be to render an ahistorical judgment. As we have noted, the political economy of the United States was drastically transformed after 1945. Economic institutions now had to cope with abundance rather than scarcity, and politicians and their associated bureaucrats were required to plan according to strategic global priorities rather than in terms of purely parochial needs. Clearly, there was little practical precedent to which decisionmakers could turn for guidance here.[6] As the menacing vision of a growing Communist bloc began to dominate the architecture of foreign policy, it was perhaps inevitable that such primary concerns would crowd out and eventually subordinate the interests of others who could not demonstrate the parallelism of their needs with the goals of this larger mission. The steel industry—long held by those in government to be only a domestic interest (a view due, no doubt, to the industry's own desire to minimize its prior international involvement with foreign cartels)—thus could not command the attention of others in attracting public favors. The American Century was in full bloom during the 1950s, and the steelmakers could hardly be expected to succeed in their push for protectionist considerations given this environment.

Nevertheless, some significant questions still need to be raised. As we have shown repeatedly in this study, the steelmakers themselves were painfully aware of their narrowing competitive gap with offshore rivals. Admittedly, they failed to undertake steps within their own control that might have ameliorated their declining condition. Yet, at the same time, these men were never able to convince the shapers of American public policy (including the public itself) of the necessity to act decisively to shore up their diminishing advantage if long-term prosperity were to be attained. Although the politics of the Truman administration were often so overtly postured that one can more easily understand why (traditionally staunch Republican) steel executives did not achieve much cooperation with government between 1945 and 1952, the case is not so clear in the crucial Eisenhower years. Indeed, according to one recent evaluation the objective of Eisenhower was to establish a "corporate commonwealth," which would "limit the role of the federal state" while concomitantly striving for "a strong and pervasive emphasis on cooperation between business and government."[7] Although the goal of a limited state seems to have been achieved in terms of the paucity of benefits conferred on steel, by no means were the Eisenhower years marked by a noninterventionist state into steel's affairs. And the outcome of this latter fact rendered almost ludicrous Eisenhower's goal of business–government cooperation, at least in this particular arena. Why was this?

The essential issue here remains the appropriate balance of business–government interaction in a democratic capitalist society. Why, where, when, and how should public interests intervene in private markets in order to shape outcomes deemed necessary to national welfare? This challenging agenda, long the concern of numerous scholars, has recently been joined by several impressive new studies. In one of the more penetrating, the historian Thomas K. McCraw points to the necessity of shaping any intervention to the particularized economic properties of the industry under examination.[8] This is most pertinent counsel to the steel case as it forces recognition of a critical factor that thwarted not only Eisenhower's conciliatory hopes for business–government peace in this sector, but those of prior administrations as well. Perhaps the most important differences between steelmakers and the government were their abiding disagreements regarding the most appropriate degree of industry concentration and the pricing policies for the firms to pursue—conflicts, it will be recalled, that had divided the two parties since 1901. American public policymakers could be distinguished from those in practically every other nation in the world by their singular pursuit of policies that favored dissolution of the larger integrated steel firms and a rigid adherence to conventional marginal cost theory as the only legitimate method by which to price steel products.

Indeed, there was almost a willful lack of creativity by domestic policymakers as how best to achieve public interest objectives from the industry except through this acutely narrowed agenda. Policymakers for steel in other lands, however, did not choose so constricted a path for their own producers—not in Europe, Japan, other industrialized states, newly industrializing countries, or nonindustrialized nations that somehow managed to erect steelmaking facilities. Perhaps the only deviation by American policymakers from their usual position was an occasional call for nationalization of the industry. But as the records of those nations that did make such a choice clearly show, this policy could hardly be beneficial in the long run to any of the principal constituencies—firms, workers, or the public at large. Eisenhower, for all his sincere concerns for a more cooperative relationship, was never able to enlist others (even in his own administration) to undertake the efforts necessary. The agenda of conflict established by U.S. Steel's Elbert H. Gary and the federal government in the first decades of the century effectively bounded the debate in the 1950s as well.

Would improved business–government–labor relations, were they achieved in the 1950s, have mattered that much? Some might argue that they would not, as steel was a mature sector facing declining secular-demand trends in an economy that was rapidly shifting to a postindustrial mix that emphasized services. Management behavior, in this context, appears as only a residual factor that would have little effect on long-term economic outcomes. Yet my own conclusion to this question differs from such a restricted analysis. Although no one can deny the diminishing intensity of steel consumption in the recent American economy, one must also acknowledge that the contemporary adjustment process has been less than satisfactory. Many have been injured, and few have benefited.

As such, one would have to ask what alternatives might have provided for a stronger resistance by the industry to emerging rivals or at least what could have been done to allow for a smoother transition to a less prosperous future. Although a number of options appear feasible in response, a fine recent study of business–government relations in America by the historian Richard H. K. Vietor suggests that properly constructed and managed industry advisory councils appear to have the highest potential for achieving the economic goals of our democratic society.[9] So-constituted mechanisms can provide a forum for the development of market-conforming means of intervention without any undue interference from the deadened hands of public bureaucrats. Yet while there is much to recommend in this proposal, it presumably would be able to function only within a larger framework of national industrial policy that legitimates such public–private interaction. Given the current conditions of the steel industry in America, it is not surprising to find that national industrial policy has been strongly recommended as a palliative to bring about recovery.[10]

If only the producers, labor, and public officials would reason together, it is suggested, then new programs could be formulated that might improve industry performance and restore economic rewards. Unfortunately, as recent history clearly testifies, such a course appears doomed to failure.[11] Proposals for an industrial policy in steel do not sufficiently address the historical framework of institutional relations in this arena nor do they appear to appreciate the deeply embedded culture of distrust that has long animated the involved parties. Although industrial policies might be efficacious to some sectors of the American economy, the steel industry is not one of them.

But while the opportunity for change may already have been lost in steel, this does not imply that nothing is to be gained from the intense examination of this one industry as has been conducted in my study. America as a nation is entering a future far less advantageous than that characterized by the American Century years of 1945–65; global hegemony, both in the political and economic spheres, has receded over the past two decades and there is little prospect that a significant reversal to this trend will occur soon, if at all. As such, the steel case provides dramatic witness to what can happen when key actors facing the circumstances of decline fail to acknowledge the power of the past to influence their current behavior—that is, when they fail to learn from the past and adjust accordingly. History, we might conclude, is too often an undervalued variable in the calculus of policy decisionmaking in America. As the cultural historian Warren I. Susman has well noted, history for most Americans is often ideology.[12] Yet by better understanding this nation's complex institutional evolution relative to business, government, and labor, our policymakers—both public and private—might be able to transcend these ideological barriers and thus improve both the quality and utility of their subsequent actions. At the least, the steel industry case should allow us to draw this lesson of hope from its otherwise turbulent past.

Notes

Abbreviations

DDE Library	Dwight D. Eisenhower Presidential Library, Abilene, KS
Hill Papers	John W. Hill Papers, Mass Communication History Center, State Historical Society of Wisconsin, Madison, WI
HST Library	Harry S. Truman Presidential Library, Independence, MO
Lamont Papers	Thomas W. Lamont Papers, Baker Library, Harvard Business School, Boston, MA
SML	Schwab Memorial Library, Bethlehem Steel Corporation, Bethlehem, PA

Preface

1. See W. Adams and J. B. Dirlam, "Big Steel, Invention, and Innovation," *Quarterly Journal of Economics* 80 (May 1966), pp. 167–189; Gardiner C. Means, *Pricing Power and the Public Interest, a Study Based on Steel* (New York: Harper & Row, 1962); Otto Eckstein and Gary Fromm, "Steel and Postwar Inflation," Study Paper No. 2, U.S. Congress, JEC, 86th Con., 1st sess. *Materials Prepared in Connection with the Study of Employment, Growth, and Price Levels* (Washington, DC: GPO, 1959); John M. Blair, "Administered Prices: A Phenomenon in Search of a Theory," *American Economic Review* 49 (May 1959), pp. 431–450; M. A. Adelman, "Steel, Administered Prices, and Inflation," *Quarterly Journal of Economics* 75 (February 1961), pp. 16–40; and Walter Adams and Joel B. Dirlam, "Steel Imports and Vertical Oligopoly Power," *American Economic Review* 54 (September 1964), pp. 626–655. For somewhat less damning commentary, see Robert W. Crandall, *The U.S. Steel Industry in Recurrent Crisis* (Washington, DC: Brookings, 1981); and Donald F. Barnett and Louis Schorsch, *Steel, Upheaval in a Basic Industry* (Cambridge, MA: Ballinger, 1983).

Some scholars have found the industry's alleged abuse of corporate power to be a significant factor in its decline; see Grant McConnell, *Private Power and American Democracy* (New York: Alfred A. Knopf, 1967), pp. 328–329; Edward Greer, "The Political Economy of U.S. Steel Prices in the Postwar Period," in Paul Zarembka (ed.), *Research in Political Economy,* Vol. 1 (Greenwich, CT: JAI Press, 1977) 1:59–86; Estes Kefauver, *In a Few Hands, Monopoly Power in America* (Baltimore: Penguin Books, 1965), pp. 104–136; William E. Scheuerman, *The Steel Crisis: The Economics and Politics of a Declining Industry* (Westport, CN: Greenwood Press, 1986); and P. A. Baran and P. M. Sweezy, *Monopoly Capital* (New York: Monthly Review Press, 1966), pp. 60, 81–88.

2. Joseph A. Schumpeter, *History of Economic Analysis* (New York: Oxford University Press, 1954), p. 12.

Chapter 1

1. See Ralph L. Nelson, *Merger Movements in American Industry, 1895–1956* (Princeton: Princeton University Press, 1959); Naomi R. Lamoreaux, *The Great Merger Movement in American Business, 1895–1904* (New York: Cambridge University Press, 1985); H. R. Seager and C. A. Gulick, Jr., *Trust and Corporation Problems* (New York: Harper & Brothers, 1929). On the formation of U.S. Steel, see U.S. Bureau of Corporations, *Report of the Commissioner of Corporations on the Steel Industry, Part I: Organization, Investment, Profits, and Position of the United States Steel Corporation* (Washington, DC: GPO, 1911).

2. See Peter Temin, *Iron and Steel in Nineteenth Century America* (Cambridge: MIT Press, 1964).

3. See C. R. Daugherty, M. G. DeChazeau, and S. S. Stratton, *The Economics of the Iron and Steel Industry,* 2 vols. (New York: McGraw-Hill, 1937); and George W. Stocking, *Basing Point Pricing and Regional Development—A Case Study of the Iron and Steel Industry* (Chapel Hill: University of North Carolina Press, 1954). On the subject of price elasticities for steel products, see the testimony and other exhibits submitted by Professor T. Yntema in U.S. Congress, TNEC, 76th Cong. 3rd sess., Hearings, *Investigation of Concentration of Economic Power,* pt. 26 (Washington, DC: GPO, 1940).

4. For more systematic treatment of these economic conditions, see the discussion in F. M. Scherer, *Industrial Market Structure and Economic Performance,* 2nd ed. (Chicago: Rand McNally, 1980), pp. 205–220. For their relevance to the steel industry at the turn of the century, see Lamoreaux, *op. cit.,* pp. 46–86.

5. See Joseph A. Schumpeter, *The Theory of Economic Development* (Cambridge: Harvard University Press, 1934). For an interesting disagreement with this approach, see George W. Stocking, *Workable Competition and Antitrust Policy* (Nashville, TN: Vanderbilt University Press, 1961).

6. For information on the new philosophy of management in the industry following the creation of U.S. Steel, see the voluminous testimony before the Stanley Committee: U.S. Congress, House Special Committee on Investigation of the United States Steel Corporation, 62nd Cong. 1st sess, Hearings, *United States Steel Corporation* (Washington, DC: GPO, 1911). In addition, see "Co-operation and Conciliation in the Steel Industry," *Iron Age* 80 (November 28, 1907), p. 1549; Melvin I. Urofsky, *Big Steel and the Wilson Administration* (Columbus: Ohio State University Press, 1969), pp. 1–36; and Ida M. Tarbell, *The Life of Elbert H. Gary, the Story of Steel* (Boston: D. Appleton, 1925).

7. See Abraham Berglund, "The United States Steel Corporation and Price Stabilization," *Quarterly Journal of Economics* 38 (November 1923), pp. 1–30; A. Berglund,

"The United States Steel Corporation and Industrial Stabilization," *Quarterly Journal of Economics* 38 (August 1924), pp. 607–630; and Edward S. Meade, "The Price Policy of the United States Steel Corporation," *Quarterly Journal of Economics* 22 (May 1908), pp. 452–466.

8. For a description of the basing-point scheme, see Stocking, *Basing Point Pricing.* For evidence of its prior usage in the industry, see Urofsky, *op. cit.,* p. 245; Stocking, *Basing Point Pricing,* pp. 50–51; and E. H. Gary, "Address of the President," in AISI, *Yearbook, 1924* (New York: AISI, 1924), pp. 240–242. For a lengthy review of the industry usage of the scheme in the early twentieth century, see the action brought against the practice by the FTC: *In the Matter of United States Steel Corporation et al.,* Docket 760, FTC, 8 FTC Decisions 1 (1924). This suit forced the steelmakers to abandon the Pittsburgh Plus system; it was soon replaced, however, with a multiple basing-point pricing scheme.

9. The AISI was formed in 1908 by Elbert H. Gary. It soon took over the duties of the American Iron and Steel Association, an industry group that had been in existence since 1855 and that did not finally expire until 1912. There is no history of the AISI currently available. For details surrounding its formation, see Joseph G. Butler, Jr., *Fifty Years of Iron and Steel* (Cleveland: Penton Press Co., 1923), p. 160; AISI, *Officers, Directors, Constitution and By-Laws* (New York: AISI, 1914), p. 3; and *Bulletin of the American Iron and Steel Association* 42 (April 15, 1908), p. 35. Also see Paul H. Tedesco, "Patriotism, Protection, and Prosperity: James Moore Swank, The American Iron and Steel Association, and the Tariff, 1873–1913" (Ph.D. diss., Boston University, 1970). For information on the early workings of the AISI, see Maurice H. Robinson, "The Gary Dinner System: An Experiment in Cooperative Price Stabilization," *Southwestern Political and Social Science Quarterly* 7 (September 1926), pp. 128–159; and testimony throughout the Stanley Committee Hearings: U.S. Congress, House, *United States Steel Corporation.*

10. For industry labor policies, see Charles A. Gulick, *Labor Policy in the United States Steel Corporation* (New York: Columbia University Press, 1924); John A. Garrity, "The United States Steel Corporation Versus Labor: The Early Years," *Labor History* 1 (Winter 1960), pp. 3–38; David Brody, *Steelworkers in America, the Nonunion Era* (Cambridge: Harvard University Press, 1960); and Gerald G. Eggert, *Steelmasters and Labor Reform, 1886–1923* (Pittsburgh: University of Pittsburgh Press, 1981).

Regarding the stabilization of international steel trade, Gary originally had hoped to form an international version of the AISI that would be devoted to the same goals as the domestic group. See "Proceedings at a Luncheon Given by the President," *Proceedings of the AISI, 1910* (New York: AISI, 1910), pp. 169–192; "An Iron and Steel Conference," *Iron Age* 87 (February 23, 1911), p. 471; "The Brussels International Conference," *Iron Age* 87 (June 22, 1911), p. 1501; "The International Iron and Steel Conference," *Iron Age* 88 (July 13, 1911), p. 74; "International Iron and Steel Conference," *Iron Age* 88 (July 20, 1911), pp. 144–147; "International Cooperation in the Steel Trade," *Iron Age* 88 (August 3, 1911), pp. 234–235; "International Iron and Steel Association," *Iron Age* 88 (August 3, 1911), pp. 252–258; and "Address of the President," in AISI, *Yearbook, 1912* (New York: AISI, 1912), pp. 17–18. Because of opposition from the Justice Department, this international body was never formed. As well, it is unclear whether or not foreign producers were that enthusiastic about joining such a body; see J. C. Carr and W. Taplin, *History of the British Steel Industry* (Cambridge, MA: Harvard University Press, 1962), pp. 250–251.

11. "Impromptu Remarks of the President," in AISI, *Yearbook, 1925* (New York: AISI, 1925), p. 222.

12. *United States v. United States Steel Corporation et al.,* 251 U.S. 417 (1920), brief filed by the Department of Justice in the Federal District Court of New Jersey on October 26, 1911, seeking dissolution of the trust into its original operating units.

13. *Ibid.* For an interesting recent analysis of this decision, see Thomas K. McCraw, "Rethinking the Trust Question," in T. K. McCraw (ed.), *Regulation in Perspective* (Cambridge: Harvard University Press, 1981), pp. 1–55.

14. Tarbell, *op. cit.*, pp. 91, 111, 251; Arundel Cotter, *The Authentic History of the United States Steel Corporation* (New York: Moody Magazine and Book Co., 1916), p. 99; Duncan Burn, *The Economic History of Steelmaking, 1867–1939* (Cambridge: Cambridge University Press, 1940), p. 285; "Trusts and Combines in Coal, Iron, and Steel," *Economist* (London) 103 (October 2, 1926), p. 532; and "The Foreign Fear of American Steel Exports," *Iron Age* 80 (November 14, 1907), pp. 1399–1400.

15. See Tedesco, *op. cit.;* Abraham Berglund and Philip G. Wright, *The Tariff on Iron and Steel* (Washington, DC: Brookings, 1929); M. R. Eiselen, *The Rise of Pennsylvania Protectionism* (Philadelphia: author, 1932); F. W. Taussig, *The Tariff History of the United States,* 8th ed. (New York: G. P. Putnam's Sons, 1931); and Edward Stanwood, *American Tariff Controversies in the Nineteenth Century,* 2 vols. (Boston: Houghton Mifflin, 1903).

16. William H. Becker, *The Dynamics of Business–Government Relations, Industry & Exports, 1893–1921* (Chicago: University of Chicago Press, 1982), pp. 1–19; Mary L. Eysenbach, *American Manufactured Exports, 1879–1914* (New York: Arno Press, 1976), p. 127 (Table 16); and Abraham Berglund, *The United States Steel Corporation* (New York: Columbia University Press, 1907), p. 86.

17. United States Steel Corporation, *Fifth Annual Report* (New York: author, 1906), p. 26. Also see Taussig, *Tariff History of the United States,* pp. 212–213.

18. Cotter, *op. cit.,* p. 115. Also see William T. Hogan, *Economic History of the Iron and Steel Industry in the United States,* 5 vols. (Lexington, MA: Lexington Books, 1971) 2:798; Burton I. Kaufman, *Efficiency and Expansion: Foreign Trade Organization in the Wilson Administration, 1913–1921* (Westport, CN: Greenwood Press, 1974), p. 4; and F. W. Taussig, *Some Aspects of the Tariff Question* (Cambridge: Harvard University Press, 1915), p. 193 (Chart V).

19. One such response was the formation of a German steel cartel; see Francis Walker, "The German Steel Syndicate," *Quarterly Journal of Economics* 20 (May 1906), pp. 363–399.

20. See citations in the second paragraph of n. 10.

21. Becker, *op. cit.,* pp. 1–19.

22. Edward N. Hurley, "Co-operation and Efficiency in Developing Our Foreign Trade" in AISI, *Yearbook, 1916* (New York: AISI, 1916), p. 192.

23. See the annual reports of U.S. Steel during these years for a listing of board and committee members. Also see "U.S. Steel: I," *Fortune* 13 (March 1936), pp. 63, 180, 186–188; Lewis Corey, *The House of Morgan* (New York: G. Howard Watt, 1930), pp. 415–453; and "Myron Charles Taylor," *Fortune* 13 (June 1936), pp. 117–120, 172ff.

24. O. H. Cheney, "America in the World Steel Markets—A Warning," *Annalist* 30, 775 (November 25, 1927), pp. 819–820. Cheney was vice president of American Exchange-Irving Trust Co. of New York.

25. See United States Steel Corporation, *Twenty-Eighth Annual Report* [for 1929] (New York: author, 1930), p. 33; Frank A. Southard, Jr., *American Industry in Europe* (Boston: Houghton Mifflin, 1931), p. 173 and app. X; Mira Wilkins, *The Maturing of Multinational Enterprise: American Business Abroad from 1914 to 1970* (Cambridge: Harvard University Press, 1974), pp. 11, 185–186, 306; Cleona Lewis, *America's Stake in International Investments* (Washington, DC: Brookings, 1938), pp. 256–258; and Mira Wilkins, *The Emergence of Multinational Enterprise: American Business Abroad from the Colonial Era to 1914* (Cambridge: Harvard University Press, 1970), p. 150. Also see Paul Tiffany, "Opportunity Denied: The Abortive Attempt to Internationalize the American Steel Industry, 1903–1929," *Business and Economic History* 16(1987).

26. Franklin D. Roosevelt's initial suspicions regarding the steel industry had deep roots. In 1913 he went to Washington in an official capacity for the first time to serve as the Wilson administration's new assistant secretary of the Navy. His supervisor was Josephus Daniels, a long-time supporter of William Jennings Bryan and an ardent progressive. Throughout his secretariat, Daniels fought a bitter battle with the steel producers over armor contracts for naval vessels; the primary focus of the conflict was a proposed government-built and government-operated armor plate factory, which the secretary hoped to use as a yardstick to measure bids submitted by the private armor producers (U.S. Steel, Bethlehem Steel, and Midvale Steel—the latter soon to be acquired by Bethlehem). Roosevelt, of course, became drawn into the argument; although he may not always have agreed precisely with Daniel's antisteel-company sentiments, nevertheless, he fully supported the government's position on price collusion in big business and, no doubt, formed an impression of steel industry competitive behavior through his Navy Department experience. See Frank Friedel, *Franklin D. Roosevelt, the Apprenticeship* (Boston: Little, Brown, 1952), pp. 157–173; Urofsky, *op. cit.*, pp. 117–151; and Benjamin F. Cooling, *Gray Steel and Blue Water Navy, the Formative Years of America's Military-Industrial Complex, 1881–1917* (Hamden, CT: Shoe String Press, Archon Books, 1979), pp. 183–212.

27. Jesse C. Moody, Jr., "The Steel Industry and the National Recovery Administration: An Experiment in Industrial Self-Government" (Ph.D. diss., University of Oklahoma, 1965), pp. 39–56, 87–88, 106–108, 112–113, 269, 278; Ellis W. Hawley, *The New Deal and the Problem of Monopoly, a Study in Economic Ambivalence* (Princeton: Princeton University Press, 1966), pp. 19–146; FTC, *Practices of the Steel Industry Under the Code,* published as U.S. Senate Document No. 159, 73rd Cong., 2nd sess. (Washington, DC: GPO, 1934); and Robert F. Himmelberg, *The Origins of the National Recovery Administration* (New York: Fordham University Press, 1976).

28. Eugene G. Grace, "Industry and the Recovery Act," *Scribner's Magazine* 95 (February 1934), p. 96.

29. See U.S. Department of Commerce, Bureau of Foreign and Domestic Commerce [J. Joseph Palmer], *Origin and Development of the Continental Steel Entente,* Trade Information Bulletin No. 484 (Washington, DC: GPO, 1927); Walter S. Tower, "The New Steel Cartel," *Foreign Affairs* 5 (January 1927), pp. 249–266; Richard A. Lauderbaugh, *American Steel Makers and the Coming of the Second World War* (Ann Arbor: University Microfilms International, UMI Research Press, 1980), pp. 134–135; and FTC, *Report on International Steel Cartels* (Washington, DC: GPO, 1948), pp. 21–22.

30. Ervin Hexner, *The International Steel Cartel* (Chapel Hill: University of North Carolina Press, 1943), pp. 79–83, 88–89; and Burn, *op. cit.*, pp. 456–457.

31. This action was legal under the provisions of the Webb-Pomerene Act of 1918 as long as the domestic producers did not conspire to limit competition within domestic markets. See "Export Company All Inclusive," *Iron Age* 121 (June 28, 1928), p. 1833; "Steel Export Assoc. Formed," *Commercial and Financial Chronicle* 126 (June 23, 1928), pp. 3859–3860; FTC, *Report on International Steel Cartels,* pp. 1–13, 19–21; and Ervin Hexner, "American Participation in the International Steel Cartel," *Southern Economic Journal* 8 (July 1941), pp. 67–69.

32. See Hawley, *op. cit.*, pp. 149–168; and Arthur M. Schlesinger, Jr., *The Coming of the New Deal* (Boston: Houghton Mifflin, 1958), pp. 253–260, 471–478.

33. For details on how these provisions of the Reciprocal Trade Agreements Act operated, see Harry C. Hawkins and Janet L. Norwood, "The Legislative Basis of United States Commercial Policy," in William B. Kelly, Jr. (ed.), *Studies in United States Commercial Policy* (Chapel Hill: University of North Carolina Press, 1963), pp. 69–103.

34. On the demise of the NRA, see Hawley, *op. cit.*, pp. 111–129; Bernard Bellush,

The Failure of the NRA (New York: W. W. Norton, 1975); and Schlesinger, Jr., *op. cit.*, pp. 152–176.

35. See Brody, *op. cit.;* and Leon Wolff, *Lockout: The Story of the Homestead Strike, 1892* (New York: Harper & Row, 1965).

36. See Garrity, "The United States Steel Corporation Versus Labor"; Gulick, *op. cit.;* and U.S. Bureau of Labor, *Report on Conditions of Employment in the Iron and Steel Industry,* 4 vols. (Washington, DC: GPO, 1911–1913).

37. See Eggert, *op. cit.*, pp. 103–174.

38. Brody, *op. cit.*, pp. 50–79.

39. See Joseph G. Rayback, *A History of American Labor,* rev. ed. (New York: Free Press, 1966), pp. 346–355, and Lloyd Ulman, *The Government of the Steel Workers' Union* (New York: John Wiley & Sons, 1962), pp. 3–39. Also see Hogan, *op. cit.*, 3:1168–1170.

40. "The Manifesto," *Fortune* 15 (May 1937), p. 91. The statement was originally distributed by the AISI to a variety of media on June 29, 1936.

41. *Ibid.*

42. Hogan, *op. cit.*, 3:1195–1201, 1211–1218; Richard Lauderbaugh, "Business, Labor, and Foreign Policy: U.S. Steel, the International Steel Cartel, and Recognition of the Steel Workers Organizing Committee," *Politics and Society* 6 (1976), p. 445; "It Happened in Steel," *Fortune* 15 (May 1937), p. 176; and Gordon A. Craig, *Germany, 1866–1945* (New York: Oxford University Press, 1978), p. 605.

43. "It Happened in Steel," *Fortune,* pp. 91–94, 176ff; and Irving Bernstein, *Turbulent Years: A History of the American Worker, 1933–1941* (Boston: Houghton Mifflin, 1970), pp. 448–473.

44. Lauderbaugh, *American Steel Makers,* pp. 159–163.

45. *Ibid.*, pp. 160–163.

46. *Ibid.*, p. 163; also see Hexner, "American Participation," pp. 54–79.

47. "I think a study of this whole pattern [of behavior by the ISC] will show that the American firms which organized the participation in the cartel were not interested so much in expanding American export sales as they were in limiting imports into the United States, so that the American steel industry internally could be better organized and protected against foreign competition." This was the summary statement by James S. Martin before the congressional committee that investigated U.S. participation in the ISC in 1950. Martin, the star witness, provided testimony and evidence that demonstrated clearly the involvement of American firms in the cartel. See U.S. Congress, House Subcommittee on Study of Monopoly Power, Committee on the Judiciary, 81st Cong. 2nd sess., Hearings, *Study of Monopoly Power* (Washington, DC: GPO, 1950), pt. 4A, "Steel," pp. 357–413 (Martin's testimony); pt. 4B, "Steel Exhibits," pp. 93–343 (exhibits and evidence used in Martin's testimony). This investigation was popularly known as the Celler Committee, named after the congressional chairman of the group, Representative Emanuel Celler (D-New York).

48. See "U.S. Steel's Rivals Start Price Slash," *NYT* (February 10, 1938), p. 3; "Steel Prices Kept for Next Quarter," *NYT* (February 18, 1938), p. 25; and Lauderbaugh, *American Steel Makers,* p. 166.

49. Hawley, pp. 387–403; and Kim McQuaid, *Big Business and Presidential Power, from FDR to Reagan* (New York: William Morrow, 1982), pp. 11–17.

50. Hawley, pp. 404–419; the quotation is on p. 412.

51. See Eliot Janeway, *The Struggle for Survival: A Chronicle of Economic Mobilization During World War II* (New Haven: Yale University Press, 1951); Bruce Catton, *The War Lords of Washington* (New York: Harcourt, Brace, 1948); and McQuaid, pp. 62–93.

52. See Hexner, *International Steel Cartel,* pp. 90–91; U.S. Congress, House, *Study*

of Monopoly Power, pt. 4A, pp. 370–371; and Lauderbaugh, *American Steel Makers,* p. 173.

53. For a comprehensive analysis of this concern, see Lauderbaugh, *ibid.,* pp. 17–107.

54. *Ibid.,* pp. 91, 95–96.

55. Irving S. Olds, "Address" in AISI, *Yearbook, 1941* (New York: AISI, 1941), p. 47.

56. Walter S. Tower, "Address of the President" in AISI, *Yearbook, 1941,* p. 37.

57. Lauderbaugh, *American Steel Makers,* pp. 174–175.

58. *Ibid.,* p. 106. Regarding the capacity-expansion controversy in general at this time, see David C. Motter, "Government Controls over the Iron and Steel Industry During World War II: Their Development, Implementation, and Economic Effect" (Ph.D. diss., Vanderbilt University, 1958), pp. 109–110, 103–185.

59. *Ibid.,* pp. 140–149. Between January 1, 1940, and June 30, 1945, some $2.7 billion was expended on the addition of 15.2 million tons of capacity to existing levels; the federal government contributed $1.07 billion of this amount, accounting for 7.9 million tons of the new capacity. Also see Gerald T. White, *Billions for Defense: Government Financing by the Defense Plant Corporation During World War II* (University: University of Alabama Press, 1980).

60. Motter, *op. cit.,* pp. 186–244.

61. *Ibid.,* p. 148: U.S. Steel accounted for 34.1% of domestic steel ingot capacity on January 1, 1940; it received 41.3% of the governmental funds provided for expansion, which resulted in U.S. Steel's control of 40% of the new government-funded capacity that was built. See pp. 123–124 on the change in attitude of U.S. Steel leaders toward expansion. Also see Lauderbaugh, *American Steel Makers,* pp. 106–107; and Hogan, 3:1205–1211.

62. See citations in n. 51.

Chapter 2

1. See Robert M. Collins, *The Business Response to Keynes, 1929–1964* (New York: Columbia University Press, 1981), pp. 77–112; Marion Clawson, *New Deal Planning, the National Resources Planning Board* (Baltimore: Johns Hopkins University Press, 1981); Otis L. Graham, *Toward a Planned Society, from Roosevelt to Nixon* (New York: Oxford University Press, 1976), pp. 69–90; Byrd L. Jones, "The Role of Keynesians in Wartime Policy and Postwar Planning, 1940–1946," *American Economic Review* 62 (May 1972), pp. 125–133. For steel industry planning needs in particular, see U.S. Congress, Senate Subcommittee on Surplus Property of the Committee on Military Affairs and the Industrial Reorganization Subcommittee of the Special Committee on Postwar Economic Policy and Planning, 79th Cong., 1st sess. Joint Hearings, *War Plants Disposal—Iron and Steel Plants* (Washington, DC: GPO, 1946); and William S. Hill, Jr., "The Business Community and National Defense: Corporate Leaders and the Military, 1943–1950" (Ph.D. diss., Stanford University, 1979), pp. 31–69, 166–179.

2. Walter S. Tower, "Address of the President," in AISI, *Yearbook, 1943* (New York: AISI, 1943), p. 37.

3. *Ibid.,* pp. 39–40.

4. Walter S. Tower, "Address of the President," in AISI, *Yearbook, 1947* (New York: AISI, 1947), pp. 635–636.

5. Walter S. Tower, "Address," in AISI, *Yearbook, 1944* (New York: AISI, 1944), p. 33.

6. For details surrounding ISC, see Chapter 1.

7. Douglas A. Fisher, *Steel in the War* (New York: U.S. Steel Corp., 1946), p. 13; and Duncan Burn, *The Steel Industry, 1939–1959* (Cambridge: Cambridge University Press, 1961), p. 132. Also see "Steel," *Iron Age* 157 (January 3, 1946), pp. 76–81.

8. See Alan Sweezy, "The Keynesians and Government Policy, 1933–1939," *American Economic Review* 62 (May 1972), pp. 116–124; Jones, "The Role of Keynesians . . . "; and Patrick D. Reagan, "The Architects of Modern American National Planning" (Ph.D. diss., Ohio State University, 1982).

9. For an informed discussion of the relative nature of iron and steel as a basic industry in the domestic economy, see Henry W. Broude, *Steel Decisions and the National Economy* (New Haven: Yale University Press, 1963), pp. 29–60.

10. Fritz M. Marx, "The Bureau of the Budget: Its Evolution and Present Role, I," *American Political Science Review* 39 (August 1945), p. 684.

11. See Clawson, *op. cit.*, pp. 176–186; Robert M. Collins, "Positive Business Responses to the New Deal: The Roots of the Committee for Economic Development, 1933–1942," *Business History Review* 52 (Autumn 1978), pp. 369–391; Edward Berkowitz and Kim McQuaid, *Creating the Welfare State* (New York: Praeger, 1980), pp. 126–128; Stephen K. Bailey, *Congress Makes a Law—The Story Behind the Employment Act of 1946* (New York: Columbia University Press, 1950), pp. 20–28; Hugh S. Norton, *The Employment Act and the Council of Economic Advisers, 1946–1976* (Columbia: University of South Carolina Press, 1977), pp. 73–104; and Herbert Stein, *The Fiscal Revolution in America* (Chicago: University of Chicago Press, 1969), pp. 131–204.

12. See *Who's Who in America, 1948–1949,* s. v. "Bean, Louis H."

13. See Oral History Interview with Louis H. Bean, p. 47, in Oral History Files, (HST Library).

14. Bean had been involved in earlier government studies of the steel industry, such as those conducted by the National Resources Planning Board in 1940 under the direction of Gardiner C. Means, who had been an associate of Bean's since they worked together in the Department of Agriculture in 1933. See Clawson, *op. cit.*, pp. 152–153; and Gladys L. Baker, *et al.*, *Century of Service—The First 100 Years of the United States Department of Agriculture* (Washington, DC: U.S. Department of Agriculture, 1963), p. 247.

15. See Louis J. Paradiso (under the direction of Gardiner C. Means), *Capital Requirements: A Study in Methods as Applied to the Iron and Steel Industry* (Washington, DC: National Resources Planning Board, 1940). Also see Hill, *op. cit.*, pp. 31–69, 166–179.

16. For example, John W. Snyder, Truman's director of the Office of War Mobilization and later his secretary of the treasury; Snyder was an old friend of Truman's who exerted a substantial degree of influence with the president, especially regarding economic issues. For examples of this as well as Snyder's conservative leanings, see Alonzo L. Hamby, *Beyond the New Deal: Harry S. Truman and American Liberalism* (New York: Columbia University Press, 1973), pp. 3–51; and Norton, *op. cit.*, p. 102.

17. "Girdler Anticipates Prosperous Future and Heavy Steel Demand," *Iron Age* 158 (October 10, 1946), p. 108.

18. See Wilfred Sykes, "The Future of the Steel Industry," in AISI, *Yearbook, 1947* (New York: AISI, 1947), pp. 68–83; and E. W. Axe and Co., Research Department, *The Postwar Outlook for the Steel Industry* (Tarrytown, NY: author, 1944), pp. 4–5.

19. Bottleneck effects imply that steel is such an important factor in relation to input needs of other sectors (e.g., automobiles, appliances, construction) that a shortage of steel will necessarily affect economic activity in those other sectors; leverage effects refer to the contention that steel is such a large and important sector in the

economy that any recessions (or growth) in steel output will affect the rate of growth in the economy as a whole. For discussion of these questions, see Charles K. Rowley, *Steel and Public Policy* (London: McGraw-Hill, 1971), pp. 176–187; and Broude, *op. cit.*, pp. 70–90.

20. See Ellis W. Hawley, *The New Deal and the Problem of Monopoly, a Study in Economic Ambivalence* (Princeton: Princeton University Press, 1966), pp. 247–269; and Harmon Ziegler, *The Politics of Small Business* (Washington, DC: Public Affairs Press, 1961), pp. 87–92.

21. For a broader discussion of the Truman administration's views on this subject, see Hamby, *op. cit.*, pp. 293–310; and Robert L. Branyan, "Antimonopoly Activities During the Truman Administration" (Ph.D. diss., University of Oklahoma, 1961).

22. H. S. Truman to Henry Wallace, February 18, 1946, President's Secretary File, Box 136, General File—Small Business Folder, HST Library. Wallace was secretary of commerce at this time and was very much interested in restoring the prominence of small businesspersons in the American economy. The Senate had established a Small Business Committee in 1940; the House in 1941.

23. See Hill, *op. cit.*, pp. 1–30.

24. It should be noted, however, that Congress had created a Smaller War Plants Corporation to assist small business to obtain a share of the military contracts being let during the war; see Ziegler, *op. cit.*, pp. 93–100. The Commerce Department had sponsored legislation in 1943 creating an assistant secretary of commerce for small business; when the Smaller War Plants Corporation was terminated in 1946, an Office of Small Business (in the Commerce Department) took up symbolic leadership in this area. Commerce remained a strong supporter of small business during Truman's administration; it planned to put "services to small business as the heart of the [department's] program under [the administration's] reorganization," and it also pushed plans to make risk capital available to this sector through federal sources. See Wallace to the president, February 11, 1946, President's Secretary File, Box 136, General File—Small Business Folder, HST Library, as well as the extensive material on the subject of Truman's support of small business in this same folder. For details surrounding the culmination of this movement, that is, creation of the Small Business Administration in 1953, see Ziegler, *op. cit.*, pp. 104–115.

25. See Bert Cochran, *Harry Truman and the Crisis Presidency* (New York: Funk & Wagnalls, 1973), p. 46.

26. Truman to Wallace, February 18, 1946, *op. cit.*

27. There was, however, a substantial amount of political debate regarding the amount of authority that the Small Business Committee would be able to wield; see Ziegler, *op. cit.*, pp. 81–84.

28. See U.S. Congress, Senate Special Committee to Study Problems of American Small Business, 80th Cong., 1st sess., Hearings, *Problems of American Small Business* (Washington, DC: GPO, 1947), 4 vols.

29. *Ibid.*, pt. 4, pp. 587–701; pts. 5–8, 11–18, 28, 30–31.

30. The United States also produced three-fifths of the world output of crude steel in that year; see Louis Lister, *Europe's Coal and Steel Communty, an Experiment in Economic Union* (New York; Twentieth Century Fund, 1960), pp. 432–433, Table 2–2.

31. For an overview of the labor turmoil in this period, see Joel Seidman, *American Labor from Defense to Reconversion* (Chicago: University of Chicago Press, 1953); and Harry A. Millis and Emily C. Brown, *From the Wagner Act to Taft-Hartley: A Study of National Labor Policy and Labor Relations* (Chicago: University of Chicago Press, 1950). Also see Don Q. Crowther and Staff, "Work Stoppages Caused by Labor–Management Disputes in 1946," *Monthly Labor Review* 64 (May 1947), pp. 780–800.

32. See "Shortage Inquiry Expected to Air Steel Capacity Controversy," *Iron Age* 160 (July 31, 1947), p. 98; and "Steel Man Denies Output Curb," *NYT* (September 26, 1947), p. 37.

33. U.S. Congress, Senate, *Problems of American Small Business,* pt. 7, p. 893.

34. A subsequent House investigating committee was formed to study black markets throughout the economy; influenced by the Senate hearings, the House probers decided to use the steel industry as a proxy for all such trade practices because the phenomenon appeared so widespread in that one sector. See U.S. Congress, House Subcommittee to Investigate Questionable Trade Practices, Committee on Public Works, 80th Cong., 2nd sess., *Investigating Questionable Trade Practices,* Interim Report (Washington, DC: GPO, 1948), Committee Print No. 34, October 14, 1948.

35. See T. Campbell, "Steel at the Crossroads," *Iron Age* 161 (January 1, 1948), pp. 130–135; on the administration's pique at the large producers in other sectors as well as steel, see Hamby, *op. cit.,* p. 136.

36. The more radical political forces were led in Congress by Senator James Murray (D-Montana), who introduced a bill that would have the RFC underwrite necessary capacity expansion and have the government operate the new mills; see "Shortage Inquiry . . . ," *Iron Age.* However, there does not apppear to have been much support for such action by the administration, although it was willing to wave the flag of governmental intervention as a means of coercing the producers to adhere to its economic policies.

37. Sykes, "Future of the Steel Industry," pp. 68–83.

38. Sykes's views on export demand were in line with other steel leaders: "Export demand following World War II may well be substantially greater than the prewar average, but it is not unreasonable to assume that the long-term demand for steel products from the United States, in view of the availability of steel from wartime developed productive facilities in foreign countries, will again revert to low levels after initial needs are met." *Ibid.,* p. 70.

39. *Ibid.,* p. 73.

40. W. Parker, "Federal Chart Makers Urge Expanding Mills; the Industry Says No," *Wall Street Journal* (June 26, 1947), p. 4.

41. U.S. Congress, Senate, *Problems of American Small Business,* pt. 8, p. 994.

42. See n. 13. It might be noted that Bean's testimony before the Wherry Committee was arranged by Senator Murray, a committee member, who obviously was opposed to industry behavior in this area. See E. J. Hardy, "Washington . . . ," *Iron Age* 163 (January 20, 1949), pp. 82–83; and Hardy, "Washinton. . .,"Iron Age 162 (August 5, 1948), pp.104–106.

43. U.S. Congress, Senate, *Problems of American Small Business,* pt. 8, pp. 1001–1002.

44. *Ibid.,* pp. 1000, 1001, 1003.

45. *Ibid.,* pt. 9, p. 1537.

46. *Ibid.,* pt. 18, pp. 2006, 2007.

47. "Steel Man Denies Output Curb," *NYT,* p. 37.

48. Reuther to the president, August 20, 1947, Official File, Box 1026, Folder OF 342 (May–December, 1947), p. 2, HST Library.

49. See U.S. Congress, Senate, *Problems of American Small Business,* pt. 16, pp. 1847–1887 (for Reuther's testimony). The testimony for the USWA was provided by Otis Brubaker, director of research for the union; although he called for action, he offered no specifics (see pt. 18, pp. 1975–1993). What is perhaps most interesting about the USWA position, however, is that it represented a drastic change from that taken only the year before. In April 1946 ILO—a worldwide amalgamation of free-trade unionists that eventually was folded into the United Nations—held its first meeting of the Industrial Committee on Iron and Steel in Cleveland, OH. The meetings were

attended by representatives of labor, government, and industry, and they were intended to find ways to improve the working conditions and social standards of ironworkers and steelworkers the world over. Clint Golden, a vice president of the USWA, was one of the leading labor representatives in attendance. He disapproved of steel-capacity expansion, fearing it "would result in a surplus of steel-production facilities and cause widespread unemployment"; others agreed, with one stating, "Mr. Golden's views represent the views of the workers' group as a whole." Yet, within a year, at the Wherry Committee Hearings, Brubaker was endorsing the call for domestic expansion. The evidence is not clear as to why labor reversed itself on this crucial issue, yet it does provide insight into the problems that would surround the formulation of an effective American policy for steel.

For sources, see ILO, Iron and Steel Committee, *Regularisation of Production and Employment at a High Level* (Geneva: author, 1947); the policies adopted by the labor, government, and industry groups are stated on pp. 69–71. For commentary on the progress of the Cleveland conference, see the coverage in the *NYT:* "ILO Delegates Approved," *NYT* (April 20, 1946), p. 14; W. H. Waggoner, "Map Better Living in Steel Industry," *NYT* (April 24, 1946), p. 20; W. H. Waggoner, "Steel Job Security Asked by ILO Group," *NYT* (April 27, 1946), p. 11; W. H. Waggoner, "Steel Industry and Workers Split on Issue of a Guaranteed Wage," *NYT* (April 28, 1946), p. 35; and W. H. Waggoner, "World Issues Vex ILO Steel Leaders," *NYT* (April 29, 1946), p. 5. The quotations in the first paragraph of this note are from *NYT,* April 27, 1946.

50. The depression rumor apparently was founded on an article written by the syndicated columnist Joseph Alsop, published in numerous newspapers on September 5, 1947. This was a "fantastic statement," according to Thomas W. Lamont, head of J. P. Morgan & Company, and he tried to counter the rumor by providing columnist David Lawrence with data that would show otherwise. See Lamont to Lawrence, September 19, 1947, and Lamont to Irving S. Olds, September 19, 1947, both in Box 229, Folder 16, Lamont Papers. Olds, of course, was the chairman of U.S. Steel Corporation at this time.

51. A number of sources may be consulted for details surrounding the steel industry's public relations efforts at this time. For example, see Douglas A. Fisher, *Steel Serves the Nation, 1901–1951, the Fifty Year Story of United States Steel* (New York: U.S. Steel Corp., 1951), pp. 93–97; "Remarks of J. Carlisle MacDonald at the Annual Meeting of Officials of United States Steel Corporation and Subsidiary Companies, January 10, 1945," Box 225, Folder 25, Lamont Papers—describes U.S. Steel's PR program for that year, as told by the firm's PR director; Edward L. Ryerson, "The Steel Industry's Public Relations Program," in AISI, *Yearbook, 1947* (New York: AISI, 1947), pp. 655–664; and J. M. Larkin, "Public Relations in Bethlehem: Its History, Organization, Methods, Results, Potentialities" (mimeo, May 20, 1948), Corporate History File, Industrial and Labor Relations Folder, SML, report by Bethlehem Steel's vice president for PR to other company officers.

52. See "Getting Better Acquainted, 1947–48," a report on public relations to the members of AISI that describes the grass roots community-relations efforts undertaken by the institute; a copy is in the AISI File, SML. For details on the role of Hill & Knowlton, see Hill Papers; George F. Hamel, "John W. Hill, Public Relations Pioneer" (Master's thesis, University of Wisconsin, 1966); and John Hill, *The Making of a Public Relations Man* (New York: David McKay, 1963).

53. Memorandum for Mr. T. W. Lamont from R.G.W., October 16, 1947, Box 229, Folder 16, Lamont Papers.

54. Bradford B. Smith, *America's Steel Capacity—What It Is, What It Does* (New York: AISI, 1948). More than 22,000 copies of this pamphlet were sent to key opinion leaders in the country; see George S. Rose, "Activities of the Institute—1948," in AISI, *Yearbook, 1949* (New York: AISI, 1949), p. 604.

55. See David F. Austin, "Supply vs. Demand—A Continuing Struggle," in AISI, *Yearbook, 1948* (New York: AISI, 1948), pp. 88–95; "Austin Answers Capacity Critics; Labels Demands 'Fantastic'," *Iron Age* 161 (June 3, 1948), p. 123; U.S. Steel Corporation, News Release, May 21, 1948, a copy is in Box 42, News Releases-AISI, 1946–1950, 1952 Folder, Hill Papers—summarizes a speech made by Dr. R. E. Zimmerman, the firm's vice president for research and technology, before the American Society for Metals on the capacity issue.

56. Hardy, "Washington . . . ," *Iron Age* 162 (August 5, 1948), p. 106.

57. "U.S. Steel's Economist in Strong Defense of His Capacity Argument," *Iron Age* 162 (August 14, 1947), p. 142.

58. See, for example, Harold J. Ruttenberg, "End the Steel Famine," *Harper's Magazine* 96 (February 1948), pp. 111–117; "Steel," *Time* (January 17, 1949), p. 77; Thomas E. Mullaney, "Steel Men Defend Industry's Output," *NYT* (January 9, 1949), p. III-1; A. A. Mol, "Does the Nation Need More Steel Capacity?" *Barron's Weekly* (January 7, 1949), p. 11. In mid-April 1949 the Cooperative League, U.S.A., sponsored a panel discussion on "Steel—The Industrial Bottleneck," which attracted considerable attention; see Hardy, "Washington . . . ," *Iron Age* 163 (April 21, 1949), pp. 96–98.

59. "Industrial News Summary," *Iron Age* 162 (December 16, 1948), p. 129; also see "Industrial News Summary," *Iron Age* 163 (January 6, 1949), p. 305.

60. In January 1949 a Mr. Epinard de Witt of Los Angeles sent Truman an elaborate and detailed plan on a "proposed World's Most Modern Self Sufficient Steel Mill"; whether they took him seriously or not, administration officials replied to de Witt and even subjected his plan to some investigations. See Official File, Box 1027, Folder OF 342 (January 1949), HST Library. Other proposals were taken far more seriously; see, for example, the extensive correspondence between the president and Representative Wright Patman (D-Texas), who was actively lobbying for federal government support to the Lone Star Steel Company, a firm located in Patman's home district. This correspondence is located in Official File, Box 1025, Folder OF 342 (1945–Feb. 1947), HST LIbrary.

61. U.S. Department of Interior, *National Resources and Foreign Aid. Report of J. A. Krug, Secretary of the Interior* (Washington, DC: author, October 9, 1947). Also see F. Belair, Jr., "Krug Finds US Resources Sufficient to Aid Europe," *NYT* (October 19, 1947), pp. 1, 44.

62. See "Steel Capacity," *NYT* (December 8, 1948), p. 49; and Hardy, "Washington . . . ," *Iron Age* 163 (January 20, 1949), p. 86.

63. Hardy, "Washington . . . ," *Iron Age* 161 (April 15, 1948), p. 104.

64. Hardy, "Washington . . . ," *Iron Age* 163 (March 10, 1949), pp. 132–134; and Charles Sawyer, *Concerns of a Conservative Democrat* (Carbondale: Southern Illinois University Press, 1968), pp. 171–207.

65. See Hardy, "Washington . . . ," *Iron Age* 163 (March 10, 1949), pp. 132–134; and Hardy, "Washington . . . ," *Iron Age* 161 (April 15, 1948), p. 104.

66. Smith, *America's Steel Capacity,* pp. 34–35.

67. U.S. Congress, Senate, Special Committee to Study Problems of American Small Business, 80th Cong., 2nd sess., *Steel Supply and Distribution Problems, Final Report* (Washington, DC: GPO, 1949), p. 29. Issued as Senate Report No. 43, February 10, 1949.

68. *Ibid.*

69. CEA, "Inter-agency Report on Steel and Essential Steelmaking Materials," mimeo, March 15, 1949; a copy is in CEA Papers, Box 22, Steel Folder, DDE Library. The quote is from p. 2 of the Introduction and Summary. The report was apparently compiled by Dr. Edgar M. Hoover, an economist at the University of Michigan on contract to the CEA; see Hardy, "Washington . . . ," *Iron Age* 161

(March 25, 1948), p. 96; and Hardy, "Washington . . . ," *Iron Age* 163 (March 10, 1949), p. 132.

70. CEA, "Inter-agency Report . . . ," p. II–12ff.

71. *Ibid.*, p. IV–1.

72. See, for example, Charles S. Russell and William J. Vaughan, *Steel Production: Processes, Products, and Residuals* (Baltimore: Johns Hopkins University Press, 1976).

73. See G. F. Sullivan, "Steel Capacity," *Iron Age* 163 (January 6, 1949), pp. 198–205, especially p. 201.

74. See, for example, G. W. Hewitt, "Iron Ore Supply for the Future," in AISI, *Yearbook, 1947* (New York: AISI, 1947), pp. 336–354; and Marvin Barloon, "The Question of Steel Capacity," *Harvard Business Review* 27 (March 1949), pp. 227–228. Experimentation was underway to find means of converting the low-grade iron ore deposits of the Mesabi region into usable materials, chiefly through pelletization of taconite deposits; yet, this was plagued by difficulty, especially water pollution problems. See Robert V. Bartless, *The Reserve Mining Controversy* (Bloomington: Indiana University Press, 1980). Also see C. C. Henning and R. W. Braund, "Present and Prospective Sources of Supply of Steelmaking Raw Materials," in AISI, *Yearbook, 1950* (New York: AISI, 1950), pp. 251–273; and William T. Hogan, *Economic History of the Iron and Steel Industry in the United States,* 5 vols. (Lexington, MA: Lexington Books, 1971), 4:1481–1490.

75. See CEA, "Inter-agency Report . . . ," pp. II–7ff.

76. See AISI, *Annual Statistical Report, 1947* (New York: author, 1948), p. 29.

77. CEA, "Inter-agency Report . . . ," pp. I–9, II–6. The search for additional sources of scrap was undertaken by the government, the AISI, and the Institute of Scrap Iron and Steel. John R. Steelman, President Truman's assistant, was closely involved with this search, and he apparently went to some lengths to aid the industry: see Official File, Boxes 1026,1027, 1030, and 1031, various folders, HST Library. One result of this effort was a joint industry–government mission to Germany to identify potential scrap supplies left over from the war: see U.S. Department of Commerce, "Report of Industry–Government Scrap Iron and Steel Mission to Germany" (Washington, DC: author, 1948), a copy is in Official File, Box 1031, Folder OF 345, HST Library.

78. Barloon, *op. cit.*, p. 233.

79. See D. I. Brown, "f.o.b. mill," *Iron Age* 163 (January 6, 1949), pp. 168–177. Also see "U.S. Steel Giving Up Base-Point Pricing," *NYT* (July 8, 1948), p. 31; "New Pricing Basis for Steel Spreads," *NYT* (July 9, 1948), p. 23; G. F. Sullivan, "The Basing Point System," *Iron Age* 161 (January 1, 1948), pp. 226–227; "Steel Heads See Government Shadow on Industry," *Iron Age* 161 (June 10, 1948), pp. 91–93; G. F. Sullivan, "Steel Men Say FTC Order Leaves the Basic Problem Unsolved," *Iron Age* 161 (June 24, 1948), pp. 119–120; and T. Campbell, "Steel Industry Poised to Plunge Into F.O.B. Mill Sales System," *Iron Age* 162 (July 8, 1948), pp. 119–120.

80. For example, see Sullivan, "Steel Capacity," *Iron Age* 163 (January 6, 1949), p. 202.

81. Hogan, *op. cit.,* 4:1564–1583. Also see Bela Gold, Gerhard Rosegger, and Myles G. Boylan, Jr., *Evaluating Technological Innovations* (Lexington, MA: Lexington Books, 1980), pp. 175–209.

82. See J. W. Kirkpatrick, "Oxygen in Open Hearth Steelmaking," in AISI, *Yearbook, 1961* (New York: AISI, 1961), pp. 207–209.

83. See Earle G. Hill, "Increasing Open Hearth Production by Use of Oxygen, Better Refractories and Control of Slag," in AISI, *Yearbook, 1949* (New York: AISI, 1949), pp. 139–141; and see Sullivan, "Steel Capacity," p. 203, where the author states: "Almost all the estimates of tremendous and immediate production increases from oxygen have so far proved too optimistic, but it took a year or so to discover the fact."

84. See David R. Dilley and David L. McBride, "Oxygen Steelmaking—Fact vs. Folklore," *Iron and Steel Engineer* 44 (October 1967), pp. 133–134. Also see Kenneth Warren, *World Steel, an Economic Geography* (New York: Crane, Russak & Co., 1975), pp. 57–59.

85. Dilley and McBride, *op. cit.*, pp. 133–134.

86. CEA, "Inter-agency Report . . . ," p. I–7.

87. Barloon, *op. cit.*, p. 229.

88. See "Department of Commerce Report on Recent Steel Price Increases, March 1948," a copy is located in Official File, Box 1026, Folder OF 342 (March–April 1948), HST Library.

89. See Sullivan, "Steel Capacity," pp. 202–203.

90. Another option would have been for Congress to impose a utilitylike regulatory system for steel. By recognizing the peculiar economic properties of steelmaking and acknowledging that traditional marginal-pricing economic theories were not appropriate for this industry, the government might have been able to provide the assistance the industry needed in financing new capacity construction. Although some moves in this direction were made in 1950, none were serious, and all, of course, were adamantly opposed by industry leaders.

Chapter 3

1. On Truman's prepresidential years, see Bert Cochran, *Harry Truman and the Crisis Presidency* (New York: Funk & Wagnalls, 1973), pp. 1–115; Alonzo L. Hamby, *Beyond the New Deal: Harry S. Truman and American Liberalism* (New York: Columbia University Press, 1973), pp. 3–51; and Robert J. Donovan, *Conflict and Crisis* (New York: W. W. Norton, 1977), pp. 3–33.

2. Myron C. Taylor to T. W. Lamont, April 18, 1945, Box 133, Folder 12, Lamont Papers.

3. Walter S. Tower, "Steel Faces the Postwar Years," in AISI, *Yearbook, 1945* (New York: AISI, 1945), p. 12.

4. *Ibid.*, pp. 17–18. Also see "Steel Men Assail Federal Control," *NYT* (May 24, 1946), p. 26.

5. See Craufurd D. Goodwin and R. Stanley Herren, "The Truman Administration: Problems and Policies Unfold," in C. D. Goodwin (ed.), *Exhortation and Controls—The Search for a Wage-Price Policy, 1945–1971* (Washington, DC: Brookings, 1975), pp. 14–15; also see Barton Bernstein, "Economic Policies of the Truman Administration," in R. S. Kirkendall (ed.), *The Truman Period as a Research Field* (Columbia: University of Missouri Press, 1967), pp. 87–149.

6. Goodwin and Herren, *op. cit.*, pp. 15–27; and Hamby, *op. cit.*, pp. 53–145.

7. Goodwin and Herren, *op. cit.*, p. 23.

8. Donovan, *op. cit.*, pp. 112–115; and Cochran, *op. cit.*, p. 200.

9. Goodwin and Herren, *op. cit.*, pp. 24–35; and Bernstein, "Economic Policies of the Truman Administration," pp. 87–149.

10. See Paul A. C. Koistinen, "Mobilizing the World War II Economy: Labor and the Industrial–Military Alliance," *Pacific Historical Review* 42 (November 1973), pp. 443–478. Also see Joseph G. Rayback, *A History of America Labor*, rev. ed. (New York: Free Press, 1966), pp. 387–390; and Kim McQuaid, *Big Business and Presidential Power, from FDR to Reagan* (New York: William Morrow, 1982), pp. 132–149.

11. Hamby, *op. cit.*, pp. 66–68.

12. See Joel Seidman, *American Labor from Defense to Reconversion* (Chicago: University of Chicago Press, 1953); Harry A. Millis and Emily C.Brown, *From the Wagner Act to Taft-Hartley: A Study of National Labor Policy and Labor Relations*

(Chicago: University of Chicago Press, 1950); and Howell J. Harris, *The Right to Manage* (Madison: University of Wisconsin Press, 1982).For a listing of strike days lost, see *The Statistical History of the United States, from Colonial Times to the Present* (New York: Basic Books, 1976), Series D 970–985, p. 179.

13. Rayback, *op. cit.*, pp. 389-391; and John L. Blackman, Jr., *Presidential Seizure in Labor Disputes* (Cambridge: Harvard University Press, 1967), pp. 30–31ff., 33–36ff.

14. See Barton J. Bernstein, "The Truman Administration and the Steel Strike of 1946," *Journal of American History* 52 (March 1966), p. 792; Richard W. Nagle, "Collective Bargaining in Basic Steel and the Federal Government, 1945–1960" (Ph.D. diss., Pennsylvania State University, 1978), pp. 38–41; and U.S. Departmentof Labor [E. Robert Livernash], *Collective Bargaining in the Basic Steel Industry—A Study of the Public Interest and the Role of Government* (Washington, DC: author, 1961), pp. 254–258.

15. Nagle, *op. cit.*, p. 27; and Bernstein, "Steel Strike of 1946," p. 792. Also see Benjamin Fairless to President Truman, November 13, 1945, and Benjamin Fairless to John W. Snyder, November 13, 1945, both in Official File, Box 1025, Folder OF 342 (1945–February 1947), HST Library.

16. Bernstein, "Steel Strike of 1946," p. 793.

17. *Ibid.*, pp. 793–794; U.S. Department of Labor, *op. cit.; Nagle, op. cit.*, pp. 46–50; Donovan, *op. cit.*, pp. 166–167; and I. W. Abel, *Collective Bargaining—Labor Relations in Steel: Then and Now* (New York: Columbia University Press, 1976), pp. 44–46.

18. Bernstein, "Steel Strike of 1946," p. 795; and Alfred Friendly, "American Industry's Grand Strategy," *Nation* 162 (January 19, 1946), pp. 62–63.

19. Nagle, *op. cit.*, pp. 51–53; and Bernstein, "Steel Strike of 1946," p. 795.

20. Truman to Martha Ellen Truman and Mary Jane Truman, January 23, 1946, in Robert H. Ferrell (ed.), *Off the Record—The Private Papers of Harry S. Truman* (New York: Harper & Row, 1980), p. 83.

21. Bernstein, "Steel Strike of 1946," pp. 796–800.

22. Nagle, *op. cit.*, pp. 58–59.

23. Goodwin and Herren, *op. cit.*, p. 28; Bernstein, "Steel Strike of 1946," pp. 801–813. On the tonnage lost during the strike, see T. E. Mullaney, "Record Year Seen in 1947 for Steel," *NYT* (January 2, 1947), pp. 31, 38; and T. Campbell, "Steel—1946–1947," *Iron Age* 159 (January 2, 1947), pp. 66–71.

24. This is not to imply, however, that steel leaders and the administration were not on speaking terms with one another. On June 20, 1946, in fact, Benjamin F. Fairless of U.S. Steel received the prestigious Medal of Merit on orders from President Truman; see Fairless to Truman, June 14, 1946, Official File, Box 1046, Folder OF 357C (January–August 1946), HST Library. Soon afterward, Fairless wrote to Thomas Lamont of J. P. Morgan, "I am happy to see our great Corporation recognized by Government and I do hope this incident might be helpful in bringing about a better relationship between Government and the Steel Corporation." See Fairless to Lamont, June 24, 1946, Box 229, Folder 7, Lamont Papers.

25. Byrd L. Jones, "The Role of Keynesians in Wartime Policy and Postwar Planning, 1940–1946," *American Economic Review* 62 (May 1972), pp. 125–133; Herbert Stein, *The Fiscal Revolution in America* (Chicago: University of Chicago Press, 1969), pp. 169–196; Karl Schriftgiesser, *Business and Public Policy* (Englewood Cliffs, NJ: Prentice-Hall, 1967), pp. 12–18; and Goodwin and Herren, *op. cit.*, pp. 9–11.

26. See B. D. Hulen, "OPA Price Controls End," *NYT* (June 30, 1946), p. 1; and Donovan, *op. cit.*, pp. 198–199, 235.

27. See Susan Hartmann, *Truman and the 80th Congress* (Columbia: University of Missouri Press, 1971), p. 19; Hamby, *op. cit.*, p. 136; and Donovan, *op. cit.*, p. 239. For background on the subject, see Addison T. Cutler, "Price Control in Steel," in

OPA, Office of Temporary Controls, *Studies in Industrial Price Control* (Washington, DC: GPO, 1947), pp. 37–85. Issued as General Publication No. 6, Historical Reports of War Administration; OPA.

28. Cutler, *op. cit.*, pp. 84–85.

29. Goodwin and Herren, *op. cit.*, pp. 40–48.

30. Hamby, *op. cit.*, pp. 136–137.

31. "Sees Violation of Law; Clark Rules on Proposed Sale of Consolidated to Columbia," *NYT* (February 24, 1947), p. 27. Also see T. E. Mullaney, "Steel Men Fight for Coast Trade," *NYT* (February 24, 1947), p. 27.

32. See E. J. Hardy, "Washington . . . ," *Iron Age* 160 (September 25, 1947), p. 94.

33. R. Porter, "Truman Again Urges Price Cuts to End Rising Threat of Recession: No Time to Reduce Taxes, He Says," *NYT* (April 22, 1947), pp. 1, 14.

34. See W. H. Waggoner, "Truman Urges Price Cuts, Fearing Inflationary Spiral," *NYT* (March 27, 1947), pp. 1, 8; R. Porter, "Force of Public Opinion Revived on Price Factor," *NYT* (April 6, 1947), p. III–1; and L. W. Moffett, "Washington . . . ," *Iron Age* 159 (April 24, 1947), pp. 74–76.

35. L. Starr, "President Asks Price Tests Before Any Coal-Steel Rises," *NYT* (July 15, 1947), pp. 1, 15.

36. CEA, *The Midyear Economic Report of the President to the Congress, July 21, 1947,* in *The Economic Reports of the President* (New York: Reynal and Hitchcock, 1948), p. 80.

37. Starr, *op. cit.*, p. 15.

38. This comes out quite clearly in correspondence between members of the U.S. Steel board of directors; see Arthur Andersen to Thomas Lamont, July 15, 1947, Box 229, Folder 14, Lamont Papers. Andersen was a J. P. Morgan partner who served on U.S. Steel's board along with Lamont; he usually sent summaries of board meetings to Lamont when the latter was unable to attend.

39. "Steel Price Rises General; U.S. Steel, Bethlehem Join Move," *NYT* (July 30, 1947), pp. 1, 13.

40. Lamont to Andersen, July 31, 1947, Box 229, Folder 15, Lamont Papers; "Prices Are Raised by 5 Steel Firms," *NYT* (August 1, 1947), p. 28.

41. T. E. Mullaney, "Unabated Output for Steel Is Seen," *NYT* (August 3, 1947), pp. III–1.

42. Lamont to E. M. Voorhees, August 1, 1947, Box 229, Folder 16, Lamont Papers. Voorhees was chairman of the Finance Committee of U.S. Steel as well as an officer and director of the firm.

43. See R. H. Fetridge, "Along the Highways and Byways of Finance," *NYT* (November 7, 1947), p. III–3; contains a brief and laudatory portrait of U.S. Steel's chairman. Stettinius had been brought in as chairman of the board in 1938 in order to improve U.S. Steel's image with both the public and Washington; prior to this time, he had served in public relations positions with General Motors. See "U.S. Steel: IV," *Fortune* 13 (June 1936), p. 114; and "Stettinius Heads U.S. Steel; Taylor Says Farewell to Stockholders," *Steel* 102 (April 4, 1938), p. 29. Stettinius, of course, left the firm in 1940 to work in the war mobilization effort for President Roosevelt; in 1944 he was promoted to secretary of state under FDR.

44. Based on information cited in Paul A. Tiffany, "The Steel Industry Responds: The Rise of Public Relations" (Working Paper, Berkeley, CA, 1980).

45. Lamont to Fairless, September 25, 1947, Box 229, Folder 16, Lamont Papers.

46. For example, U.S. Steel officials were concerned about governmental complaints regarding the firm's involvement in the Geneva steelworks in Utah. This mill had been built during World War II with public funds, but it was operated and managed by U.S. Steel under contract. The plant was located in Utah owing to local access

to raw materials and because of fears that a site on the West Coast (which was the location of the mill's markets) would be subject to enemy attack. When the war ended, U.S. Steel was unsure about its plans for Geneva; the government wanted operations to continue in order to help ease the steel shortage. But the firm was concerned with the poor location of the plant; its distance from coastal markets would add significantly to costs. If any U.S. Steel facilities were to serve coastal demand, they should be built in either Los Angeles or San Francisco, reasoned company planners. But governmental pressures to keep Geneva operating were strong; moreover, Henry J. Kaiser wanted to take over the mill and merge it into his Kaiser Steel Company. As a result, U.S. Steel reluctantly agreed to buy the Geneva works from the government and keep it operational; it paid $40 million for the plant, which had been built at a cost of $200 million. Thus concern with public pressures was a major factor in shaping the firm's domestic strategy for competition in the booming West Coast market. For sources, see "Pacific Coast Study," an internal report for U.S. Steel made in 1945, located in Box 230–1 of the Lamont Papers; "Statement Relative to Past and Possible Future Participation of United States Steel Corporation in Pacific Coast Steel Markets," from Irving S. Olds to Board of Directors, July 16, 1945, Box 228, Folder 29, Lamont Papers; and Lamont to Olds, July 28, 1945, Box 228, Folder 30, Lamont Papers. In addition, see U.S. Congress, Senate Subcommittee on Surplus Property of the Committee on Military Affairs and the Industrial Reorganization Subcommittee of the Special Committee on Postwar Economic Policy and Planning, 79th Cong., 1st sess., Joint Hearings, *War Plants Disposal—Iron and Steel Plants* (Washington, DC: GPO, 1946).

47. See R. Hendershot, "Policy Dangerous," *New York World-Telegram* (September 15,1947), copy is in Box 229, Folder 16, Lamont Papers.

48. See AISI, *Annual Statistical Report, 1947* (New York: author, 1948); and William T. Hogan, *Economic History of the Iron and Steel Industry in the United States,* 5 vols. (Lexington, MA: Lexington Books, 1971), 4:1654–1656.

49. See *The Outlook for Key Commodities, Resources for Freedom, a Report to the President by the President's Materials Policy Commission,* 2 vols. (Washington, DC: GPO, 1952) 2:17. This is the report of the Paley Commission, which investigated the U.S. position in critical raw materials in the postwar period.

50. Campbell, "Steel," *Iron Age* 157 (January 3, 1946), p. 78; and Hendershot, "Policy Dangerous,"

51. Hendershot, *ibid.* This sentiment about prices is also a prevalent theme in the correspondence between Lamont and other U.S. Steel officials in 1946 and 1947; see Lamont papers, *passim.*

52. U.S. Bureau of Labor Statistics figures show that the price of finished steel went up by 71% from 1939 to September 1948, whereas over the same interval the price of "all commodities rose by 118%"; see Marvin Barloon, "The Question of Steel Capacity," *Harvard Business Review* 27 (March 1949), p. 229.

53. Lamont to C. G. McGehee, July 15, 1947, Box 229, Folder 14, Lamont Papers. McGehee was an industrialist from Florida.

54. See McQuaid, *op. cit.;* and Robert M. Collins, *The Business Reponse to Keynes, 1929–1964* (New York: Columbia Unversity Press, 1981).

55. This private attempt at jawboning was reported in the media by columnist Stewart Alsop on September 5, 1947, in his nationally syndicated column. On September 10 Nourse issued a press release denying the statement. See Goodwin and Herren, *op. cit.*, pp. 44–45, fn. 113; and "News Release by Edwin G. Nourse," September 10, 1947, Box 229, Folder 16, Lamont Papers. It should be noted, of course, that this incident did not end the practice of CEA officials meeting privately with industry leaders to discuss positions and develop recommendations for public policy; the CEA continued to hold private talks with steel leaders in 1947 as well as in later years.

56. See S. A. Tower, "FTC Charges Plot Curbs Steel Supply," *NYT* (November 17, 1947), p. 1; the FTC suit filed at this time amended the earlier one filed on August 14 by adding more defendants.

57. See Chapter 1, n. 8 for citations.

58. For details of these developments, see George W. Stocking, *Basing Point Pricing and Regional Development—A Case Study of the Iron and Steel Industry* (Chapel Hill: University of North Carolina Press, 1954).

59. T. Campbell, "Federal Trade Commission Blasts at Industry on Price Setup," *Iron Age* 160 (August 21, 1947), p. 113.

60. This was the Wherry Committee, chaired by Senator Kenneth S. Wherry (R-Nebraska); its formal name was the U.S. Senate Special Committee to Study Problems of American Small Business. Also see J. A. Loftus, "Debate on Steel: Should Industry's Capacity Be Expanded?" *NYT* (September 28, 1947), p. IV–8.

61. See Chapter 2 for details surrounding the capacity-expansion controversy.

62. See "Memorandum for Mr. T. W. Lamont" from R.G.W., October 16, 1947, Box 229, Folder 16, Lamont Papers.

63. See Donovan, *op. cit.*, pp. 239–245; Harry S. Truman, *Year of Decisions* (Garden City, NY: Doubleday, 1955), pp. 223–224, 502–505.

64. See Hamby, *op. cit.*, pp. 195–265.

65. Hartmann, *op. cit.*, pp. 102, 113.

66. *Ibid.*, p. 122.

67. On the anti-inflation program, see Goodwin and Herren, *op. cit.*, pp. 46–47, and *ibid.* On steel's voluntary allocation program, see Hartmann, pp. 122–126; and Goodwin and Herren, *op. cit.*, pp. 46–48. Also see E. J. Hardy, "Steel Industry Views Commerce's Plans for Voluntary Allocation," *Iron Age* 161 (January 8, 1948), pp. 95–96; T. Campbell, "Steel Industry to Cooperate on Voluntary Allocation Plans," *Iron Age* 161 (January 22, 1948), p. 99; and G. Baker, "Draft Bill Sleeper Provide for Surprise Govt. Steel Allocation," *Iron Age* 162 (July 1, 1948), p. 126.

68. Goodwin and Herren, *op. cit.*, p. 48.

69. For background on the development of the system, see U.S. Department of Commerce [G. Lyle Belsley], National Production Authority, Historical Reports on Defense Production, Report No. 19, *Consultation with Industry—History of the Office of Industry Advisory Committees of the National Production Authority* (Washington, DC: author, 1953), copy is in the National Archives, Record Group 277, National Production Authority, Office of the Executive Secretary, Historical Reports on Defense Production, Box. 7.

70. John D. Morris, "Harriman Says Only Steel Faces Control at Present," *NYT* (November 27, 1947), pp. 1, 28; and "Steel Men Called to Draft Controls," *NYT* (December 23, 1947), p. 37.

71. U.S. Department of Commerce, *op. cit.*, p. 5.

72. For critical commentary on industry's advisory committees as a mechanism for business–government cooperation, see Grant McConnell, *Private Power and American Democracy* (New York: Alfred A. Knopf, 1967), pp. 255–280.

73. Hardy, "Washington . . . ," *Iron Age* 161 (January 15, 1948), p. 98.

74. S. A. Tower, "Executives Defend Rise in Steel Price as Forced by Costs," *NYT* (March 3, 1948), p. 1, 2.

75. "Steel Men Called by Congress Group to Explain Prices," *NYT* (February 22, 1948), p. 1.

76. *Ibid.* Also see A. Leviero, "President Orders FBI Investigation in Steel Price Rise," *NYT* (February 25, 1948), pp. 1, 17; and "Industrial News Summary," *Iron Age* 161 (March 4, 1948), p. 3.

77. "Department of Commerce Report on Recent Steel Price Increases, March 1948"; CEA to the President, March 10, 1948; and Tom C. Clark, Attorney General,

to the President (n.d.): all released by Charles G. Ross, secretary to the president, on March 12, 1948, copies in Official File, Box 1026, Folder OF 342 (March–April 1948), HST Library. Also see G. Baker, " 'Don't Let It Happen Again,' Says Government to Steel Industry," *Iron Age* 161 (March 18, 1948), p. 106.

78. CEA to the President, March 10, 1948, p. 3.

79. U.S. Congress, Joint Committee on the Economic Report, 80th Cong., 2nd sess., Hearings, *Increases in Steel Prices* (Washington, DC: GPO, 1948).

80. Hardy, "Washington . . . ," *Iron Age* 161 (March 11, 1948), p. 156.

81. U.S. Congress, Joint Committee, *op. cit.,* p. 2.

82. See Paul A. Tiffany, "Corporate Management of the External Environment: Bethlehem Steel, Ivy Lee, and the Origins of Public Relations in the American Steel Industry," in *Essays in Economic and Business History* 5 (1987), pp. 1–18.

83. In response to criticism from government, "businessmen had become cautious, if not timid," said Secretary of Commerce Charles Sawyer. See Charles Sawyer, *Concerns of a Conservative Democrat* (Carbondale: Southern Illinois University Press, 1968), p. 176.

84. *Federal Trade Commission v. Cement Institute,* 333 U.S. 683 (1948). Details surrounding this case are discussed in Stocking, *op. cit.,* pp. 144–189.

85. See "U.S. Steel Giving Up Base-Point Pricing," *NYT* (July 8, 1948), p. 31; "New Pricing Basis for Steel Spreads," *NYT* (July 9, 1948), p. 23; G. F. Sullivan, "The Basing Point System" *Iron Age* 161 (January 1, 1948), pp. 226–227; Hardy, "Washington . . . ," *Iron Age* 161 (May 27, 1948), pp. 104–106; "Steel Heads See Government Shadow on Industry," *Iron Age* 161 (June 10, 1948), pp. 91–93; G. F. Sullivan, "Steel Men Say FTC Order Leaves the Basic Problem Unsolved," *Iron Age* 161 (June 24, 1948), pp. 119–120; G. F. Sullivan, "Chaos Seen If Multiple Basing Point Sales System Is Outlawed," *Iron Age* 161 (May 6, 1948), pp. 123–124; B. Lloyd, "Big Steel Consuming Points Would Become Bigger Steel Producers," *Iron Age* 161 (May 6, 1948), p. 124; T. Campbell, "Steel Industry Poised to Plunge Into F.O.B. Mill Sales System," *Iron Age* 162 (July 8, 1948), p. 119–120; B. Packard, "Spectacular Steel Developments Marred by Distribution Rhubarbs," *Iron Age* 162 (September 2, 1948), pp. 119–121. Although the Supreme Court decision on the Cement Institute case would obviously have brought about major changes in steel distribution policies, industry critics believed that the particular plan chosen by U.S. Steel—which resulted in higher prices to many customers—was part of a conspiracy to raise prices and that the Court's decision could have been implemented in a less costly manner had the Steel Corporation so desired: see Earl Latham, "The Politics of Basing Point Legislation," *Law and Contemporary Problems* 15 (Spring 1950), pp. 274–275, 291.

86. Walter S. Tower, "Steel Problems and Prospects," in AISI, *Yearbook, 1948* (New York: AISI, 1948), p. 74.

87. A great deal of legislative intrigue surrounded S. 1008, the Senate's basing-point bill. It might be noted that the bill's sponsor was Senator Joseph O'Mahoney (D-Wyoming), who had long been one of the steel industry's sharpest critics. See Latham, "Politics of Basing Point Legislation" for details. Also see *Congressional Record,* vol. 95, pt. 5, 81st Cong., 1st sess. (May 31, 1949), pp. 7027–7031; and *Congressional Record,* vol. 95, pt. 9, 81st Cong., 1st sess. (August 12, 1949), pp. 11355–11356: details what one senator called, "One of the best organized, one of the most heavily financed, and one of the most adroitly deceptive [lobbying efforts] that has ever been addressed to the Congress of the United States," that is, industry pressure to obtain passage of S. 1008.

88. Speech delivered by President Truman in Louisville, Kentucky, September 30, 1948, pp. 8, 9, copy is in the Charles S. Murphy Papers, Box 23, Monopolies Folder, HST Library.

89. See "Truman Vetoes Freight Absorption Bill," *Iron Age* 165 (June 22, 1950),

p. 101; and T. C. Campbell, "Truman's Veto—A Blow to Industry," *Iron Age* 165 (June 22, 1950), p. 101.

90. See U.S. Department of Labor, *op. cit.,* pp. 254–261.

91. Goodwin and Herren, *op. cit.,* p. 60.

92. T. Campbell, "Details on U.S. Steel Price Cuts Astound Other Steel Firms," *Iron Age* 161 (May 6, 1948), p. 125B.

93. U.S. Department of Labor, *op. cit.,* p. 259.

94. Campbell, "Details on U.S. Steel . . . ," p. 125B.

95. FTC, *Monopolistic Practices and Small Business* (Washington, DC: GPO, 1952), pp. 53–54.

96. See "U.S. Steel Yields in Inflation Fight; to Lift Pay, Prices," *NYT* (July 17, 1948), pp. 1, 2; "Price Rise of $9.34 a Ton Is Announced by U.S. Steel," *NYT* (July 21, 1948), p. 1; and T. Campbell, "U.S. Steel Leadership Prepared to Answer Challenges of Critics," *Iron Age* 162 (July 29, 1948), pp. 119–120.

97. "Prices Drop for Third Month," *NYT* (January 28, 1949), pp. 1, 25; Campbell, "U.S. Steel Leadership . . . ," p. 119; G. F. Sullivan, "Criticism of Industry for Recent Price Rises Surprises No One," *Iron Age* 162 (August 5, 1948), pp. 119–120.

98. Hardy, "Washington . . . ," *Iron Age* 162 (October 21, 1948), p. 94.

99. G. Baker, "Administration Poised to Wield Axe or Broom on Steel Industry," *Iron Age* 162 (December 16, 1948), p. 131.

100. *Ibid.* On Taft-Hartley, see Donovan, *op. cit.,* pp. 299–304.

101. Baker, "Administration Poised to Wield Axe . . . ," pp. 132.

102. *Congressional Record,* vol. 95, pt. 1, 81st Cong., 1st sess. (January 5, 1949), p. 74.

103. *Ibid.,* p. 75.

104. *Ibid.* Also see A. Leviero, "Truman Asks Increased Taxes, . . . Authority to Build Steel Mills," *NYT* (January 6, 1949), p. 1; and H. W. Cloke, "Control of Steel Held Truman Aim," *NYT* (January 6, 1949), p. 3.

105. J. S. Lawrence, "Nationalization," *Iron Age* 163 (January 27, 1949), p. 1. Also see Hardy, "Washington . . . ," *Iron Age* 163 (January 20, 1949), p. 82.

106. "Threat of Nationalization Seen for Steel," *Iron Age* 163 (January 27, 1949), p. 111. Some others also shared these beliefs (or fears). One congresswoman, for example, thought nationalization might occur within five years; see "U.S. Business Shift Like British Seen," *NYT* (December 22, 1948), p. 15.

107. See George W. Ross, *The Nationalisation of Steel* (London: Macgibbon & Kee, 1965); and Doug McEachern, *A Class Against Itself, Power and the Nationalisation of the British Steel Industry* (Cambridge: Cambridge University Press, 1980).

108. "Business Reacts Mildly to President's State of Union Message," *Iron Age* 163 (January 20, 1949), p. 101; and G. Baker, "Some White House Proposals Face Stiff Fight in Congress—Especially Steel," *Iron Age* 163 (January 13, 1949), p. 101.

109. This statement was made by Henry J. Kaiser, the West Coast industrialist and steelmaker; "Business Reacts Mildly . . . ,", p. 102.

110. "Steel Building Plan Explained by Truman," *NYT* (January 8, 1949), p. 7.

111. See Baker, "Some White House Proposals . . . ," p. 101; and C. Hurd, "Economic Group Backs President," *NYT* (March 2, 1949), p. 4. Also see U.S. Congress, Joint Committee on the Economic Report, 81st Cong., 1st sess. *Joint Economic Report* (Washington, DC: GPO, 1949), pp. 35–36 for comment. Issued as Senate Report No. 88, March 1, 1949.

112. Charles Sawyer, "Business and the American Way of Life," in AISI, *Yearbook, 1949* (New York: AISI, 1949), p. 128; and "Business Freedom Urged by Sawyer," *NYT* (May 27, 1949), p. 33. Although Sawyer's statement is obviously not conclusive evidence of the administration's intentions regarding steel producers (and the secretary himself was probably more conservative toward business interests than many of Tru-

man's other advisors), this sentiment, nevertheless, can be regarded as an important indication of the administration's views on the topic.

113. See Hamby, *op. cit.*, pp. 331–332; "Fair Deal Defeat by Coalition Seen," *NYT* (March 25, 1949), p. 3; and Baker, "Administration Poised to Wield Axe . . . ," p. 132.

114. Nagle, *op. cit.*, pp. 72–112.

115. T. E. Mullaney, "Steel Men Defend Industry's Output: Faster Increase of Capacity Injudicious They Say, in View of Abnormal Demand," *NYT* (January 9, 1949), p. III–1.

116. "Steel Ingot Capacity Rises to Record 96,000,000 Tons Per Year with 2,700,000 Tons More Scheduled to Be Added," AISI News Release, January 6, 1949, copy is in Box 42, News Releases, AISI—1946–1950, 1952 Folder, Hill Papers.

117. United States Steel Corporation, News Release, January 3, 1949, p. 4. Box 42, Hill Papers.

Chapter 4

1. See, for example, the comments of Eugene Grace, head of Bethlehem Steel Corporation: "Grace Urges Curb on Steel Capacity," *NYT* (August 1, 1947), p. 23; and T. E. Mullaney, "Unabated Output for Steel is Seen," *NYT* (August 3, 1947), p. III–1. Grace stated that European demand should not be a factor in the decision to expand domestic capacity.

2. The steel subcommittee, headed by Senator Edward Martin (R-Pennsylvania), was an arm of the Wherry Committee that investigated small business problems resulting from product shortages in major industries. See U.S. Congress, Senate Special Committee to Study Problems of American Small Business, 80th Cong., 1st sess., Hearings, *Problems of American Small Business* (Washington, DC: GPO, 1947), 4 vols. Part 15 of the Hearings testimony focused on the steel export licensing program, see pp. 1699–2026. Also see pt. 5, pp. 738–757 for details on how this program was operated. Other than this, the foreign environment of steel was not discussed in the hearings except in random passing references.

3. The most comprehensive analyses of domestic steel in the international context came from the House Select Committee on Foreign Aid, known as the Herter Committee in honor of its chairman, Representative Christian Herter (R-Massachusetts). This group was given the responsibility of studying means to implement the Marshall Plan of foreign aid to Europe. It issued two reports on the steel requirements of Europe through 1952 and how they might be met by domestic exports. Throughout this analysis, however, no concept of a coordinated American policy for steel was ever broached; the focus was on the amount of steel needed in Europe and the share that should be provided by American producers. See U.S. Congress, House Select Committee on Foreign Aid, 80th Cong., 1st sess., *United States Steel Requirements and Availabilities as They Affect European Needs for Interim Aid,* Preliminary Report 6 (Washington, DC: GPO, 1947). Issued as House Report No. 1150, November 25, 1947. Also see U.S. Congress, House Select Committee on Foreign Aid, 80th Cong., 2nd sess., *The Role of Steel in the European Recovery Program,* Supplement to Preliminary Report 6 (Washington, DC: GPO, 1948). The only other major study of the international role of steel was U.S. Tariff Commission, *Iron and Steel,* War Changes in Industry Series Report No. 15, Prepared in Response to Requests from the Committee on Finance of the U.S. Senate and the Committee on Ways and Means of the House of Representatives (Washington, DC: GPO, 1946). Although comprehensive, this study did not address the need for the United States to develop any long-term policies in this area. Finally, it might also be noted that in 1948 the FTC studied international cartels in steel, though this was hardly meant to underscore the need for national policy

in this sector. Rather, the study represented the efforts of the FTC to call judicial attention to what it viewed as potentially illegal behavior by the steelmakers in their dealings with foreign rivals. See FTC, *Report on International Steel Cartels* (Washington, DC: GPO, 1948).

4. This was obviously discussed by some in the industry at this time; J. R. Hight in his "European Letter . . . ," *Iron Age* 158 (September 26, 1946), pp. 96–98, notes just such interest. Also see "Says Imports Will Be a Greater Benefit to Nations Than Exports," *Iron Age* 157 (February 21, 1946), p. 105, which reports on an address delivered by U.S. Steel's president of its export subsidiary.

5. See "Steel Export Tonnage Was Nearly 36 Million for 5-Yr. War Period," *Iron Age* 157 (June 27, 1946), p. 148; regarding the level of exports during the 1930s, see William T. Hogan, *Economic History of the Iron and Steel Industry in the United States,* 5 vols. (Lexington, MA: Lexington Books, 1971), 3:1425–1435.

6. "Steel Exports Hold Own in Ratio to Production: Quality Steels in Fore," *Iron Age* 158 (July 4, 1946), p. 116.

7. J. Anthony, "Foreign Funds for Purchase of American Steel Seen Unlimited," *Iron Age* 158 (July 11, 1946), p. 95.

8. U.S. Tariff Commission, *op. cit.,* pp. 35–36. Also see L. W. Moffett, "Washington . . . ," *Iron Age* 157 (April 4, 1946), pp. 96–98; and "Study Prognosticates Declining Demand for American Steel in Europe," *Iron Age* 157 (April 4, 1946), p. 164.

9. U.S. Tariff Commission, *op. cit.,* p. 31.

10. J. R. Hight, "Foreign Reconstruction," *Iron Age* 157 (January 3, 1946), p. 160.

11. The literature surrounding the origins of the Cold War, with which the German question was, of course, intimately bound up, is both large and growing. See, for example, John L. Gaddis, *The United States and the Origins of the Cold War, 1941–1947* (New York: Columbia University Press, 1972); and Daniel Yergin, *Shattered Peace, the Origins of the Cold War and the National Security State* (Boston: Houghton Mifflin, 1977), pp. 303-335. Also consult the extensive bibliographies contained in both of these works.

12. For discussion, see Thomas G. Paterson, "The Economic Cold War: American Business and Economic Foreign Policy, 1945–1950" (Ph.D. diss., University of California, Berkeley, 1968), chap. 8, "Economic Containment of Europe and the Rehabilitation of Germany," especially p. 423. Also see Bruce R. Kuklick, "American Foreign Policy, Economic Policy and Germany, 1939–1946" (Ph.D. diss., University of Pennsylvania, 1968).

13. Henry Morgenthau, Jr., *Germany Is Our Problem* (New York: Harper & Brothers, 1945), quotation is in unpaged statement preceding the table of contents.

14. Herbert Feis, *Churchill, Roosevelt, Stalin; the War They Waged and the Peace They Sought* (Princeton: Princeton University Press, 1957), p. 370. Kuklick, *op. cit.,* p. 82, quotes Roosevelt, "As far as I'm concerned, I'd put Germany back as an agricultural country."

15. See Paterson, *op. cit.,* pp. 1–55, 425; Yergin, *op. cit.,* p. 95.

16. See Yergin, *op. cit.,* pp. 87–105; Robert J. Donovan, *Conflict and Crisis* (New York: W. W. Norton, 1977), pp. 72–79; Paterson, *op. cit.,* pp. 427–429; Bert Cochran, *Harry Truman and the Crisis Presidency* (New York: Funk & Wagnalls, 1973), pp. 158–162. Cochran notes that JCS 1067 went through seven drafts before Truman approved of it, indicative of the president's uncertainty on this matter (see Cochran, pp. 164–165). Also see Carl J. Friedrich, *American Experiences in Military Government in World War II* (New York: Rinehart, 1948), pp. 381–402, for a copy of the text of the JCS 1067 directive; and John Gimbel, *The American Occupation of Germany, Politics and the Military, 1945–1949* (Stanford: Stanford University Press, 1968).

17. See Diane S. Clemens, *Yalta* (New York: Oxford University Press, 1970); George C. Herring, Jr., *Aid to Russia 1941–1946: Strategy, Diplomacy, the Origins of*

the Cold War (New York: Columbia University Press, 1973); and John H. Backer, *The Decision to Divide Germany* (Durham, NC: Duke University Press, 1978).

18. Harry S. Truman, *Year of Decisions* (Garden City, NY: Doubleday, 1955), p. 235.

19. Backer, *op. cit.,* pp. 46–101; Yergin, *op. cit.,* pp. 111–119.

20. See Henry C. Wallich, *Mainsprings of the German Revival* (New Haven: Yale University Press, 1955), pp. 199–203, 369–371; Gimbel, *op. cit.,* pp. 19–51; Backer, *op. cit.,* pp. 88–116; and Robert A. Pollard, *Economic Security and the Origins of the Cold War, 1945–1950* (New York: Columbia University Press, 1985), p. 95, fn. 42.

21. Lucius D. Clay, *Decision in Germany* (Garden City, NY: Doubleday, 1950), p. 19.

22. See *Who's Who in America, 1946–1947,* s.v. "Draper, William H., Jr."

23. Draper quoted in James S. Martin, *All Honorable Men* (Boston: Little, Brown, 1950), p. 203.

24. *Ibid.,* pp. 164, 173.

25. U.S. Congress, Senate, *Problems of American Small Business,* pt. 18, p. 2020.

26. Clay, *op. cit.,* p. 331. Also see Paterson, *op. cit.,* p. 428, fn. 125.

27. J. R. Hight, "European Letter . . . ," *Iron Age* 157 (May 23, 1946), p. 82. Also see David Anderson, "Luxembourg Firm Is Under Inquiry," *NYT* (August 29, 1946), p. 37; J. R. Hight, "World Steel Production," *Iron Age* 159 (January 2, 1947), p. 89; "French and Luxemburg Steel Men Agitating for New Steel Cartel," *Iron Age* 159 (January 16, 1947), p. 112; J. Raymond, "'Bizonia' Breaks Up Cartel in Banking," *NYT* (August 4, 1948), p. 6; E. A. Morrow, "Germans Are Ordered to Dissolve New Cartel for Steel Scrap Export," *NYT* (August 18, 1948), p. 6; H. Callendar, "French-German Tie: Custom Union Held U.S. Aim in Europe," *NYT* (November 13, 1949), pp. 1, 42; M. L. Hoffman, "European Cartels Revived to Block Freeing of Trade," *NYT* (November 21, 1949), pp. 1, 10; and "Global Letter—Review of World Markets," *Iron Age* 165 (April 27, 1950), p. 30.

28. Hight, "European Letter . . . ," *Iron Age* (May 23, 1946), p. 82.

29. D. Anderson, "Luxembourg Seen as German's Prey," *NYT* (June 10, 1946), p. 35; E. J. Hardy, "Washington . . . ," *Iron Age* 164 (July 7, 1949), pp. 112–114; F. Belair, Jr., "Delay Looms on ERP Action Despite Appeal for Speed," *NYT* (January 8, 1948), pp. 1, 18; E. J. Hardy, "The Federal View—This Week in Washington," *Iron Age* 164 (December 15, 1949), p. 48. Also see FTC, *op. cit.*

30. Byrnes quoted in Paterson, *op. cit.,* p. 395.

31. See Michael J. Hogan, "The Search for a 'Creative Peace': The United States, European Unity, and the Origins of the Marshall Plan," *Diplomatic History* 6 (Summer 1982), pp. 267–285.

32. See Paterson, *op. cit.,* pp. 399, 437. John R. Steelman, an assistant to President Truman, stated that even though most of these individuals were Republicans, they were chosen by Truman because they were the most competent candidates for the job to be done; see "Steelman's Memoirs," President's Personal File, Box 2, "Memoirs File," John R. Steelman Folder, Oral History Interview, pp. 28–29, HST Library.

33. J. R. Hight, "European Aid Program Would Boom American Steel Rate Next Year," *Iron Age* 160 (July 17, 1947), p. 95.

34. Ernest T. Weir, president of National Steel, for example, was one of those in the industry opposing aid. See "Says US Can't Fill World Steel Needs," *NYT* (August 2, 1947), p. 17; and Paterson, *op. cit.,* p. 399, fn. 12. It might be noted, however, that by 1949 Weir had begun to change his mind on this subject; see "Weir Expects Steel Output to Shrink to 65 Pct of Capacity in Fourth Quarter," *Iron Age* 164 (July 28, 1949), p. 103.

35. See T. E. Mullaney, "No Exports' Rush Planned by Steel," *NYT* (April 6, 1947), p. III–1. Also see "Global Letter—Review of World Markets," *Iron Age* 164

(August 25, 1949), p. 138; and Anthony, "Foreign Funds for Purchase of American Steel . . . ," pp. 95–96. For more generalized background information, see Kenneth Warren, *World Steel, an Economic Geography* (New York: Crane, Russak & Co., 1975).

36. Pressures from local consumers were manifested in Senate Hearings of the Wherry Committee, which considered small business concerns, see U.S. Congress, Senate, *Problems of American Small Business,* pts. 5, 15; "Commerce Department Asked for More Stringent Export Controls," *Iron Age* 160 (July 24, 1947), p. 102.

37. Truman to George C. Marshall, December 23, 1947, Official File, Box 1026, Folder OF 342, (May–December 1947) HST Library.

38. See E. J. Hardy, "State Dept. Recommends Revision of European Steel Requests," *Iron Age* 161 (January 15, 1948), pp. 120–121. The Commerce Department plan was voluntary, nevertheless, it was guided by such tight administrative restrictions that compliance was assured, see "Export Steel," *Iron Age* 161 (January 8, 1948), p. 93; and "First Quarter Export Quotas Announced," *Iron Age* 161 (February 12, 1948), p. 118.

39. B. Packard, "Iron and Steel Exports Will Be Reduced 2 Million Tons This Year," *Iron Age* 162 (December 9, 1948), p. 147.

40. Secretary of Commerce W. Averell Harriman took this view, and accordingly argued for an increase in German steel production to relieve export pressures on American steelmakers. See Yergin, *op. cit.,* p. 309. Also see Kuklick, *op. cit.,* pp. 332–333; and Gimbel, *op. cit.,* pp. 147–185.

41. "Europe Is Speeding Comeback in Steel," *NYT* (April 21, 1948), p. 41.

42. A. M. Anderson to T. W. Lamont, February 18, 1947, Box 229, Folder 12, Lamont Papers.

43. "Ruhr Steel and the Future," *Economist* (London) 157 (December 3, 1949), p. 1241.

44. See Duncan Burn, *The Steel Industry, 1939–1959* (Cambridge: Cambridge University Press, 1961), p. 139, Table 8. Burn's figures are in long tons, while Table 4.3 of this book shows short tons; thus the difference.

45. Wallich, *op. cit.,* p. 344.

46. See "German Dismantling Tangle Persists But Recommendations Call for More Steel," *Iron Age* 162 (December 30, 1948), p. 73. Also see "German Steel Future Becomes Political Dynamite," *Iron Age* 157 (February 7, 1946), p. 123; Hardy, "State Dept. Recommends Revision . . . ," p. 121; "International Body Will Control Ruhr's Coal, Coke and Steel," *Iron Age* 162 (July 1, 1948), p. 154; "Germans Balk on Steel," *NYT* (July 16, 1948), p. 4; "European Letter . . . ," *Iron Age* 162 (December 16, 1948), pp. 124–126; and "A British View of German Steel," *Economist* (London, 157 (August 27, 1949), p. 472. Burn, *op. cit.,* pp. 398–403, offers interesting comment on these developments, as does William Diebold, Jr., *The Schuman Plan, a Study in Economic Cooperation, 1950–1959* (New York: Praeger, 1959), pp. 28–37; and John C. Campbell, *The United States in World Affairs, 1945–1947* (New York: Harper & Brothers, 1947), pp. 179–182. In the United States there was considerable agitation to apply domestic antitrust statutes to German industry, see Steuart L. Pittman, "The Foreign Aid Programs and the United States Government's Antitrust Policy," in Kingman Brewster, Jr., *Antitrust and American Business Abroad* (New York: McGraw-Hill, 1958), app. A, pp. 459–473.

47. J. Raymond, "Ruhr Output Rise Spurred by West," *NYT* (June 6, 1948), p. 8. There was, of course, much controversy in the United States over what to do about Germany's future, and in the immediate postwar years no general consensus emerged. The Marshall Plan thus represented, in many respects, a compromise among the contending positions. See Gimbel, *op. cit.,* pp. 147–185; and Yergin, *op. cit.,* pp. 317–321.

48. See the testimony of James S. Martin in U.S. Congress, House Subcommittee

on Study of Monopoly Power, Committee on the Judiciary, 81st Cong., 2nd sess, Hearings, *Study of Monopoly Power,* (Washington, DC: GPO, 1950), pt. 4A, "Steel," pp. 405–406.

49. E. A. Morrow, "Mills to Regain Mines," *NYT* (November 18, 1948), p. 3. The reference to Britain involves that nation's demands that the German iron and steel industry be broken up and that the industry be nationalized; see Burn, *op. cit.,* pp. 400–401.

50. Backer, *op. cit.,* pp. 88–116; Gimbel, *op. cit.,* pp. 163–186; Kuklick, *op. cit.,* pp. 332–334; and Paterson, *op. cit.,* p. 429.

51. Such as W. Averell Harriman. See Paterson, *op. cit.,* p. 430. Harriman, of course, was most concerned with rebuilding German industry, a view strongly shared in by Herbert Hoover, President Truman's special consultant on European recovery. Together, they constantly urged reindustrialization in Germany. See Louis P. Lochner, *Herbert Hoover and Germany* (New York: Macmillan, 1960); Joan Hoff Wilson, *Herbert Hoover, Forgotten Progressive* (Boston: Little, Brown, 1975), pp. 256–260; and Yergin, *op. cit.,* pp. 309, 319–320.

52. See Paterson, *ibid.,* pp. 432–433, especially p. 433, fn. 142. For a record of U.S. diplomatic discussion of this problem, see U.S. Department of State, *Foreign Relations of the United States, 1948,* vol. 2 (Washington, DC: GPO, 1948), 448–455, 492–494, 703–866.

53. *Ibid.*

54. Gimbel, *op. cit.,* pp. 182–184. On the report, see ECA, Industrial Advisory Committee, *Report on Plants Scheduled for Removal as Reparations from the Three Western Zones of Germany* (Washington, DC: author, 1949), pp. 2–3. In May and June 1948, the United States Steel Export Corporation (U.S. Steel's export subsidiary) made a study of the German steel industry for the ECA, the Army, and the Joint Congressional Committee, see ECA, *ibid.,* p. 3; and Gimbel, *op. cit.,* pp. 182–184. This report was updated in late 1948 for the purposes of the IAC, and it is included in the ECA study as "Recommendations for Retention of Steel Capacity in Western Germany," following p. 38 of the report. It might be noted that the recommendations of the Humphrey Committee were a foregone conclusion, in that the group was created solely to give ECA Administrator Hoffman some leverage with Congress in his battle to save German industry from falling into foreign hands; see U.S. Department of State, *Foreign Relations of the United States, 1948,* vol. 2, 792–794; and U.S. Department of State, *Foreign Relations of the United States, 1949,* vol. 3 (Washington, DC: GPO, 1974), pp. 546–642.

55. ECA, *Report on Plants,* p. 1. Also see "Three Power Agreement Drastically Alters Plant Dismantling in Germany," *Iron Age* 163 (April 21, 1949), p. 121.

56. See Burn, *op. cit.,* pp. 398–403; and Diebold, *op. cit.,* p. 362ff.

57. Wallich, *op. cit.,* p. 380.

58. Kuklick, *op. cit.,* pp. 332–334; and Manuel Gottlieb, *The German Peace Settlement and the Berlin Crisis* (New York: Paine-Whitman, 1960), pp. 92–94, 124–127. Also see D. Ansboro, "Outlines Permanent Control Plan for German Industry," *Iron Age* 157 (January 3, 1946), pp. 197–198.

59. "Ruhr Steel and the Future," *Economist,* p. 1241.

60. Diebold, *op. cit.,* pp. 32–37; and C. Daniel, "French Give Plan for Ruhr Control," *NYT* (November 18, 1948), p. 3.

61. Burn, *op. cit.,* pp. 400–401. Also see F. Belair, Jr., "Revitalized Ruhr Termed the Heart of Marshall Plan," *NYT* (August 12,1947), p. 1; Morrow, "Mills to Regain Mines," *NYT,* p. 3; E. J. Hardy, "Washington . . . ," *Iron Age* 162 (December 23, 1948), pp. 78–80; and J. Raymond, "12 Germans Named to Run Ruhr Steel," *NYT* (February 26, 1949), p. 5.

62. Burn, *op. cit.,* pp. 448–452; and Diebold, *op. cit.,* pp. 356ff. Also see Graham

D. Taylor, "The Rise and Fall of Antitrust in Occupied Germany, 1945–1948," *Prologue* 11 (Spring 1979), pp. 23–39.

63. The formation of European administrative agencies to work with the United States in reviving Europe's economy were greatly influenced by perceptions of Soviet intentions. Thus, the ECE, a unit of the United Nations, was originally involved in formulating plans for European steel industry recovery. However, because of Allied fears surrounding possible Soviet obstructionism within the ECE (Russia, of course, was a UN Security Council member), the western democracies, led by the United States, created the OEEC to deal with European industrial reconstruction. This agency was not affiliated with the UN and was thus outside the bounds of Soviet influence. The OEEC was a refinement of the CEEC, which had been formed in July 1947. See W. O. Henderson, *The Genesis of the Common Market* (Chicago: Quadrangle, 1962), p. 132; and Kuklick, *op. cit.,* pp. 335–371. Also see M. L. Hoffman, "18 Nations to Join in Steel Survey," *NYT* (November 29, 1947), p. 3; and Lincoln Gordon, "The Organization for European Economic Cooperation," *International Organization* 10 (February 1956), pp. 1–11.

64. ECA, *European Recovery Program, Iron and Steel Commodity Study* (Washington, DC: author, 1949), p. 2. The conclusions regarding excess capacity were the same as those reached by the ECE in *European Steel Trends in the Setting of the World Market* (New York: author, November, 1949), pp. 67, 72. Much of the excess capacity apparently was expected to develop because of the building programs undertaken by the smaller nations of Europe that had traditionally been importers of iron and steel, see Burn, *op. cit.,* p. 145; "Steel Prices and Plans," *Economist* (London) 156 (April 2, 1949), pp. 619–620; and "Global Letter—Review of World Markets," *Iron Age* 165 (January 12, 1950), pp. 26, 110.

65. "Grace Calls Subsidy of Shipping Vital," *NYT* (October 28, 1949), p. 47.

66. "Weir Expects Steel Output to Shrink . . . ," *Iron Age,* p. 115.

67. See "ECA Studies Proposal to Lengthen Iron, Steel Investment Program," *Iron Age* 164 (July 28, 1949), p. 115.

68. See Robert A. Brady, *The Rationalization Movement in German Industry* (Berkeley: University of California Press, 1933); Charles S. Maier, "Between Taylorism and Technocracy: European Ideologies and the Vision of Industrial Productivity in the 1920s," *Journal of Contemporary History* 5 (1970), pp. 27–61, specially pp. 55–56; and Judith A. Merkle, *Management and Ideology* (Berkeley: University of California Press, 1980), pp. 172–207, especially pp. 196–199.

69. "West Coast Export Men See European Competition," *Iron Age* 163 (April 7, 1949), p. 120.

70. ECA, *European Recovery Program, op. cit.,* p. 11.

71. "A British View of German Steel," *Economist.* Also see Louis Lister, *Europe's Coal and Steel Community, an Experiment in Economic Union,* (New York: Twentieth Century Fund, 1960), p. 8; and Burn, *op. cit.,* pp. 382–385.

72. This was a continuation of concerns that had developed during the early 1940s with the influence of cartels on the outbreak of World War II. See, for example, U.S. Congress, Senate, Subcommittee on War Mobilization, Committee on Military Affairs, 78th Cong., 2nd sess, *Economic and Political Aspects of International Cartels,* Monograph No. 1 (Washington, DC: GPO, 1944); this report was written by Corwin D. Edwards for the Kilgore Committee, the chief congressional investigator of cartels during the war. Also see Corwin D. Edwards (ed.), *A Cartel Policy for the United Nations* (New York: Columbia University Press, 1945); G. W. Stocking ad M. W. Watkins, *Monopoly and Free Enterprise* (New York: Twentieth Century Fund, 1951); and K. L. Mayall, *International Cartels, Economic and Political Aspects* (Rutland, VT: Charles E. Tuttle, 1951).

73. U.S. Congress, House, *Study of Monopoly Power,* p. 359. Also see "Steel Cartel Seen Riddling Our Policy," *NYT* (April 25, 1950), pp. 39, 41.

74. U.S. Congress, House, *Study of Monopoly Power,* p. 407.

75. *Ibid.,* pp. 659–673. Also see "Fairless Denies Pact," *NYT* (April 25, 1950), p. 41; and "Head of U.S. Steel Cites Competition," *NYT* (April 26, 1950), p. 43.

76. B. Packard, "Steel Exporters See Rough Sledding Ahead," *Iron Age* 165 (March 2, 1950), p. 116.

77. "Global Letter—Review of World Markets," *Iron Age* 165 (March 16, 1950), p. 26.

78. The possibility of a resumption of the prewar steel cartel had occurred to many prior to the formation of the ECSC. Both French and German private industrialists were known to favor this proposal; see Anne T. Moore, "France and the Schuman Plan, 1948–1953" (Ph.D. diss., University of North Carolina, 1963), pp. 56ff. Marshall Plan officials, however, especially the Chief Administrator Paul G. Hoffman, opposed renewal; see Callendar, "French-German Tie . . . ," *NYT,* pp. 1, 42. Diebold, in the most comprehensive study of the Schuman Plan written in English, also indicates that domestic interests must surely have been aware of the growing sentiment for some form of control over the Continental industry; yet, he essentially concludes that the Schuman Plan itself was launched independently by purely European forces and that American interests played no part in the decision (see Diebold, *op. cit.,* pp. 44–45). Another contemporary observer reaches the same conclusion; "The Schuman Plan as it now stands, does not contemplate a restoration of the pre-war steel cartel, even in a semi-public form. . . . The problem will be to break up the habits of collusion and cartelization." William N. Parker, "The Schuman Plan—A Preliminary Prediction," *International Organization* 6 (August 1952), p. 390. American diplomats were unaware of the negotiations for the plan, see the extensive discussions in U.S. Department of State, *Foreign Relations of the United States, 1950,* vol. 3 (Washington, DC: GPO, 1977) pp. 691–767. Dean Acheson, Truman's secretary of state, noted on May 9, 1950, "We have not been consulted or involved in the proposals [for the Schuman Plan], if made, in any way" (*ibid.,* p. 692).

79. Diebold, *op. cit.,* quotation is on p. 1. Also see "Franco-German Steel-Coal Pool Sought," *Iron Age* 165 (May 18, 1950), pp. 106–107; and Diebold, *op. cit.,* pp. 41–43.

80. For the history of events leading up to formation of the ECSC, see Moore, *op. cit.,* Diebold, *op. cit.,* and Lister, *op. cit.*

81. Diebold, *op. cit.,* pp. 48–60.

82. "Franco-German Steel-Coal Pool Sought," p. 107.

83. See J. R. Hight, "World Steel Production," p. 89.

84. It should be noted, however, that Japan had been a major customer of American steelmakers during the 1920s. During that decade the nation began to build up its own capacity for production, and by the 1930s dependence on imports had been significantly reduced. Japan, however, was not a force in world markets during this period—production was geared to domestic needs rather than exports. See Takafusa Nakamura, *Economic Growth in Prewar Japan,* trans. R. A. Feldman (New Haven: Yale University Press, 1983).

85. See William Lockwood, *The Economic Development of Japan* (Princeton: Princeton University Press, 1954), pp. 3–37.

86. U.S. Tariff Commission, *op. cit.,* p. 169. (However, see n. 91, Chapter 6).

87. The onset of the Cold War stimulated American interest in the productive capability of Soviet steelmakers. See J. R. Hight, "Two Worlds and Steel," *Iron Age* 161 (January 1, 1948), p. 147; Hight, "World Steel Production," p. 95; and J. Anthony, "World Production," *Iron Age* 157 (January 3, 1947), p. 158.

88. See E. F. Denison and W. K. Chung, "Economic Growth and Its Sources," in H. Patrick and H. Rosovsky (eds.), *Asia's New Giant—How the Japanese Economy Works* (Washington, DC: Brookings, 1976), p. 77. Also see Takafusa Nakamura, *The Postwar Japanese Economy, Its Development and Structure*, trans. J. Kaminski (Tokyo: University of Tokyo Press,1981), pp. 3–48; and Chalmers Johnson, *MITI and the Japanese Miracle, the Growth of Industrial Policy, 1925–1975* (Stanford: Stanford University Press, 1982). The steelmaker's attitude toward Japan only reflected, of course, the larger pattern of opinion in the United States. Eisenhower's Secretary of State, John Foster Dulles, would note as late as 1954 that "there was little future for Japanese products in the United States . . . ," and that "Japan should not expect to find a big U.S. market because the Japanese don't make the things we want." See U.S. Department of State, *Foreign Relations of the United States, 1952–1954,* vol. 14, pt. 2 (Washington, DC: GPO, 1985), pp. 1693, 1725.

89. See W. G. Beasley, *The Modern History of Japan,* 2nd ed. (New York: Praeger, 1974), pp. 280–281; Yergin, *op. cit.,* pp. 70, n250; Nakamura, *Postwar Japanese Economy,* pp. 30–40; and Alonzo L. Hamby, *Beyond the New Deal: Harry S. Truman and American Liberalism* (New York: Columbia University Press, 1973), p. 57.

90. See Nakamura, *Postwar Japanese Economy,* p. 31; William S. Borden, *The Pacific Alliance: United States Foreign Economic Policy and Japanese Trade Recovery, 1947–1955* (Madison: University of Wisconsin Press, 1984), pp. 65ff; T. J. Hamilton, "Reparation Policy for Japan Held Up," *NYT* (May 27, 1946), p. 4; "Pauley Blueprints War-Shorn Japan," *NYT* (November 29, 1946), p. 16; and Kiyoshi Kawahito, *The Japanese Steel Industry* (New York: Praeger, 1972), pp. 5–7. Statistical sources on Japanese steel production during the war are inconsistent; the figures cited are drawn from Ken-ichi Iida, *History of Steel in Japan* (Tokyo: Nippon Steel Corporation, 1973), p. 29.

91. See E. Brewin, "Details Rise and Fall of Japanese Industrial Empire," *Iron Age* 158 (October 17, 1946), p. 124; and Borden, *ibid.,* pp. 61–102.

92. Borden, *op. cit.,* p. 77ff.

93. Kawahito, *op. cit.,* p. 8; Denison and Chung, *op. cit.,* p. 73; and Borden, *op. cit.,* pp. 63–64.

94. Lockwood, *op. cit.,* pp. 64–77; Kawahito, *op. cit.,* pp. 3–17; Iida, *op. cit.,* pp. 21–32; Jerome B. Cohen, *Japan's Economy in War and Reconstruction* (Minneapolis: Univeristy of Minnesota Press, 1949). Also see "Estimate Japanese Potential Scrap at 7 Million Tons; 5 to 6-Year Supply," *Iron Age* 163 (April 14, 1949), p. 132; A. F. Matthews, "Japan's Yawata Steel Plant Survives the War," *Iron Age* 157 (May 23, 1946), pp. 104–105; Brewin, "Details Rise and Fall," pp. 123–124; and Anthony, "World Production," p. 159.

95. Beasley, *op. cit.,* pp. 280–291; Kawahito, *op. cit.,* pp. 8–17; Friedrich, *op. cit.,* pp. 340–354; and "Japan's Steel Industry Makes Strong Effort to Restore Prewar Output," *Iron Age* 162 (December 9, 1948), pp. 161, 164.

96. See Borden, *op. cit.;* Johnson, *op. cit.;* Nakamura, *Postwar Japanese Economy, op. cit.;* and Iida, *op. cit.,* p. 30.

97. Kawahito, *op. cit.,* pp. 8–9; and Iida, *op. cit.,* p. 30.

98. Kawahito, *op. cit.,* pp. 8–9.

99. Johnson, *op. cit.,* pp. 198–241; and Kawahito, *op. cit.,* pp. 11–12.

100. "Japan's Steel Industry Makes Stronger Effort," p. 161; and Iida, *op. cit.,* p. 32.

101. Johnson, *op. cit.,* pp. 207–211; and Iida, *op. cit.,* p. 30.

102. Leonard H. Lynn, *How Japan Innovates, a Comparison with the U.S. in the Case of Oxygen Steelmaking* (Boulder, CO: Westview Press, 1982), pp. 43–118; Nakamura, *Postwar Japanese Economy,* p. 70; and Kawahito, *op. cit.,* pp. 24–25, 28.

103. S. D. Smoke, "World Steel," *Iron Age* 163 (January 6, 1949), p. 196.

104. See Eleanor M. Hadley, *Antitrust in Japan* (Princeton: Princeton University Press, 1970); Richard Caves and Masu Uekusa, "Industrial Organization," in H. Patrick and H. Rosovsky (eds.), *Asia's New Giant—How the Japanese Economy Works* (Washington, DC: Brookings, 1976), p. 483; and Iida, *op. cit.*, p. 32.

Chapter 5

1. See Craufurd D. Goodwin and R. Stanley Herren, "The Truman Administration: Problems and Policies Unfold," in C. D. Goodwin (ed.), *Exhortation and Controls—The Search for a Wage-Price Policy, 1945–1971* (Washington, D.C.: Brookings, 1975), pp. 63–65; Karl Schriftgiesser, *Business and Public Policy* (Englewood Cliffs, NJ: Prentice-Hall, 1967), pp. 34–37; and G. F. Sullivan, "Steel Companies Unwilling to Start a Price War at This Time," *Iron Age* 163 (June 30, 1949), p. 99.

2. Richard W. Nagle, "Collective Bargaining in Basic Steel and the Federal Government, 1945–1960" (Ph.D diss., Pennsylvania State University, 1978), pp. 73–74; and "Spotlighting 1949," *Iron Age* 165 (January 5, 1950), p. 154.

3. J. L. Loftus, "U.S. Steel Rejects 4th Round Pay Raise; Strike Threatened," *NYT* (July 7, 1949), pp. 1, 19; quotation is on p. 19.

4. Nagle, *op. cit.*, pp. 78–80.

5. L. Stark, "Injunction Voted into Labor Bill by Senate, 50–40," *NYT* (June 29, 1949), pp. 1, 3.

6. This was the Steel Industry Board, appointed on July 15, 1949. See U.S. Department of Labor [E. Robert Livernash], *Collective Bargaining in the Basic Steel Industry—A Study of the Public Interest and the Role of Government* (Washington, DC: author, 1961), pp. 261–264; and Nagle, *op. cit.*, pp. 78–81.

7. These statements were made by Clarence B. Randall, president of Inland Steel. Randall had recently returned to his firm after several years with the ECA in Europe. See Nagle, *op. cit.*, pp. 86–87.

8. The industry's economic position was formulated by Dr. Jules Bachman, a professor from New York University who had been retained by the steelmakers. See Nagle, *op. cit.*, p. 88.

9. The union relied on its own personnel in testimony before the board. The quotations in the text are drawn from Nagle, *op. cit.*, pp. 92, 100, 106.

10. Steel Industry Board, *Report to the President of the United States on the Labor Dispute in the Basic Steel Industry* (Washington, DC: GPO, 1949). This was submitted on September 10, 1949.

11. See "Steel Board Report Surprises Industry," *Iron Age* 164 (September 15, 1949), pp. 117, 158, and T. Campbell, "The Steel Board's Recommendations," *Iron Age* 164 (September 15, 1949), p. 7.

12. A. H. Raskin, "U.S. Steel Rejects Findings of President's Fact Board as 'Revolutionary Doctrine'," *NYT* (September 15, 1949), pp. 1, 23.

13. J. Delaney, "Steel Strike Centered About a Principle," *Iron Age* 164 (October 6, 1949), pp. 147–148; and U.S. Department of Labor, *op. cit.*, p. 268.

14. U.S. Department of Labor, *ibid.*, pp. 264–269; "U.S. Steel and Union Agree; Steel Industry 95 Pct Signed," *Iron Age* 164 (November 17, 1949), p. 114; and Nagle, *op. cit.*, pp. 105–109.

15. Nagle, *op. cit.*, p. 102.

16. The most significant labor contract signed in this period was that between Ford Motor Co. and the UAWA on September 29, 1949. See Nagle, *op. cit.*, p. 108.

17. See T. Campbell, "Steel and the Public," *Iron Age* 163 (January 6, 1949), pp. 178–183; George S. Rose, "Activities of the Institute—1948," in AISI, *Yearbook, 1949* (New York: AISI, 1949), pp. 605–606; T. E. Mullaney, "Steel Is Opening 'Build-Up'

Drive," *NYT* (January 23, 1949), p. III–6; "Steel's Community Relations," *NYT* (July 9, 1948), p. 23; and "Employee Program in Steel Advanced," *NYT* (August 9, 1948), p. 22.

18. U.S. Steel's archives are not open to outside researchers, hence, it is impossible to survey primary corporate sources to determine why the firm acted as it did on this matter.

19. This, of course, is only a surmise. See n. 18.

20. J. B. Delaney, "Pension Pacts to Put Pressure on Prices," *Iron Age* 164 (November 17, 1949), p. 113.

21. Extras are those additional costs incurred by producers to customize orders and involve such jobs as specialized cutting, different finishes, unusual chemical compositions of products, and so on. The price increases announced by U.S. Steel averaged approximately 4% more than existing levels. See G. F. Sullivan, "Steel Extras Revised on Cost Basis," *Iron Age* 164 (December 22, 1949), pp. 87–88; "Ernie Weir Answers Steel Price Critics," *Iron Age* 164 (December 29, 1949), p. 69; and T. E. Mullaney, "US Steel's Prices Increased $4 a Ton; An Inquiry Is Slated," *NYT* (December 16, 1949), p. 1.

22. U.S. Congress, Joint Committee on the Economic Report, 81st Cong. 2nd sess. *Basic Data Relating to Steel Prices* (Washington, DC: GPO, 1950), "Letter of Transmittal," p. v.

23. E. J. Hardy, "The Federal View—This Week in Washington," *Iron Age* 164 (December 29, 1949), p. 36.

24. U.S. Congress, Joint Committee on the Economic Report, 81st Cong., 2nd sess., Hearings, *December 1949 Steel Price Increases* (Washington, DC: GPO, 1950). Also see G. Baker, "The Steel Hearings—A Political Flop," *Iron Age* 165 (February 2, 1950), pp. 107–108.

25. U.S. Cong., Report of the Joint Committee on the Economic Report, 81st Cong., 2nd sess., *December 1949 Steel Price Increases* (Washington, DC: GPO, 1950). Issued as Senate Report No. 1371, March 27, 1950. Also see Hardy, "The Federal View—This Week in Washington," *Iron Age* 165 (March 9, 1950), pp. 48–50; Hardy, "The Federal View—This Week in Washington," *Iron Age* 165 (March 23, 1950), pp. 54–56; "O'Mahoney Urges New Steel Industry Probe,"*Iron Age* 165 (March 30, 1950), pp. 111–112; and "Steel Men Say O'Mahoney Ignores Cost Factors," *Iron Age* 165 (April 6, 1950), pp. 109–110.

26. U.S. Cong., Report, . . . *December 1949 Steel Price Increases,* p. 7.

27. *Ibid.,* p. 29.

28. *Ibid.,* p. 30.

29. See "Steel Price Rise Is Unwarranted, Marjority of Joint Committee Holds," *NYT* (March 8, 1950), pp. 39, 44; "Report on Steel Prices," *NYT* (March 11, 1950), p. 14; "'Big Steel' Capacity 32% of Total; Taft Proposes Congress Inquiry," *NYT* (March 16, 1950), p. 49; J. D. Morris, "Steel Price Hearings by U.S. Urged Before Any New Rises," *NYT* (March 27, 1950), pp. 1, 13; and "Steel Politics Attacked," *NYT* (March 28, 1950), p. 12.

30. See Irving S. Olds to Harry S. Truman, February 9, 1950, Official File, Box 155Y, Folder OF 852 (United States Steel Corp.), HST Library. Attached to the letter is a copy of Benjamin F. Fairless's testimony before the JEC; soon after, Olds sent Truman a copy of a recent speech by Fairless, "Guilty Before Trial," in which Fairless denied that U.S. Steel was in violation of the antitrust laws.

31. See U.S. Congress, House Subcommittee on Study of Monopoly Power, Committee on the Judiciary, 81st Cong., 2nd sess., Hearings, *Study of Monopoly Power* (Washington, DC: GPO, 1950). pt. 4A, "Steel."

32. See citations in Chapter 4, nn. 73–78.

33. The accusations against the industry appear to have sunk from sight following the Celler Committee Hearings and James S. Martin's sensational testimony. There is no further reference to the matter in the press. The reasons for this remain conjectural.

34. See FTC, *Report of the Federal Trade Commission on the Concentration of Productive Facilities, 1947—Total Manufacturing and 26 Selected Industries* (Washington, DC: GPO, 1950), pp. 23–25; and FTC, *A List of 1,000 Large Manufacturing Companies, Thier Subsidiaries and Affiliates, 1948* (Washington, DC: GPO, 1951).

35. Hardy, "The Federal View—This Week in Washington," *Iron Age* 164 (December 29, 1949), p. 38.

36. U.S. Congress, House, *Study of Monopoly Power,* pt. 4A, "Steel", p. 117. Also see "Economist' Harmony Sour to Steel's Weir," *Iron Age* 165 (May 11, 1950), p. 86; G. Baker, "Steel Inquisition Hints Justice Dept. Suit," *Iron Age* 165 (May 4, 1950), p. 107; Hardy, "The Federal View—This Week in Washington," *Iron Age* 165 (May 11, 1950), p. 60; and G. Baker, "Celler Readies Antitrust Traps for Steel," *Iron Age* 165 (May 18, 1950), pp. 107–108.

37. U.S. Congress, House, *Study of Monopoly Power,* pt. 4A, "Steel,"pp. 465, 466.

38. Hardy, "The Federal View—This Week in Washington" *Iron Age* 165 (May 11, 1950), p. 60.

39. There were, however, other results from Celler's work. The most noteworthy was eventual congressional passage (in 1950) of the Celler-Kefauver Act, which amended Section 7 of the Clayton Act of 1914 regarding corporate mergers. In general, this amendment made it more difficult for firms to merge. See, for example, Peter Asch, *Economic Theory and the Antitrust Dilemma* (New York: John Wiley & Sons, 1970), pp. 291–292.

40. See Hardy, "The Federal View—This Week in Washington," *Iron Age* 167 (January 4, 1951), p. 259. It should be noted, however, that the absence of available primary evidence on this subject makes more conclusive statements impossible.

41. See Burton I. Kaufman, *The Oil Cartel Case: A Documentary Study of Antitrust Activity in the Cold War Era* (Westport, CT: Greenwood Press, 1978).

42. Particularly Representative Celler and Senator Kefauver; each was able to utilize his position as chair of an antimonopoly subcommittee to continue the investigation of large corporations and their effect on economic activity. See Theodore P. Kovaleff, *Business and Government During the Eisenhower Administration, a Study of the Antitrust Policy of the Antitrust Division of the Justice Department* (Athens: Ohio University Press, 1980).

43. U.S. Congress, House, *Study of Monopoly Power,* pt. 4A, "Steel," pp. 791, 794. Mr. Levi was Edward H. Levi, Counsel, Subcommittee on Study of Monopoly Power, Committee on the Judiciary, U.S. Congress, House (81st Cong., 2nd sess.).

44. See Hardy, "The Federal View—This Week in Washington." *Iron Age* 165 (June 8, 1950), p. 69. The House Small Business Committee, chaired by Wright Patman (D-Texas), had recently undertaken several investigations of steel industry activities. See U.S. Congress, House, Select Committee on Small Business, 81st Cong. 1st sess., Hearings, *Patterns of Steel Distribution in the States of Arkansas, Louisiana, Oklahoma, and Texas* (Washington, DC: GPO, 1949); and U.S. Congress, House, Select Committee on Small Business, 81st Cong., 2nd sess., Hearings, *Steel, Acquisitions, Mergers, and Expansion of 12 Major Companies, 1900 to 1950* (Washington, DC: GPO, 1950).

45. See "Steel Plans 4 Million-Ton Capacity Boost," *Iron Age* 165 (June 8, 1950), p. 91; T. Metaxas, "Steel to Add 6,383,000 Ton Capacity by '52," *Iron Age* 166 (July 27, 1950), pp. 80–81; Hardy, "The Federal View—This Week in Washington," *Iron Age* 166 (October 12, 1950), p. 149. Also see J. Delaney, "Capacity for Steelmaking Rises in 1950," *Iron Age* 166 (December 21, 1950), p. 95.

46. U.S. Department of Commerce, News Release, October 2, 1950, a copy is attached to Charles Sawyer to the president, September 30, 1950, President's Secretary File, Box 136, General File—Steel Folder, HST Library. Symington's news conference was held on October 23, 1950. See W. C. Bryant, "U.S. Advocates of More Plant Expansion Seen to Be Carrying the Day," *Wall Street Journal* (October 24, 1950), p. 1. On the new capacity goals, see C. M. White, "Capital Requirements and Productivity," in AISI, *Yearbook, 1951* (New York: AISI, 1951), p. 24.

47. See Maeva Marcus, *Truman and the Steel Seizure Case* (New York: Columbia University Press, 1977), pp. 3–6.

48. See T. E. Lloyd, "Industry's Vital Need—Liberalized Depreciation," *Iron Age* 161 (January 1, 1948), pp. 167–177; T. E. Mullaney, "Depreciation Seen Problem in Steel," *NYT* (February 20, 1949), pp. III–1; G. F. Sullivan, "Steel Company Earnings Rose 32 pct. in 1948 on Sales of $8.5 Billion," *Iron Age* 163 (April 7, 1949), p. 146; "Weir Says Steel Capacity Surpasses Needs," *Iron Age* 165 (April 6, 1950), p. 144; G. Hardy, "Accelerated Plant Amortization Likely," *Iron Age* 166 (August 24, 1950), p. 83; and "Washington Moves on Faster Depreciation, " *Iron Age* 166 (October 5, 1950), p. 105.

49. "Industry Gets 5-Year Write-Offs," *Iron Age* 166 (October 19, 1950), p. 95, and Hardy, "The Federal View—This Week in Washington," *Iron Age* 166 (November 30, 1950), p. 61.

50. "Weir Says Steel Capacity Surpasses Needs," p. 114.

51. See J. Delaney, "Steel Capacity in '52 Set at Over 120 Million Tons," *Iron Age* 167 (February 1, 1951), pp. 121–122; "Background Memo—1950 Steel Expansion and Production," February 1951, AISI News Release for Writers/Editors. A copy is in SML, Associations File, AISI–Economics Folder. Also see G. Baker, "Steel Expansion Gets Green Light," *Iron Age* 168 (October 4, 1951), p. 199; "Iron Age Summary—Annual Review," *Iron Age* 167 (January 4, 1951), p. 15; E. H. Collins, "Facts on Steel Production,"*NYT* (October 23, 1950), p. 34. Administration officials hoped that a 120 million ton capacity level would be reached, see "Truman Asks Heavy Tax Rise . . . ," *NYT* (January 13, 1951), p. 1.

52. See "Flock of Regional Firms Besiege U.S. for Help in Building New Mills," *Wall Street Jounal* (February 26, 1951), pp. 1, 2,; "Yolo: It Plans a Mysterious Million Tons of Steel for the West—With Aid," *Wall Street Journal* (February 28, 1951), pp. 1, 12; "Gibralter's Mill Would Serve Detroit's Auto Plants—If RFC Helps," *Wall Street Journal* (March 5, 1951), pp. 1, 12. Also see R. Vickers and M. Karmin, "Two RFC Steel Loans Are in Default," *Wall Street Journal* (June 8, 1954), p. 1.

53. See "Lone Star Steel's New Plant to Produce 350,000 Tons of Pipe," *Iron Age* 167 (Febrary 1, 1951), p. 122. This plant located in Daingerfield, Texas, was originally built with governmental funds for World War II use. After the war it was purchased from the War Assets Administration by a group of local Texas investors who had the political support of their congressman, Wright Patman. Patman was instrumental in aiding this venture from the start; see Patman to the president, January 21, 1947; and Truman to Patman, Janauary 23, 1947; both in Official File, Box 1025, (1945–February 1947) Folder OF 342, HST Library. On the New London mill, see B. Packard, "New England Mill Start Seen in Six Months," *Iron Age* 167 (January 18, 1951), p. 79; "Tax Write-Off Set for Steel Plant: O.D.M. Grants Benefits for New $26,400,000 Mill Projected in New England Area," *NYT* (October 31, 1953), p. 22; and Henry W. Broude, *Steel Decisions and the National Economy* (New Haven: Yale University Press, 1963), pp. 110–121. The proposed New England mill, it should be noted, was never built.

54. See "Wilson Puts the Brakes on Granting Certificates of Necessity for Expansion of Steel Industry," *Iron Age* 167 (March 8, 1951), pp. 110–111; W. C. Bryant,

"Planners of New Mills May Lose Federal Aid If Fund Raising Lags," *Wall Street Journal* (March 30, 1951), pp. 1, 12; and "Steel Men Worry on Over Expansion: Leaders Say Capacity May Be Pushed to Excessive Heights by Defense Hysteria," *NYT* (February 4, 1951), p. F-1. In addition, see "Fast Tax Write-Offs—Hasty Approval Under Fire," *Iron Age* 167 (April 12, 1951), p. 119; U.S. Department of Commerce, National Production Authority, Historical Reports on Defense Production, Report No. 28, *Iron and Steel—History of the Iron and Steel Divison of the National Production Authority* (Washington, DC: author, 1953); and U.S. Secretary of the Treasury, *Final Report on the Reconstruction Finance Corporation* (Washington, DC: GPO, 1959).

55. See the citations in n. 53. Also see Hardy, "The Federal View—This Week in Washington," *Iron Age* 167 (March 29, 1951), p. 81.

56. See "Eastern Steel Mill," *NYT* (April 26, 1950), p. 43. As noted, reports had circulated earlier regarding this proposal; see "Bowles Asks Survey on Steel Mill Site," *NYT* (April 9, 1950), p. 32.

57. See "U.S. Steel Eastern Mill Moves a Step Closer," *Iron Age* 165 (January 5, 1950), pp. 329–330; "U.S. Steel Verifies Eastern Mill Plan," *Iron Age* 165 (May 4, 1950), p. 110; and R. A. Smith, "Bethlehem Steel and the Intruder," *Fortune* 47 (March 1953), pp. 100–105ff.

58. T. C. Campbell, "U.S. Steel to Begin Eastern Mill in 1951," *Iron Age* 165 (February 2, 1950), p. 109.

59. "Bethlehem Adding Sheet Capacity in East," *Iron Age* 165 (February 2, 1950), p. 117; Smith, "Bethlehem Steel" pp. 100–105ff. J. Delaney, "National's Eastern Mill Rivals U.S. Steel's, " *Iron Age* 167 (January 11, 1951), p. 78. National Steel's proposed mill, however, was never built; see "No Plans for Eastern Mill Site," *Iron Age* 169 (May 22, 1952), p. 78.

60. See J. Delaney, "Steelmen Seek Key to Future Market Trends," *Iron Age* 167 (May 10, 1951), p. 110; J. B. Delaney, "Steel: J&L Growth Defies Prophets," *Iron Age* 168 (November 8, 1951), pp. 88–89; and "Steel Men Worry on Over-Expansion . . . ," *NYT*, p. F-1.

61. H. G. Batcheller, "Battlefronts of Steel," in AISI, *Yearbook, 1950* (New York: AISI, 1950), pp. 48–49, 55, 57.

62. Goodwin and Herren, *op. cit.,* pp. 71–76.

63. *Ibid.,* p. 76; and Marcus, *op cit.,* pp. 11–12.

64. For more detailed organization chart, see "Harrison Heads DPA as Straw Boss to Wilson," *Iron Age* 167 (January 11, 1951), p. 81. Also see Marcus, *op cit.,* pp. 9–16 for discussion.

65. See "Two Top Steel Firms Raise Pay, Prices," *NYT* (December 1, 1950), p. 19; "New Steel Prices Cover Wage Rises," *NYT* (December 1, 1950), pp. 39, 43; "Built-In Inflation," *NYT* (December 2, 1950), p. 12; "Steel Industry Alarmed," *NYT* (December 20, 1950), p. 25; and Hardy, "The Federal View—This Week in Washington" *Iron Age* 167 (January 4, 1951), p. 259.

66. J. Delaney, "Steel Profits, Taxes, Sales Reach New Highs," *Iron Age* 167 (April 12, 1951), pp. 126ff.

67. See Kiyoshi Kawahito, *The Japanese Steel Industry* (New York: Praeger, 1972), pp. 17–22. He notes, "The Korean War boom lasted for a short time, but it is still widely considered to have provided the biggest 'break' in the postwar development of the Japanese steel industry" (p. 21). The surge in demand brought on by the war also affected others in the industry; in the U. S., traditional steel importers formed a new trade association to protect themselves from the many neophytes called into the market by the rising demand. This was the American Institute of Imported Steel, formed in January of 1951 (see "Steel Importers Seek Price Freeze Relief," *Iron Age* 167 [February 8, 1951], p. 108). Moreover, the same boom allowed other nations pre-

viously dependent on imports for much of their steel to begin home production (the impetus stemming from the fact that the boom drained off supply and left them without adequate steel for their needs); see T. Metaxas, "Renaissance of Steel in South America," *Iron Age* 166 (September 28, 1950), pp. 92–93. Owing to similar causes, traditional U. S. export markets in Canada were being lost to others, chiefly British steelmakers who saw an opportunity to penetrate North American customers. See "British Bid for Canadian Steel Market," *Iron Age* 165 (January 12, 1950), p. 73; and "Canada Turn to U.K. for Needed Steel, " *Iron Age* 166 (July 6, 1950), p. 112. American attitudes toward German steel-production capabilities also underwent revision because of the stretched demand situation; see T. Metaxas, "German Steel—A Weapon for the Allies," *Iron Age* 166 (December 7, 1950), pp. 138–139. On the other hand, not all agreed that increased German output was in the best interests of America, see T. Prittie, "Allies Snub German Bid to Lift Output Though Shortages Plague West," *Wall Street Journal* (December 5, 1951), p. 1. It remained until 1952 before any significant let up in demand occurred, see J. C. Potter, "Importers Report End of Post-Korea Buying Spree by U.S. Industry," *Wall Street Journal* (February 14, 1952), p. 1

It is worth noting that the Korean War boom reversed a world wide recession in the steel industry that had been in effect since the spring of 1949. It was this downturn that, no doubt, inspired (at least partially) the formation of the Schuman Plan in Europe, just as it is probable that the increased level of steel demand following the outbreak of the Korean War also affected the early development of the plan.

68. The federal price controls had been weakened by congressional passage of the Capehart amendment in July 1951, which allowed firms to pass on costs they had incurred between June 1950 and July 1951; see Marcus, *op. cit.*, pp. 27–29; and Goodwin and Herren, *op. cit.*, pp. 81–82. the steelmakers' costs went up faster than they could pass them on, thus causing the drop in earnings for the year.

69. See G. Baker, "Steel Wage Rise Gets Solid Backing," *Iron Age* 168 (October 11, 1951), p. 75.

70. Steelman to the president, October 19, 1951, President's Secretary File, Box 136, General File—Steel Folder, HST Library. Steelman had long been involved in labor mediation duties for the government and, in fact, had been offered earlier the position of secretary of labor by Truman.

71. See Alonzo L. Hamby, *Beyond the New Deal: Harry S. Truman and American Liberalism* (New York: Columbia University Press, 1973), pp. 403–458; Bert Cochran, *Harry Truman and the Crisis Presidency* (New York: Funk & Wagnalls, 1973), pp. 311–352; and Marcus, *op. cit.*, pp. 35–37. In April 1951, Truman fired General Douglas MacArthur, commander of the Korean forces.

72. Truman's attorney general, J. Howard McGrath, for example, was forced to resign in April 1952 because of alleged corruption in his department. See Robert J. Donovan, *Tumultuous Years, the Presidency of Harry S. Truman, 1949–1953* (New York: W. W. Norton, 1982), pp. 162–170, 114,–127, 332–339, 365–370; and Hamby, *op. cit.*, pp. 459–480.

73. Donovan, *ibid.* Also see Alan F. Westin, *The Anatomy of a Constitutional Law Case* (New York: Macmillan, 1958), pp. 1–2.

74. Truman asserts that Secretary of Defense Robert A. Lovett had for months been telling him that any stoppage of the flow of steel from the mills to the battlefield would prove disastrous to the war effort. See Harry S. Truman, *Years of Trial and Hope* (Garden City, NY: Doubleday, 1956), pp. 466–467. This position is strongly corroborated by John R. Steelman, see "Steelman's memoirs," December 8, 1954, p. 1, in Official File, President's Personal File, Box 2, Memoirs File, John R. Steelman Folder, HST Library.

75. See n. 70. Truman's message is penciled in at the bottom of the memo.

76. This represented a new basic contract for the steelworkers; the existing contract, written in 1947, had been amended in 1948, 1949, and 1950. See Nagle, *op. cit.*, pp. 124–125.

77. Quoted in Grant McConnell, *The Steel Seizure of 1952* (Indianapolis, IN: Bobbs-Merrill, 1960), p. 6.

78. CEA to the president, RE: Forthcoming Steel Negotiations, November 15, 1951, President's Secretary File, Box 136, General File, Steel Folder, HST Library.

79. *Ibid.*, p. 1.

80. CEA, "Price-Wage Situation in Steel," November 14, 1951, David Lloyd Papers, Box 5, Steel No. 1 Folder, HST Library.

81. See Marcus, *op. cit.*, p. 59.

82. *Public Papers of the Presidents of the United States, Harry S. Truman, 1951* (Washington, DC: GPO, 1962), p. 651. Truman made the statement on December 22, 1951.

83. The WSB was composed of an equal number of public, labor, and industry representatives. The approval of the recommendations was unanimous from the first two groups and was unanimously opposed by the third. See Nagle, *op. cit.*, pp. 139–146; and Marcus, *op. cit.*, pp. 63–73.

84. Marcus, ibid., p. 63.

85. *Ibid.*, p. 65; and Nagle, *op. cit.*, pp. 145–146

86. Nagle, *ibid.*, pp. 143–144.

87. See Table 7.7 for comparative data.

88. Marcus, *op. cit.*, p. 65; and Nagle, *op. cit.*, pp. 140–141.

89. Nagle, *ibid.*, pp. 149–150.

90. See "Steelman's Memoirs," p. 5; and "Memorandum," March 31, 1952, Neustadt to Murphy, Charles S. Murphy Papers, Box 28, Steel Strike April 1–15, 1952 Folder, HST Library. In this latter memo, Richard Neustadt (a staff assistant to Murphy, the latter serving as special counsel to the president) specifically recalled the events of the 1946 steel strike episode.

91. "Steelman's Memoirs," p. 14.

92. Marcus, *op. cit.*, pp. 69–72.

93. *Ibid.*, p. 73; and Hamby, *op. cit.*, p. 455. Also see Goodwin and Herren, *op. cit.*, p. 85.

94. Truman, *Years of Trial and Hope,* pp. 488–492.

95. "Steelman's Memoirs," *passim*; and Marcus, *op. cit.*, p. 74.

96. See "Memorandum," Neustadt to Murphy, HST Library; and Marcus, *op. cit.*, pp. 75–79.

97. Donovan, *Tumultuous Years,* pp. 386–387.

98. These observations about the Taft-Hartley Act were made by Truman in his personal review of the draft of "Steelman's Memoirs"; they are penciled in the margins of the draft. See "Steelman's Memoirs," p. 2. The book Truman referred to was Fred A. Hartley, *Our New National Labor Policy; the Taft-Hartley Act and the Next Steps* (New York: Funk & Wagnalls, 1948).

99. Marcus, *op. cit.*, pp. 79–80.

100. See Charles Sawyer, *Concerns of a Conservative Democrat* (Carbondale: Southern Illinois University Press, 1968), p. 257.

101. See Executive Order No. 10340, issued April 8, 1952, by the president, "Directing the Secretary of Commerce to Take Possession of and Operate the Plants and Facilities of Certain Steel Companies." A copy is in the Official File, Box 897, Folder OF 272 Steel Companies, HST Library. The quotations in the text are found on p. 2 of the order.

102. See Marcus, *op. cit.;* Westin, *op. cit.*; McConnell, *op. cit.*; and Richard E. Neustadt, *Presidential Power* (New York: John Wiley & Sons, 1980).

103. Quoted in Marcus, *op. cit.,* p. 253.

104. *Ibid.,* quotations on pp. 79, 253–254, 254, respectively.

105. *Ibid.*; and Westin, *op. cit.* There is an extensive bibliography surrounding this case in the law jounals; the most important works are cited in the above two studies.

106. See W. V. Packard, "Steel: The Whipping Boy Fights Back," *Iron Age* 169 (February 14, 1952), pp. 83–84; "Steel Concerns Plan Hot Aid Campaign to Court Public Support," *Wall Street Journal* (April 11, 1952), p. 1; Edward L. Ryerson, "Management Thinking and Public Relations," address delivered September 1952 in which the speaker (chair of the AISI's Public Relations Committee) credits PR for the margin of victory in the industry's steel-seizure-case triumph (copy is in AISI Vertical Files, Box P-13, Eleutherian Mills-Hagley Foundation Historical Library, Wilmington, DE); and George F. Hamel, "John W. Hill, Public Relations Pioneer" (Master's thesis, University of Wisconsin, Madison, 1966), pp. 33–34.

107. See Nagle, *op. cit.,* p. 179.

108. These assertions are made in "Steelman's Memoirs." He claims that it was Clarence Randall, president of Inland Steel, who was most instrumental in ratcheting up the controversy through his introduction of personal invective into the debate. See "Steelman's Memoirs," pp. 3, 6, 11, 13. The characterization of Truman's actions as a "sad mistake" was made by Randall in an address delivered in Washington on April 25, 1952, see Clarence B. Randall, "Seizure. . . . The New Push-Button Warfare on Business," p. 12, Vertical File, Steel Industry and Trade Section, HST Library.

109. "Steelman's Memoirs," pp. 20–21; Nagle *op. cit.,* pp. 164–192; and Marcus, *op. cit.,* pp. 102–227. Also see *Youngstown Sheet & Tube Co. v. Sawyer,* 343 U.S. 579, 589 (1952); and J. A. Loftus, "Supreme Court Voids Steel Seizure, 6 to 3; Holds Truman Usurped Powers of Congress; Workers Again Strike as Mills are Returned," *NYT* (June 3, 1952), pp. 1, 23.

110. Nagle, *op. cit.,* p. 195. Also see J. B. Delaney, "Steel: Union Tries Split and Conquer," *Iron Age* 169 (June 26, 1952), p. 59.

111. "Steelman's Memoirs," pp. 3–4.

112. A. Stevens, "Truman Summons Steel Disputants as Defense Bogs," *NYT* (July 24, 1952), p. 1. Also see A. H. Raskin, "Steel Lack Shuts Army's Top Plant Producing Shells," *NYT* (July 23, 1952), pp. 1, 40.

113. Nagle, *op. cit.,* pp. 200–202; A. H. Raskin, "Steel Strike is Settled with Increases in Pay and Prices," *NYT* (July 25, 1952, pp. 1, 6.

114. See J. B. Delaney, "Steel: Strike's End Stirs Hopes for New Era," *Iron Age* 170 (July 31, 1952), p. 37.

115. Barton Bernstein, "Economic Policies of the Truman Administration," in R. S. Kirkendall (ed.), *The Truman Period as a Research Field* (Columbia: University of Missouri Press, 1967), p. 273.

116. See, for example, Phillip E. Stebbins, "Truman and the Seizure of Steel: A Failure in Communication," *Historian* 34 (November 1971), pp. 1–21; and Chong-do Hah and R. M. Lindquist, "The 1952 Steel Seizure Revisited: A Systematic Study in Presidential Decision Making," *Administrative Science Quarterly* 20 (December 1975), pp. 587–605. It was often noted by critics that if Truman had seriously believed the situation to be so critical, he could have invoked the Taft-Hartley Act to end the impasse. Moreover, as the president had never formally declared the nation to be at war in Korea, some might claim that no real emergency existed.

117. See Steelman to the president, August 1, 1952, and Steelman to the president, August 2, 1952, both President's Secretary File, Box 136, General File-Steel Folder, HST Library. In his memoirs, however, Truman stressed that the maintenance of adequate steel supplies for the war effort was of the highest priority and that this factor was not given sufficient attention in the Supreme Court decision. See Truman, *Years of Trial and Hope,* pp. 476–478.

Chapter 6

1. Arthur B. Homer, "What Does the Future Hold?" Address delivered to the Bethlehem Steel Management Group Meeting, Bethlehem, PA, February 2, 1953, p. 2. A copy is in SML, Arthur B. Homer File, Bethlehem Steel Corporation, Bethlehem, PA. Also see G. H. Baker, "Industry Sees Good Capitol Relations," *Iron Age* 170 (November 13, 1952), p. 117.

2. Arthur B. Homer, "The Doorway to Stabilized Progress," in AISI, *Yearbook, 1955* (New York: AISI, 1955), pp. 85, 86.

3. See Claude Robinson, "Current Public Opinion Trends and Their Meaning for Business Decision," in AISI, *Yearbook, 1955* (New York: AISI, 1955), pp. 95–136; and Donald M. K. Smith, "Public Opinion Survey Results," in AISI, *Yearbook, 1955* (New York: AISI, 1955), p. 178.

4. See Robert W. Crandall, *The U.S. Steel Industry in Recurrent Crisis* (Washington, DC: Brookings, 1981), pp. 18–40.

5. See Richard N. Gardner, *Sterling-Dollar Diplomacy,* expanded ed. (New York: McGraw-Hill, 1969), pp. 158–160; and Robert A. Pastor, *Congress and the Politics of U.S. Foreign Economic Policy, 1929–1976* (Berkeley: University of California Press, 1980), pp. 93–96.

6. *Public Papers of the Presidents of the United States, Harry S. Truman, 1946* (Washington, DC, GPO, 1962), p. 354. Truman made the statement on July 23, 1946.

7. Among the many sources on America's postwar economic expansionism, see, for example, William Appleman Williams, *The Tragedy of American Diplomacy,* rev. ed. (New York: Delta, 1962), pp. 269–276; John L. Gaddis, *The United States and the Origins of the Cold War, 1941–1947* (New York: Columbia University Press, 1972), pp. 316–352; Robert A. Pollard, *Economic Security and the Origins of the Cold War, 1945–1950* (New York: Columbia University Press, 1985); and Thomas G. Paterson, "The Economic Cold War: American Business and Economic Foreign Policy, 1945–1950" (Ph.D diss., University of California, Berkeley, 1968), pp. 357–394.

8. Pastor, *op. cit.,* p. 95.

9. See, for example, William A. Brown, Jr., *The United States and the Restoration of World Trade* (Washington, DC: Brookings, 1950), pp. 47–63; Gardner, *op. cit.*, pp. 158–160; Paterson, *op. cit.,* pp. 62–65.

10. See William Diebold, Jr., "The End of the I.T.O.," in *Essays in International Finance,* No. 16, Princeton University, Department of Economics and Social Institutions (October 1952), pp. 1–37; Pastor, *op. cit.,* pp. 97–99; Kenneth W. Dam, *The GATT, Law and International Economic Organization* (Chicago: University of Chicago Press, 1970), pp. 10–16; and John W. Evans, *The Kennedy Round in American Trade Policy* (Cambridge: Harvard University Press, 1971), pp. 8–11.

11. See John M. Leddy and Janet L. Norwood, "The Escape Clause and Peril Points Under the Trade Agreements Program," in William B. Kelly, Jr. (ed.), *Studies in United States Commercial Policy* (Chapel Hill: University of North Carolina Press, 1963), pp. 124–173. Also see Sumner Welles, "Pressure Groups and Foreign Policy," *Atlantic Monthly* 180 (November 1947), pp. 63–67.

12. This was particularly true of textile interests that were lured to the South by the promise of low-cost, nonunion labor. By 1955 this industry would become an important participant in the protectionist coalition. See Raymond A. Bauer, Ithiel de Sola Pool, and Lewis A. Dexter, *American Business and Public Policy, the Politics of Foreign Trade* (New York: Atherton Press, 1963), pp. 62–65. Also see Pastor, *op. cit.,* p. 96 and p. 97, Table 4.

13. Statement by the president, News Release, June 16, 1951, Presidential Papers, Official File, Box 901, 275A (June 1951) Folder, HST Library.

14. Gardner, *op. cit.,* p. 373.

15. Richard E. Neustadt to Charles Murphy, July 7, 1952, Official File, Box 901, 275A (August 1952–1953) Folder, HST Library.

16. *Ibid.* Also see George M. Elsey to Charles Murphy, June 25, 1952, Official File, Box 901, 275A (August 1952–1953) Folder, HST Library.

17. Public Advisory Board for Mutual Security, *A Trade and Tariff Policy in the National Interest* (Washington, DC: GPO, 1953).

18. Walter S. Tower, "Steel Faces the Postwar Years," in AISI, *Yearbook, 1945* (New York: AISI, 1945), p. 19.

19. See the testimony of George W. Wolf, president of the U.S. Steel Export Company (a U.S. Steel Corporation subsidiary), before the U.S. Congress, House Subcommittee on Foreign Trade and Shipping, Special Committee on Post-War Economic Policy and Planning; 78th Cong., 2nd sess.; and 79th Cong., 1st sess., Hearings, *Post-War Economic Policy and Planning,* pt. 4, Problems of Foreign Trade and Shipping (Washington, DC: GPO, 1945), pp. 927–941. Wolf read his testimony from a booklet, "International Trade," that his firm had recently published; it was highly supportive of liberalized free-trade measures. Also see R. H. Johnston, "Export," *Iron Age* 157 (January 3, 1946), p. 167; L. W. Moffett, "Washington . . . ," *Iron Age* 158 (October 17, 1946), pp. 80–82; "Largest Reciprocal Trade Agreement Reflects Tariff Cuts," *Iron Age* 158 (November 21, 1946), pp. 113–114; and L. W. Moffett, "Washington . . . ," *Iron Age* 158 (December 5, 1946), pp. 96–98.

20. For annual summaries of AISI activities in this area, see George S. Rose, "Activities of the Institute," in AISI, *Yearbook, 1947, . . . 1948, . . . 1949,* and *. . . 1951* (New York: AISI); see, pp. 701, 679, 610, and 503, respectively. Also see L. W. Moffett, "Washington . . . ," *Iron Age* 158 (December 26, 1946), p. 92; "Institute Protests High Tariff Rates for Raw Materials," *Iron Age* 159 (January 16, 1947), p. 115; E. J. Hardy, "Washington . . . ," *Iron Age* 161 (April 1, 1948), p. 104; "Steel Duties Put at 7% of Market," *NYT* (December 6, 1949), p. 52; T. Metaxas, "Tariffs: Do you Want Them Low or High," *Iron Age* 170 (November 27, 1952), pp. 37–38. Regarding the Committee for Reciprocity Information, see U. S. Department of State, Office of Public Affairs, *The United States Reciprocal Trade Agreements Program and the Proposed International Trade Organization, A Summary of Background Information* (Washington, DC: author, March, 1948), pp. 5–6.

21. "Tariff Battle Breaking in Congress," *Steel* 120 (February 3, 1947), pp. 65–66.

22. See "Present Tariff Rates on Steel Products Are Nominal, Afford Little Real 'Protection,'" *Steel Facts* 99 (December 1949), pp. 1–2; and "Little Effect Expected on Tariffs at Torquay," *Steel Facts* 103 (August 1950), p. 4. The latter citation, reporting on a tariff brief filed by the AISI with the Committee for Reciprocity Information, specifically refers to the reconstruction in Europe as a potential threat to American steelmakers. (*Steel Facts* was a publication of the AISI.)

23. The president was considered the leader of the internationalist wing of the Republican party; see Bauer, Pool, and Dexter, *op. cit.,* p. 28–30. For a detailed analysis of Eisenhower's complete record on foreign trade issues, see Burton I. Kaufman, *Trade & Aid, Eisenhower's Foreign Economic Policy, 1953–1961* (Baltimore: Johns Hopkins University Press, 1982). The president often expressed his commitment to free trade in private correspondence. See, for example, Eisenhower to E. E. Hazlett, July 20, 1954, p. 2, Papers of Dwight D. Eisenhower, Name Series, Box 18, Hazlett (1954) Folder, DDE Library. Hazlett was a boyhood friend of Eisenhower's, whom the president often confided in through long letters.

24. *Public Papers of the Presidents of the United States: Dwight D. Eisenhower, 1953* (Washington, DC: GPO, 1960), p. 15.

25. Eisenhower to E. E. Hazlett, August 3, 1956, Paper of Dwight D. Eisenhower, Name Series, Box 18, Hazlett (January 1956–November 1958) Folder, DDE Library.

26. Dwight D. Eisenhower, *Mandate for Change, 1953–1956* (Garden City, NY: Doubleday, 1963), p. 195.

27. Examples of such behavior would include the president's response to McCarthyism and racial integration. Although recent revisionist treatments of Eisenhower attempt to picture him as an activist, dynamic leader (e.g., Fred I. Greenstein, *The Hidden-Hand Presidency, Eisenhower as Leader* [New York: Basic Books, 1982]), there are, nevertheless, some areas in which his tentativeness and ambivalence seemed to dominate. Foreign economic policymaking was apparently one such area; see Kaufman, *Trade & Aid,* pp. 8–9; and Robert Griffith, "Dwight D. Eisenhower and the Corporate Commonwealth," *American Historical Review* 87 (February 1982), pp. 87–122.

28. Kaufman, *Trade & Aid,* p. 94. Also see Raymond Vernon, "Trade Policy in Crisis," in *Essays In International Finance,* No. 29, in Princeton University, International Finance Section (March 1958), for a prerevisionist evaluation of Eisenhower's leadership in foreign trade matters.

29. See Bauer, Pool, and Dexter, *op. cit.,* pp. 30–39; Pastor, *op. cit.,* pp. 101–102; and Kaufman, *Trade & Aid,* pp. 16–17.

30. Eisenhower to the Hon. Joseph W. Martin, Jr., Speaker of the House, May 1, 1953, Official File, Box 587, 116-M—Commission on Foreign Economic Policy, 1953 Folder, DDE Library. Also see "Statement by the President upon Signing the Trade Agreements Extension Act of 1953," *Public Papers, Eisenhower, 1953,* pp. 558–559.

31. See Robert Sheehan, "Clarence Randall: Statesman from Steel," *Fortune* 49 (January 1954), pp. 120–122, 132ff.; and Richard Harkness and Gladys Harkness, "Private Life of a Steel Boss," *Saturday Evening Post* (December 6, 1952), pp. 32–33ff.

32. Clarence B. Randall, *A Creed for Free Enterprise* (Boston: Little, Brown, 1952). On the book's best-seller status, see Kaufman, *Trade & Aid,* p. 19. For Randall's commentary on the nature of the competitive environment in industrial Europe, see pp. 150–153 of his book.

33. Kaufman, *Trade & Aid,* p. 20.

34. "Statement of the American Iron and Steel Institute Before the Commission on Foreign Economic Policy," December 2, 1953. A copy is in CFEP, Records, 1953–1954 File, Box 19, Hearings-Presentations, AISI Folder, DDE Library. For press coverage of this filing, see "Foreign Trade," *Iron Age* 172 (December 17, 1953), p. 95.

35. There are no available AISI internal records that might illuminate any big firm–small firm debate concerning the AISI statement; according to AISI officials, the institute's library was closed in 1970 and its contents discarded.

36. "Statement of the AISI . . . ," p. 3.

37. *Ibid.* For another view of the industry's foreign trade position, see Percy W. Bidwell, *What the Tariff Means to American Industries* (New York: Harper & Brothers, 1956), pp. 162–176. It might be noted that European producers claimed their exports of steel into America would not increase if U.S. tariff rates were lowered; see "OEEC Statement on US Tariffs on Steel," January 2, 1953, Public Advisory Board for Mutual Security, Box 70, Studies—Trade Policy Studies—RB-73–RB-79 Folder, CFEP, DDE Library.

38. "Statement of the AISI . . . ," p. 4.

39. *Ibid.,* p. 5.

40. *Ibid.*

41. *Ibid.,* p. 6.

42. See G. H. Baker, "Industry: Cooperates with Capitol," *Iron Age* 171 (June 18, 1953), p. 93.

43. For details of this oil industry venture, see Richard H. K. Vietor, *Energy Policy in America Since 1945* (New York: Cambridge University Press, 1984), pp. 37–42.

44. See U. S. Department of Commerce [G. Lyle Belsley], National Production Authority, Historical Reports on Defense Production, Report No. 19, *Consultation with Industry—History of the Office of Industry Advisory Committees of the National Production Authority* (Washington, DC: author, 1953); and Report No. 28, *Iron and Steel—History of the Iron and Steel Division of the National Production Authority* (Washington, DC: author, 1953). Copies of these reports are in National Archives, Record Group 277, National Production Authority, Office of the Executive Secretary, Historical Reports on Defense Production, Boxes 7 and 11.

45. C. E. Egan, "Sheaffer Resigns as Commerce Aide; Reputed Astin Foe," *NYT* (September 19, 1953), pp. 1, 6. Also see J. R. Lee, "US to Revamp Industry Consultant Methods," *Journal of Commerce* (December 16, 1953), pp. 1, 8; and U.S. Congress, House Committee on the Judiciary, Antitrust Subcommittee, 84th Cong., 1st sess., Hearings, *WOC's and Government Advisory Groups* (Washington, DC: GPO, 1955), pt. 1, p. 596, and pt. 3, p. 592. The quotation is from a letter of June 12, 1953, from Assistant Attorney General Stanley Barnes to Stephen F. Dunn, general counsel of the Department of Commerce; it is reproduced in the Hearings testimony, pt. 1, p. 596. It should be noted that in October 1953 the Commerce Department established a watered-down version of these earlier proposals in the Business and Defense Services Administration, a unit within the Commerce Department. Although the groups were retained through 1957, they were constantly held in suspicion by some members of Congress and the Justice Department. See Michael D. Reagan, "The Business and Defense Services Administration, 1953–1957," *Western Political Quarterly* 14 (June 1961), pp. 569–586, especially pp. 575–578.

46. Some smaller industries, such as watch manufacturing, lead and zinc, bicycle manufacturing, and certain agricultural commodities were able to obtain tariff protection from the Eisenhower administration, see Kaufman, *Trade & Aid,* pp. 37–46, 76–80; and Linda A. Cahn, "National Power and International Regimes: The United States Commodity Policies 1930–1980" (Ph.D diss., Stanford University, 1980). But larger and more nationally significant industries, such as textiles and petroleum, were also winning protectionist favors from the government, see Burton I. Kaufman, *The Oil Cartel Case: A Documentary Study of Antitrust Activity in the Cold War Era* (Westport, CT: Greenwood Press, 1978); Robert Engler, *The Politics of Oil* (Chicago: University of Chicago Press, 1961); and Bauer, Pool, and Dexter, *op. cit., passim.*

47. CFEP, *Report to the President and the Congress* (Washington, DC: author, January 23, 1954). Also see CFEP, *Minority Report* (Washington, DC: author, January 30, 1954).

48. See Bauer, Pool, and Dexter, *op. cit.,* pp. 40–49; and Kaufman, *Trade & Aid,* pp. 17–26.

49. On Randall's lobbying activities, see Minutes of Cabinet Meeting, February 26, 1954; and Maxwell Rabb to John Foster Dulles, March 2, 1954, "Resume of Presentation to the Cabinet by Clarence B. Randall"; both are in Cabinet Series, Box 3, Cabinet Meeting of February 26, 1954 Folder, DDE Library. Also see Arthur Burns to Gabriel Hauge, February 13, 1954, "Comments on Recommendations of Randall Commission", Arthur F. Burns Papers, Box 7, White House Correspondence—Hauge, 1953–1956 Folder, DDE Library; and "Legislative Leadership Meeting," March 29, 1954, Supplementary Notes, in Legislative Meeting File, Box, 1, Legislative Meetings 1954 (March–April) Folder, DDE Library. In addition, see R. M. Stroupe, "Foreign Trade: Ike Seeks Streamlining," *Iron Age* 173 (April 15, 1954), p. 77; and G. H. Baker, "Washington News," *Iron Age* 173 (May 20, May 27, June 17, 1954), pp. 115, 97, and 125, respectively. For Eisenhower's personal views on the report, see Eisenhower, *Mandate for Change,* pp. 292–294. Much of the opposition views held by the more protectionist-minded members of Congress were captured in the *Minority Report* cited in n. 47.

50. See Pastor, *op. cit.*, p. 102, and Bauer, Pool, and Dexter, *op. cit.*, pp. 50–58.

51. See Kaufman, *Trade & Aid*, pp. 41–43; Pastor, *op. cit.*, pp. 102–103; and Bauer, Pool, and Dexter, *op. cit.*, pp. 59–73.

52. Kaufman, *Trade & Aid*, pp. 113–132; Pastor, *op. cit.*, pp. 73–74.

53. Kaufman, *Trade & Aid*, pp. 43–44. Also see Gabriel Hauge to Arthur Burns, May 26, 1955, "The Organization for Trade Cooperation and Its Relation to the GATT," Arthur F. Burns Papers, Box 7, White House Correspondence—Hauge, 1953–1956 Folder, DDE Library.

54. Bauer, Pool and Dexter, *op. cit.*, p. 54; and Kaufman, *Trade & Aid*, pp. 39–40. Japan, in serious economic trouble in early and mid-1950s, was finally admitted to GATT in 1955, see William S. Borden, *The Pacific Alliance: United States Foreign Economic Policy and Japanese Trade Recovery, 1947–1955* (Madison: University of Wisconsin Press, 1984), pp. 176–187.

55. Kaufman, *Trade & Aid, passim;* and David A. Baldwin, *Foreign Aid and American Foreign Policy* (New York: Praeger, 1966).

56. See Eisenhower to Dodge, December 1, 1954, Official File, Box 592, 116-EE—CFEP Folder, DDE Library. On the organization of this function within the administration, see Gabriel Hauge, "White House Staff work,"*Looking Ahead*, supp. No. 2 (1955), pp. 3–4. (*Looking Ahead* is a journal published by the National Planning Association.) Also see Kaufman, *Trade & Aid*, pp. 36–37.

57. "For the President: The Development and Coordination of Foreign Economic Policy," p. III–2, Administrative Series, Box 10, CFEP Folder, DDE Library (n.d.). Also see, "Proposal for a New United States Foreign Economic Policy," Dodge Series—Subject Subseries, Box 2, Economic Policy Folder, DDE Library (n.d.).

58. Randall to the president, January 3, 1961, p. 3, Administrative Series, Box 32, Clarence Randall Folder, DDE Library.

59. See, for example, Robert A. Divine, *Eisenhower and the Cold War* (New York: Oxford University Press, 1981); and Herbert S. Parmet, *Eisenhower and the American Crusades* (New York: Macmillan, 1972).

60. See citations in n. 46.

61. "For the President: The Development and Coordination of Foreign Economic Policy," p. III–5, DDE Library.

62. See AISI, *Competitive Challenge to Steel* (New York: AISI, 1961), p. 7.

63. The slogan "Trade, not aid" had its origins in press accounts of the administration policy on this subject. See "Aid or Trade? A Crisis Ahead," *Business Week* (August 16, 1952), pp. 152–153; and "Trade, Not Aid," *Business Week* (December 13, 1952), p. 172. Also see Thomas V. DiBacco, "American Business and Foreign Aid: The Eisenhower Years," *Business History Review* 41 (Spring 1967), pp. 21–35.

64. See Kenneth Warren, *World Steel: An Economic Geography* (New York: Crane, Russak & Co., 1975), pp. 266–304; and "Steel: Ex-Importers Roll Their Own," *Iron Age* 174 (December 2, 1954), pp. 77–78. Also see John B. Parrish, "Iron and Steel in the Balance of World Power," *Journal of Political Economy* 64 (October 1956), pp. 369–388 for an interesting discussion of the role of the steel industry in the international balance of power.

65. Thus, for example, Eisenhower was eager to provide loan funds for the Tata Iron and Steel Co. of India, a privately owned mill, whereas the Soviet Union assisted in the construction of a state-owned steel mill in that country. See the relevant documents in DDE Diary Series, Box 9, DDE Personal Diary, 1-1-55/11-10-55 Folder, DDE Library. Also see G. J. McManus, "Inside India's Steel Expansion," *Iron Age* 183 (April 23, 1959), pp. 70–71; William B. Ewald, Jr., *Eisenhower the President—Crucial Days, 1951–1960* (Englewood Cliffs, NJ: Prentice-Hall, 1981), p. 235; and William A. Johnson, *The Steel Industry of India* (Cambridge: Harvard University Press, 1966), p. 21. This rivalry with the Soviet Union was exacerbated in the mid-1950s by a strategic

change in Soviet foreign policy whereby that country began to emphasize economic rather than military aid, see Kaufman, *Trade & Aid,* pp. 58–73.

66. See Richard B. Mancke, "Iron Ore and Steel: A Case Study of the Economic Causes and Consequences of Vertical Integration," *Journal of Industrial Economics* 20 (July 1972), pp. 220–229; Crandall, *op. cit.,* p. 20; and Gerald Manners, *The Changing World Market for Iron Ore, 1950–1980, an Economic Geography* (Baltimore: Johns Hopkins University Press, 1971).

67. Warren, *op. cit.,* p. 267.

68. *Ibid.,* pp. 266ff. Also see FTC Bureau of Economics, *Staff Report on the United States Steel Industry and Its International Rivals* (Washington, DC: author, November, 1977); Lennart Friden, *Instability in the International Steel Market,* trans. Roger Tanner (Stockholm: Beckmans, 1972); and CIA, National Foreign Assessment Center, *The Burgeoning LDC Steel Industry: More Problems for Major Steel Producers* (Washington, DC: author, 1979).

69. U.S. Department of Commerce, Bureau of Defense Services Administration, *International Iron and Steel* 2 (August 1956), p. 4, and Table B, pp. 16–19.

70. U.S. Congress, Senate Committee on Finance, 90th Cong., 1st sess. (Washington, DC: GPO, 1967), p. 32. Committee Print, *Steel Imports.*

71. One source estimated that between 1945 and 1962, U.S. funds channeled through AID and targeted for foreign steel mill construction totaled approximately $2 billion. See Manners, *op. cit.,* p. 99.

72. William Diebold, Jr., *The Schuman Plan, a Study in Economic Cooperation, 1950–1959* (New York: Praeger, 1959), pp. 543–551; and Duncan Burn, *The Steel Industry 1939–1959, a Study in Competition and Planning* (Cambridge: Cambridge University Press, 1961), pp. 147–148.

73. See "Freight: Seek Steel Export Rate Cut," *Iron Age* 173 (May 13, 1954), p. 83; and Diebold, *The Schuman Plan;* pp. 542–543.

74. Diebold, *ibid.,* p. 546.

75. Clarence B. Randall, "A Steel Man Looks at the Schuman Plan," *Atlantic Monthly* 186 (October 1950), p. 37. Also see Clarence B. Randall, "European Steel: Monopoly in the Making," *Atlantic Monthly* 188 (October 1951), pp. 34–38.

76. Diebold, *The Schuman Plan,* p. 557 and p. 562.

77. "Letter to Chairmen of Senate Foreign Relations and House Foreign Affairs Committees Concerning the European Coal and Steel Community," June 15, 1953, *Public Papers, Eisenhower, 1953,* p. 443.

78. For details, see Eisenhower, *Mandate for Change,* pp. 398–405.

79. "Letter to Chairmen . . . ," *Public Papers, Eisenhower, 1953,* p. 444.

80. This remark, made in congressional hearings, is quoted in Comptroller General of the United States, GAO, *Report to the Congress of the United States, Examination of 100 Million Dollar Loan to ECSC, June 30, 1957* (Washington, DC: author, March 13, 1958), p. 11. Also see "Steel," *Iron Age* 173 (May 13, 1954), p. 82.

81. One reason for the domestic steel industry's lack of opposition was that the major part of the 100 million loan funds would be devoted to ECSC investments in coal mining facilities, not steelmaking operations. See Diebold, *The Schuman Plan,* pp. 320–326; 544–545.

82. Clarence B. Randall, "Steel: The World's Guinea Pig," *Atlantic Monthly* 190 (December 1952), p. 33. Also see Diebold, *The Schuman Plan,* pp. 16–20, for a contrasting view.

83. See Louis Lister, *Europe's Coal and Steel Community, an Experiment in Economic Union* (New York: Twentieth Century Fund, 1960), pp. 211–213; Diebold, *The Schuman Plan,* pp. 254–286; and Burn, *op. cit.,* pp. 146–154.

84. See Burn, *ibid.,* pp. 434–447. Also see "Steel: Europe Cuts Exports Prices," *Iron Age* 170 (September 11, 1952), p. 92; K. M. Bennett, "Steel: Nibble at Foreign

Offers," *Iron Age* 170 (July 3, 1952), p. 75; "Steel Pool: How Deep Are Cracks?" *Iron Age* (January 22, 1953), p. 46; "Steel: Europe Needs More Customers," *Iron Age* 173 (March 4, 1954), p. 99; and "Europe: Upturn in Steel Market," *Iron Age* 173 (April 1, 1954), p. 77.

85. Quoted in Diebold, *The Schuman Plan,* p. 256.

86. See Martin J. Rosen, "The Brussels Entente: Export Combination in the World Steel Market," *University of Pennsylvania Law Review* 106 (June 1958), p. 1083.

87. Diebold, *The Schuman Plan,* p. 492; Burn, *op. cit.,* p. 343; and Rosen, *op. cit.,* pp. 1088–1089. It is interesting to note that when an ECSC delegation visited the United States in June 1953 seeking American loans, one of its members claimed before a congressional committee,"As Mr. Monnet pointed out, in the steel market [i.e., the ECSC] there is complete free competition; with nobody fixing prices anymore," See U.S. Congress, Senate Committee on Foreign Relations, 83rd Cong., 1st sess., Hearings, *European Coal and Steel Community* (Washington, DC: GPO, 1953), p. 31.

88. For American and British reaction, see Diebold, *The Schuman Plan,* p. 499, n. 36, and "Steel: Denationalized, U.K. Confident," *Iron Age* 172 (July 16, 1953), p. 81. As noted previously, U.S. producers—at least the larger ones—did not offer much comment of any kind on the formation and early operations of the ECSC. See Franklin R. Root, *The European Coal and Steel Community,* Part II, (College Park: University of Maryland, Bureau of Business and Economic Research, June 1956). This study consists of an analysis of interviews with executives from six large steel firms on the subject of the ECSC. The firms considered are U.S. Steel, Bethlehem, National, Republic, Inland, and Jones & Laughlin—that is, the largest integrated producers in the domestic industry. However, it might be noted that the American producers did at one time appear to demonstrate concern with events in Europe: This was through the AISI "Statement" submitted to the Randall Commission in December 1953 (see n. 34 for full citation). The institute, it will be recalled, recommended that the government commit itself to an industrial policy for internationalist firms.

For details surrounding the ECSC share of U.S. steel trade, see U.S. Department of Commerce, Bureau of Defense Services Administration, *International Iron and Steel* 1 (June 1955), p. 19, Table 5 (however, covers, only the years 1950–54). The only exception to this pattern of stability in foreign markets occurred during the Korean War when imports rose to historic highs because of the acute domestic shortages and the 1952 steelworkers' strike. Yet some officials foresaw future problem with imports because of the competitive changes taking place abroad. "American imports of iron and steel products have been reaching relatively high levels in recent years," stated the AISI in 1955; meanwhile, the Department of Commerce inaugurated publication of a new quarterly report titled *International Iron and Steel* that same year. In the first edition, the department noted, "The steel industry in the U.S. is becoming more international in its interests and concerns and, therefore, in need of more information about the changing pattern of world steel production and consumption." See AISI, *Charting Steel's Progress,* 1955 ed. (New York: author, 1956), p. 47; and U.S. Department of Commerce, Bureau of Defense Services Administration, *International Iron and Steel* 1 (March 1955), p. *2.*

89. See "Steel: State Dept. Protests Export Pricing by Steel Pool," *Iron Age* 172 (November 19, 1953), p. 113; and "Steel Pool: Who's Denting It?" *Iron Age* 172 (December 24, 1953), pp. 22–23. Also see Rosen, *op. cit.,* p. 1094 and p. 1096, fn. 109; Diebold, *op. cit.,* p. 492; and "Europe's First Anti-Cartel Law," *Metal Bulletin* (London) (June 21, 1955), pp. 18–19.

90. See Rosen, *op. cit.,* pp. 1096–1099; Diebold, *ibid.,* p. 492–501 and 559–561; and Lister, *op. cit.,* p. 216–219.

91. See Diebold, *The Schuman Plan,* pp. 497–499; and Rosen, *op. cit.,* p. 1099. Also see "Toward Free Competition in Europe—The High Authority Versus Cartels,"

ECSC High Authority Bulletin 6 (April 1955), pp. 1–5. The article claims that the HA prohibited the Brussels Entente from developing into an effective cartel. Nevertheless, the entente did not cease to exist. Although governmental pressures were instrumental in reducing the cartel's effectiveness after 1955, what appears to have been a more compelling brake was the global steel recession of 1958–62. The entente was simply unable to enforce its strictures during these years. But by late 1962, with steel demand improving, the group was once again able to achieve results. Morever, Japanese steelmakers also formed an export cartel in the early 1960s . According to one U.S. Senator, this latter group joined with European producers to split the lucrative American market between themselves, thus avoiding head-to-head competition. This action was precedent setting for the Japanese in that they had never previously participated in any international steel agreements. However, it should be noted that Japanese steel officials had monitored cartel activities in the global steel industry since at least 1926, the year in which the ISC was formed; see Leonard H. Lynn and Timothy McKeown, "The Development of Trade Associations in the U.S. and Japan," unpublished working paper, Carnegie-Mellon University, Department of Social and Decision Sciences, College of Humanities and Social Sciences (1987), pp. 24–26. Also see "Brussels Convention Is an Export Cartel," *Iron Age* 190 (November 22, 1962), p. 13; "EEC Goes After 'Bad' Cartels," *Business Week* (April 21, 1962), pp. 110–116; U.S. Department of State, Bureau of Intelligence and Research [Corwin D. Edwards], *Cartelization in Western Europe* (Washington, DC: author, June 1964), pp. 61–82; "Cartels Fence in European Steel," *Business Week* (September 3, 1966), pp. 94–98; "Japanese Ministry of Trade Authorizes Export Steel Cartel," *American Metal Market* (June 7, 1966), p. 1; "Japan–Europe Steel Cartel for Sale in U.S. Charged," *Wall Street Journal* (May 7, 1963), p. 3; and U.S. Congress, Senate, *Steel Imports,* pp. 328–330.

92. See Warren, *op. cit.,* p. 156; Burn, *op. cit.,* pp. 390ff; and Norman J. G. Pounds and William N. Parker, *Coal and Steel in Western Europe* (Bloomington: Indiana University Press, 1957), pp. 345–364.

93. See "Estimated World Steel Capacity and Output," *Iron Age* 170 (September 4, 1952), p. 95. Also see Lister, *op. cit.,* pp. 79–88, 466; and Diebold, *The Schuman Plan,* pp. 314–349.

94. See "Europe: Steel Merger Tempo Quickens," *Iron Age* 176 (July 7, 1955), p. 55; Pounds and Parker, *op. cit.,* pp. 345–350; and Diebold, *The Schuman Plan,* pp. 72–75, 356–378.

95. See Lister, *op. cit.,* p. 168. Symbolic of this development was the renaissance of the Krupp empire in Germany. Headed by Alfried Krupp von Bohlen und Halbach, the firm had been that nation's second-largest steel producer prior to the war. In 1945, however, Krupp was convicted of war crimes for his complicity with the Nazi regime; this eventually resulted in a 12-year prison term for the industrialist. As part of the denazification and decartelization drive that the Allies imposed on Germany in the immediate wake of the war, Krupp's empire was ordered broken up and sold off. Yet (as we saw earlier), American officials began to reconsider their earlier plans for Germany's future following a reassessment of Soviet motives. By 1951 Krupp was out of prison and once again in charge of his reconstituted holdings. A work force of nearly 85,000 people brought in $700 million in worldwide revenues by 1955, providing the Krupp firm with control over 17% of Germany's steel output—an amount almost equal to its market share in 1930. See J. Bell, "The Comeback of Krupp," *Fortune* 53 (February 1956), pp. 101–108, 200ff; H. Gifford, Jr., "Krupp: His Interests Are World-Wide," *Iron Age* 175 (April 7, 1955), p. 84; Lister *op. cit.,* pp. 152, 157–160. For longer treatments, see William Manchester, *The Arms of Krupp, 1587–1968* (Boston: Little, Brown, 1968), and Peter Batty, *The House of Krupp* (London: Secker and Warburg, 1966).

96. In "Britain: Denationalize Steel," *Iron Age* 175 (April 21, 1955), p. 55, it was

reported that 60% of the industry was again in private hands by the end of 1954. Also see Burn, *op. cit.*, pp. 366ff; Doug McEachern, *A Class Against Itself, Power and the Nationalisation of the British Steel Industry* (Cambridge: Cambridge University Press, 1980), pp. 103–120, and B. S. Keeling and A.E.G. Wright, *The Development of the Modern British Steel Industry* (London: Longmans, 1964), pp. 176–180.

97. On British reasons for not joining the ECSC, see Diebold, *The Schuman Plan*, pp. 48–60. Also see "Steel: British Shun Schuman Ties," *Iron Age* 173 (April 22, 1954), p. 123.

98. F. H. Harley, "Great Britain: Boosts Steel Program, " *Iron Age* 173 (March 17, 1955), p. 67. Also see Burn, *op. cit.*, p. 358.

99. See Burn, *op. cit.*, pp. 537–684; Keeling and Wright, *op. cit.*, pp. 160–200; M. J. Layton, "Some Economic Aspects of the European Steel Industry," in AISI, *Yearbook, 1953* (New York: AISI, 1953), pp. 173–179; Charles K. Rowley, *Steel and Public Policy* (London: McGraw-Hill, 1971), pp. 98–99; and McEachern, *op. cit.*, pp. 121–155.

100. Kiyoshi Kawahito, *The Japanese Steel Industry* (New York: Praeger, 1972), p. 21. Also see "Schuman Plan: Japanese Work on Plans to Combat Steel Pool Competition," *Iron Age* 171 (April 23, 1953), p. 82.

101. Kawahito, *The Japances Steel Industry*, pp. 35–47. Also see Leonard H. Lynn, *How Japan Innovates, a Comparison with the U.S. in the Case of Oxygen Steelmaking* (Boulder, CO: Westview Press, 1982), pp. 43–118; Chalmers Johnson, *MITI and the Japanese Miracle, the Growth of Industrial Policy, 1925–1975* (Stanford: Stanford University Press, 1982), pp. 207–211, 220, 226; and Takafusa Nakamura, *The Postwar Japanese Economy, Its Development and Structure*, trans. J. Kaminski (Tokyo: University of Tokyo Press, 1981), p. 70.

102. Kawahito, *op. cit.*, p. 43; and Joseph W. Ford, *The Steel Import Problem* (New York: Fordham University Press, 1961), pp. 18–27. For data on output by all countries from 1900–1959, see the foldout chart, end-piece in Burn, *op. cit.*

103. *Ibid.*, these rates were computed from the data found in the Burn foldout chart.

104. Note the relationship between national power and steel production as described by Parrish, *op. cit.* Also see Manners, *op. cit.*, pp. 83–105.

Chapter 7

1. President to Edgar Eisenhower, November 8, 1954, Papers of Dwight D. Eisenhower, Name Series, Box 11, Edgar Eisenhower–1954 Folder, DDE Library.

2. Walter S. Tower, "Subjugation or Liberation?" in AISI, *Yearbook 1952* (New York: AISI, 1952), p. 16.

3. Quoted in W. V. Packard, "Steel: Leaders See Hopeful Future," *Iron Age* 171 (June 4, 1953), p. 88.

4. As one example of the competitive effect of a temporarily sluggish market, in late 1953 the larger steelmakers were forced to absorb freight costs for customers in order to maintain sales levels; see J. B. Delaney, "Freight: Growing Steel Sales Factor," *Iron Age* 173 (Januay 7, 1954), p. 248, and J. B. Delaney, "Steel: Secret's Out, Competition Here," *Iron Age* 172 (October 8, 1953), p. 213. For evidence of industry support for changes in the depreciation laws, see n. 48, Chapter 5.

5. For an overview of this steel expansion abroad, see Kenneth Warren, *World Steel, an Economic Geography* (New York: Crane, Russak & Co., 1975); and Gerald Manners, *The Changing World Market for Iron Ore, 1950–1980, an Economic Geography* (Baltimore: Johns Hopkins University Press, 1971), pp. 9–26.

6. Henry W. Broude, *Steel Decisions and the National Economy* (New Haven: Yale University Press, 1963), p. 258.

7. *Ibid.,* for evidence of these changing attitudes, pp. 252–253.

8. For details surrounding the certificate-of-necessity program, see n. 54, Chapter 5. Also see J. B. Delaney, "Steel Forges Ahead Without Fast-Tax," *Iron Age* 179 (January 17, 1957), pp. 31–32.

9. For a schedule of the certificates granted in the first phase of the program, see "Wilson Puts the Brakes on Granting Certificates of Necessity for Expansion of Steel Industry," *Iron Age* 167 (March 8, 1951), pp. 110–111; and U.S. Department of Commerce, National Production Authority, Historical Reports on Defense Production, Report No. 28, *Iron and Steel—History of the Iron and Steel Division of the National Production Authority* (Washington, DC: author, 1953).

10. The critique came from the Defense Production Administration. See J. B. Delaney, "Steel: What Kinds of Capacity?" *Iron Age* 171 (February 5, 1953), p. 97.

11. Gilbert Burck, "The Private Strategy of Bethlehem Steel," *Fortune* 65 (April 1962), p. 112.

12. See "Iron and Steel—The Present," *Iron Age* 175 (June, 1955), p. D–10. Also see AISI, *Annual Statistical Report, 1955* (New York: author, 1956), "Annual Steel Capacity by Type of Furnance," p. 53.

13. R. D. Raddant, "Steel: Build First U.S. Oxygen Units," *Iron Age* 173 (February 11, 1954), pp. 68–69.

14. For details surrounding the BOF program, see Leonard H. Lynn, *How Japan Innovates, a Comparison with the U.S. in the Case of Oxygen Steelmaking* (Boulder, CO: Westview Press, 1982), pp. 28, 150–154, 157–163.

15. See J. W. Kirkpatrick, "Oxygen in Open Hearth Steelmaking," in AISI, *Yearbook, 1961* (New York: AISI, 1961), pp. 199–232; and David R. Dilley and David L. McBride, "Oxygen Steelmaking—Fact vs. Folklore," *Iron and Steel Engineer* 44 (October 1967), pp. 131–152. Also see W. G. Patton, "U.S. Company Reports on Oxygen Steel Process," *Iron Age* 175 (March 31, 1955), pp. 79–80; "J&L Starts Up Oxygen Steel Unit," *Iron Age* 180 (December 12, 1957), p. 87; G. J. McManus, "Low Capital Cost Spurs Swing to Oxygen Steel," *Iron Age* 181 (February 6, 1958), pp. 55–58; and G. J. McManus, "Oxygen Steel Passes Its Tests; More, Bigger Units Planned," *Iron Age* 184 (September 24, 1959), pp. 67–68.

16. Most noteworthy of these critiques is Walter Adams and J. B. Dirlam, "Big Steel, Invention, and Innovation," *Quarterly Journal of Economics* 80 (May 1966), pp. 167–189. Also see "Is Steel Technology Lagging?" *Iron Age* 180 (November 7, 1957), p. 82; "How Steel Figures Oxygen Costs," *Iron Age* 180 (October 3, 1957); p. 57, and J. McDonald, "Steel Is Rebuilding for a New Era," *Fortune* 74 (October 1966), p. 135. The inspiration for much of this critique of industrial technological leadership stemmed from a congressional inquiry into the matter; see U.S. Congress, Senate Subcommittee on Antitrust and Monopoly, Committee on the Judiciary, 85th Cong., 1st sess., Hearings, *Administered Prices,* pt. 3, "Steel" (Washington DC: GPO, 1958).

17. For reviews of this literature, see Bela Gold, Gerhard Rosegger, and Myles G. Boylan, Jr., *Evaluating Technological Innovations* (Lexington, MA: Lexington Books, 1980), pp. 119–259; and David H. Ciscel, "The Decline of the United States Steel Industry: A Study of Market Entropy" (Ph.D diss., University of Houston, 1971), pp. 7–34.

18. Adams and Dirlam, *op. cit.,* p. 186.

19. The organizational-failures framework is most forcefully expounded by Oliver E. Williamson, *Markets and Hierarchies* (New York: Free Press, 1975).

20. Industry arguments opposing early adoption of BOF are best summarized in Dilley and McBride, "Oxygen Steelmaking . . . . ," pp. 131–152.; and M. W. Lightner and D. L. McBride, "Basic Open-Hearth Steelmaking in the U.S.A.," *Journal of the*

Iron and Steel Institute 189 (July 1958), pp. 205–216. Also see William T. Hogan, *Economic History of the Iron and Steel Industry in the United States,* 5 vols. (Lexington, MA: Lexington Books, 1971), 4:1519–1526; and the comments of Lynn, *op. cit.,* pp. 15–39. There were several replies to the Adams and Direlam critique that should also be noted: Reuben E. Slesinger, "Steel Imports and Vertical Oligopoly Power: A Comment,"*American Economic Review* 56 (March 1966), pp. 152–155; A. K. McAdams, "Big Steel, Invention, and Innovation: Reconsidered," *Quarterly Journal of Economics* 81 (August 1967), pp. 457–474; and G. Rosegger, "Steel Imports and Vertical Oligopoly Power: Comment," *American Economic Review* 57 (September 1967), pp. 913–917. On "technological momentum," see Thomas P. Hughes, "Technological Momentum in History: Hydrogenation in Germany, 1898–1933," *Past & Present* 44 (August 1969), pp. 106–132.

21. See the evidence in support of this observation in Edwin Mansfield, *Industrial Research and Technological Innovation* (New York: W. W. Norton, 1968), pp. 83–108.

22. See T. M. Rohan, "Steel: Will Spend $800 Million in '55," *Iron Age* 175 (March 24, 1955), pp.51–53; "Steel: Expansion and Prices," *Iron Age* 176 (November 3, 1955), p. 58; and J. B. Delaney, "Steel: Expansion Wheels Are Turning," *Iron Age* 176 (November 24, 1955), pp. 42–43.

Although the Bethlehem mill was the only such fully integrated facility to actually be built, there were nevertheless recurrent rumors throughout the 1950s that a modern steelmaking complex would be constructed somewhere in the western United States. See T. M. Rohan, "Steel: New Integrated Mill in West?." *Iron Age* 173 (June 10, 1954), pp. 77–78; T. M. Rohan, "West: A New Fairless Works?" *Iron Age* 178 (October 4, 1956), pp. 56–57; R. R. Kay, "Steel Expansion Moving Into High," *Iron Age* 179 (January 10, 1957), p. 43; R. R. Kay, "Will Hawaii Get a Steel Mill?" *Iron Age* 179 (March 14, 1957), p. 111; R. R. Kay, "Kaiser Will Double Capacity," *Iron Age* 179 (April 18, 1957), p. 90; R. R. Kay, "Why Arizona Wants a Steel Mill," *Iron Age* 181 (February 20, 1958), p. 67; "U.S. Steel Plans Expansion in West," *Iron Age* 185 (June 9, 1960, p. 71. (Though the Bethlehem mill at Burns Harbor, IN, did not start construction until 1962, the company had signaled its intentions in 1956 when it purchased the property on which the complex would be located, see Kenneth Warren, *The American Steel Industry, 1850–1970, a Geographical Interpretation* [London: Oxford University Press, 1973], p. 315).

23. For data concerning the 1960s, see Robert W. Crandall, *The U.S. Steel Industry in Recurrent Crisis* (Washington, DC: Brookings, 1981), pp. 24–25, and D. F. Barnett and L. Schorsch, *Steel, Upheaval in a Basic Industry* (Cambridge, MA: Ballinger, 1983), pp. 37–75.

24. Arthur B. Homer, "Steel's Future—A Producer's Point of View." Address delivered to the Eastern Regional Conference of Financial Analysts Societies, New York City, November 24, 1952. A copy is in SML, Arthur B. Homer File, Bethlehem Steel Corporation, Bethlehem, PA. Also see "Steel Shipments Related to Gross National Product," *Steel Facts* 119 (April 1953), p. 6.

25. Benjamin Fairless, "America and Steel—Growing Together," in AISI *Yearbook, 1957* (New York: AISI, 1957), p. 54. Also see J. B. Delaney, "Will Steel Sales Rally in 1957?" *Iron Age* 179 (May 30, 1957), pp. 56–57.

26. See R. D. Raddant, "Steel: It's an Expanding Business," *Iron Age* 178 (September 20, 1956), pp. 51–53; G. J. McManus, "Steel Expansion Plans Stress Economy as Costs Mount," *Iron Age* 180 (September 19, 1957), pp. 99–102; and Manners, *op. cit.,* pp. 63–82.

27. See Broude, *op. cit.,* pp. 173–262. Also see T. Campbell, "Steel: Fast Tax Writeoff Is Out," *Iron Age* 176 (November 17, 1955), p. 75; G. J. McManus, "National Plans Chicago Mill," *Iron Age* 183 (February 26, 1959), p. 47; and K. M. Bennett, "Steel: Upswing in Midwest Capacity," *Iron Age* (May 22, 1952), p. 78.

28. On the rivalry with the Soviet Union, see W. V. Packard, "Steel: Reds Gain Faster Than Free World," *Iron Age* 172 (December 31, 1953), pp. 21–22; G. H. Baker, "Congress Stays Wary on Red Trade," *Iron Age* 173 (April 15, 1954), p. 87; W. V. Packard, "Reds: Narrow Steel Ingot Gap," *Iron Age* 174 (December 30, 1954), pp. 17–18; "Armaments: Eat Up Red Steel," *Iron Age* 176 (November 24, 1955), pp. 44; J. B. Delaney, "Free World Widens Steel Lead Over Reds," *Iron Age* 176 (December 29, 1955), pp. 23–24; T. Campbell, "Steel: Are We Losing Ground to Reds?" *Iron Age* 177 (February 16, 1956), pp. 64–65; "Red-Bloc Divides Export Markets," *Iron Age* 177 (July 7, 1956), p. 80; J. B. Delaney, "Free World Holds 3-to-1 Edge in Steel," *Iron Age* 178 (December 27, 1956) pp. 23–24; "West Still Leads in Steel Output but Reds Gain Technically," *Iron Age* 180 (December 26, 1957), pp. 19–21; G. F. Sullivan, "The Russian Steel Industry," *Iron Age* 182 (September 4, 1958), pp. 89–100; F. J. Starin, "New Look in Economic Warfare," *Iron Age* 182 (December 11, 1958), pp. 94–95. In the summer of 1958, a delegation of American steel officials visited Russia under State Department auspices as part of a trade mission exchange, see Francis M. Rich, "Steelmaking in the Soviet Union," in AISI, *Yearbook, 1959* (New York: AISI, 1959), pp. 45–72, 41–44.

Regarding Defense Department pressure to expand the domestic industry, see T. Campbell, "Steel: Why More Capacity Is Coming," *Iron Age* 175 (June 23, 1955), p. 53; Henry T. Simmons, "Government Planners Likely to Ask New Tax Aid to Spur Expansion," *Wall Street Journal* (June 8, 1955), p. 1; N. R. Regeimbal, "Fast Tax: Steel Makes a New Pitch," *Iron Age* 178 (September 6, 1958), p. 56; G. H. Baker, "Fight Looms on Fast Write-Offs," *Iron Age* 179 (May 9, 1957), p. 89; and "Steel: 150 Million Tons by 1960?" *Iron Age* 176 (September 15, 1955), pp. 54–55.

29. N. R. Regeimbal, "Steel: Congress Probes Sales," *Iron Age* 178 (September 20, 1956), p. 55; and T. Campbell, "Steel: No Short Cuts Out of the Woods," *Iron Age* 178 (September 13, 1956), pp. 54–55.

30. See Richard W. Nagle, "Collective Bargaining in Basic Steel and the Federal Government, 1945–1960" (Ph.d. diss., Pennsylvania State University, 1978), pp. 195, 201.

31. See H. Scott Gordon, "The Eisenhower Administration: The Doctrine of Shared Responsibility," in Craufurd D. Goodiwin (ed.), *Exhortation and Controls—The Search for a Wage-Price Policy, 1945–1971* (Washington, DC: Brookings, 1975), p. 97.

32. J. B. Delaney, "Steel: Price Rise Will Follow Wage Hike," *Iron Age* 171 (June 18, 1953), pp. 91–92; and W. V. Packard, "Steel: Pass on Price Hikes Reluctantly," *Iron Age* 171 (June 25, 1953), p. 65.

33. Gardiner C. Means, *Pricing Power and the Public Interest, a Study Based on Steel* (New York: Harper & Row, 1962), p. 113.

34. Otto Eckstein and Gary Fromm, "Steel and Postwar Inflation," Study Paper No. 2, U.S. Congress, JEC, 86th Cong., 1st sess., *Materials Prepared in Connection with the Study of Employment, Growth, and Price Levels* (Washington, DC: GPO, 1959). This analysis was subsequently critiqued in US. Department of Labor [E. Robert Livernash], *Collective Bargaining in the Basic Steel Industry—A Study of the Public Interest and the Role of Government* (Washington, DC: author, 1961), pp. 179–196.

35. The most well-publicized investigation was undertaken by the Kefauver Committee. See U.S. Congress, Senate, *Administered Prices*, pts. 2, 3, 4, "Steel."

36. See, for example, John M. Blair, "Administered Prices: A Phenomenon in Search of a Theory," *American Economic Review* 49 (May 1959), pp. 431–450. For a current updating of this position, see Edward Greer, "The Political Economy of U.S. Steel Prices in the Postwar Period," in Paul Zarembka (ed.), *Research in Political Economy*, Vol. 1 (Greenwich, CT: JAI Press, 1977), 1:59–86.

37. Walter Adams, "The Steel Industry," in W. Adams (ed.), *The Structure of*

American Industry, 5th ed. (New York: Macmillan, 1977), p. 110. Most economists, of course, tended to dismiss industry arguments about high fixed costs and price inelasticities and, instead, recommended that steelmakers should simply price at the margin, as did manufacturers of most other commodities.

38. See Leonard W. Weiss, *Case Studies in American Industry*, 3rd ed. (New York: John Wiley & Sons, 1980), p. 191; Adams, *op. cit.;* Leonard W. Weiss, *Economics and American Industry* (New York: John Wiley & Sons, 1961), pp. 293–299; and George J. Stigler and James K. Kindahl, *The Behavior of Industrial Prices* (New York: Columbia University Press, 1970), pp. 71–74. There were, of course, occasional exceptions to this leadership role, see R. D. Raddant, "Steel: The Price Pot Is Bubbling" *Iron Age* 174 (December 24, 1954), pp. 21–22, which describes a price cut initiated by National Steel in the late 1954 recession.

39. Concerning the alleged positive relationship between steel-price increases and inflation, however, there appears at best only mixed evidence to support such a hypothesis, See Charles K. Rowley, *Steel and Public Policy* (London: McGraw-Hill, 1971), pp. 151–175; and Richard Mancke, "The Determinants of Steel Prices in the United States: 1947–1965," *Journal of Industrial Economics* 16 (April 1968), pp. 147–160.

40. See Rowley, *op. cit.*, pp. 68–71, for a discussion of steel substitutes and their effect on steel markets.

41. It must also be noted that the industry had traditionally relied heavily on internally generated funds for expansion capital. Debt was avoided to a large extent, as was dilution of equity through new stock offerings. Between 1946 and 1956 only $1.5 billion was raised by the industry through the capital markets, whereas more than $8.5 billion was committed to capital expenditures—a ratio of less than 20%. See U.S. Steel Corporation, *Steel and Inflation, Fact vs. Fiction* (New York: author, 1958), pp. 181–185. The industry attributed its conservative financial strategy to fears of absorbing fixed debt or dividend requirements when it was uncertain of its ability to increase profits sufficient to service these requirements. See Eldon S. Hendriksen, *Capital Expenditures in the Steel Industry, 1900–1953* (New York: Arno Press, 1978). Others, however, disagree with the industry's stated reasons: They find the level of industry profits to have been significant in the period under discussion and, in fact, conclude that pricing policies were geared to increasing profits simply for profits' sake (i.e., to enrich shareholders). See Means, *op. cit.*, pp. 112–150 for a summary of this argument.

42. See "Steel, Story of a Shortage," (no author shown, 1952), pp. 15–16. A copy of this pamphlet may be found in the Vertical File, Steel Industry and Trade Section, HST Library.

43. See Harold G. Vatter, *The U.S. Economy in the 1950s, an Economic History* (New York: W. W. Norton, 1963), p. 155; and Benjamin F. Fairless, "Steel's Depreciation Problem," in AISI, *Yearbook, 1956* (New York: AISI, 1956), pp. 46–67.

44. See Chapters 4 and 6 for background.

45. For example, Means. *op. cit.* Means also accuses U.S. Steel of pursuing profits because of a new incentive bonus system for top management created in 1951, see pp. 155–157. It might also be noted that an incentive system for management at Bethlehem Steel also produced high salaries. In 1956, for example, Bethlehem executives comprised eleven of the eighteen highest paid corporate officers in the nation because of bonuses stemming from the firm's incentive plan. See "How Top Salaries Weathered the Recession," *Business Week* (June 13, 1959), pp. 46–66; and "End of an Era." *Forbes* (November 15, 1957), pp. 110, 113. Bethlehem's executive compensation system was described as "disgusting" by Labor Department officials who reviewed the plan in 1959; see WCW to the Under Secretary, August 10, 1959, James P. Mitchell Papers, Box 92, 1959 Steel Strike (August 1–15) (1) Folder, DDE Library.

46. "Steel Capacity," *Iron Age* 177 (February 2, 1956), p. 47.

47. See Ernest T. Weir, "The Importance of the Steel Industry," in AISI, *Yearbook, 1954* (New York: AISI, 1954), pp. 84–85; Delaney, "Expansion Wheels are Turning," pp. 42–43; Arthur B. Homer, "The Doorway to Stabilized Progress," in AISI, *Yearbook, 1955* (New York: AISI, 1955), pp. 87–88; T. Cambell, "Steel: Hold Onto Your Hat in '56," *Iron Age* 177 (January 12, 1956), p. 35; "Earnings: They're the Tops," *Iron Age* 177 (February 9, 1956), p. 48; T. Campbell, "Steel: Expansion Money Is a Problem," *Iron Age* 177 (March 22, 1956), pp. 38–39; and T. Dimond, "This New Round of Steel Expansion," *Harvard Business Review* 34 (May–June 1956), pp. 85–93.

48. Arthur B. Homer, "Meeting the Nation's Need for Steel." Address delivered to the Investment Bankers Association, Hollywood, Florida, November 27, 1956, pp. 5–6. A copy is in SML, Homer File, Bethlehem Steel Corporation.

49. Again, however, it must be noted that this view was not accepted by all observers. See Means, *op. cit.*, pp. 153–155; and Vatter, *op. cit.*, pp. 153–158.

50. As previously noted, the Defense Department was pressuring the industry to obtain accelerated depreciation benefits for new investment; for evidence, see the citations listed in n. 28. Believing that the certificate-of-necessity program might be reinstated, producers applied for nearly $500 million of such credits in the first half of 1955 alone; the program, however, was never revived. See Campbell, "Fast Tax Writeoff Is Out,", p. 75; "Steel: 150 Million Tons by 1960?" *Iron Age* 176 (September 15, 1955), p. 94; Delaney, "Steel Forges Ahead," pp. 31–33; Simmons, "Government Planners Likely to Ask New Tax Aid . . ., " p. 1; and N. R. Regeimbal, "Tax-Am: Closed to Steel, Aluminum," *Iron Age* 176 (September 29, 1955), pp. 22–23.

51. These figures were consistently quoted by all sources during the 1950s. See, for example, "Steel, Story of a Shortage," p. 16; and "Steel: What Will Price Probe Show," *Iron Age* 177 (May 17, 1956), p. 57. Also see Manners, *op. cit.*, pp. 24–25.

52. Bethlehem Steel Corporation, "A Bethlehem History, with Reference to the Reasonableness of Executive Compensation" (Bethlehem, PA: author, n.d.), p. 34. A copy is in SML. Also see Burck, "Private Strategy of Bethlehem Steel," p. 112.

53. "Bethlehem History," p. 35.

54. G. Burck, "The Transformation of U.S. Steel," *Fortune* 53 (January 1956), pp. 88–95, 198ff, and G. J. McManus, "Why 'Big Steel' Is Successful," *Iron Age* 182 (August 14, 1958), pp. 40–41. In a ranking of the managerial skills of the firms in the industry, one business journal placed U.S. Steel at the top, see "Steel," *Forbes* (January 1, 1955), pp. 21–26.

55. "Steel, Story of a Shortage," *op. cit.*, pp. 15–16.

56. Roger M. Blough, "My Side of the Steel Price Story," *Look* 27 (January 29, 1963), p. 23.

57. See Gordon, *op. cit.*, pp. 96–98, and Dwight D. Eisenhower, *Mandate for Change, 1953–1956* (Garden City, NY: Doubleday, 1963),pp. 124–127.

58. Gordon, *op. cit.*, pp. 95–134; Vatter *op. cit.*, pp. 14–16; Herbert Stein, *The Fiscal Revolution in America* (Chicago: Universtiy of Chicago Press, 1969), pp. 281–371; Alvin H. Hansen, *The American Economy* (New York: McGraw-Hill, 1957), p. 45; and G. L. Bach, *Inflation: A Study in Economics, Ethics, and Politics* (Providence, RI: Brown University Press, 1958), pp. 38–45.

59. *Public Papers of the Presidents of the United States, Dwight D. Eisenhower, 1956* (Washington, DC: GPO, 1958), p. 663. Also see Eisenhower, *Mandate for Change*, p. 489.

60. Gordon, *op. cit.*, pp. 102–106. Also see Hugh S. Norton, *The Employment Act and the Council of Economic Advisers, 1946–1976* (Columbia: University of South Carolina Press, 1977), pp. 131–164; and Kim McQuaid, *Big Business and Presidential Power, from FDR to Reagan* (New York: William Morrow, 1982), pp. 185–198.

61. Gordon, *op. cit.*, pp. 106–109; G. L. Bach, *Making Monetary and Fiscal Policy*

(Washington, DC: Brookings, 1971); and F. M. Scherer, *Industrial Market Structure and Economic Performance,* 2nd ed. (Chicago: Rand McNally, 1980), pp. 349–362.

62. U.S. Congress, Senate, *Administered Prices, op. cit.,* These hearings would continue through 1961, covering twenty seven published parts before completion.

63. See n. 32 for 1953 citations. Also J. B. Delaney, "Steel: Price Hike Averages $3.24 a Ton," *Iron Age* 174 (July 8, 1954), pp. 47–48; "Competition Cushions Steel Price Hikes," *Iron Age* 174 (July 15, 1954), p. 59; J. B. Delaney, "What Steel Price Hike Means to You," *Iron Age* 176 (July 14, 1955), pp. 59–60; "Prices: Watch Out for Inflation," *Iron Age* 176 (July 21, 1955), p. 51; J. B. Delaney, "How New Extras Affect Your Steel Bill," *Iron Age* 177 (February 9, 1956), pp. 47–48; J. B. Delaney, "Steel: Prices Are a Touchy Subject," *Iron Age* 177 (May 3, 1956), p. 53.

64. W. Gutman, "Steel Stocks: They Look Good," *Iron Age* 176 (December 1, 1955), p. 64. Also see J. B. Delaney, "Steel: Strike Cut Profits 22 Pct," *Iron Age* 171 (April 9, 1953), pp. 80–81; J. B. Delaney, "Steel: Profits Rise 38 Pct in '53," *Iron Age* 173 (April 8, 1954), p. 102; W. G. Brookfield, "Steel: Modernization Eased '54 Pains," *Iron Age* 175 (April 7, 1955), p. 92; G. J. McManus, "Steel: 1955 Was a Profitable Year," *Iron Age* 177 (April 5, 1956), p. 70; and "1956 Was One of Steel's Best," *Iron Age* 179 (March 28, 1957), p. 78.

65. "Memorandum of the Conversation Between the President and T. S. Repplier," August 3, 1955, Ann Whitman Diary, Box 6, August 1955 Folder, DDE Library. Quotation is on p. 3.

66. G. H. Baker, "Ask Steel Price Probe," *Iron Age* 176 (November 17, 1955), p. 93.

67. U.S. Department of Labor, *Collective Bargaining,* pp. 284–287.

68. *Ibid.,* pp. 286–287.

69. Eisenhower diary entry for April 8, 1954, p. 1, DDE Diary Series, Box 4, January–November 1954 Folder, DDE Library. McDonald was an occasional visitor to the White House; on these occasions he and the president exchanged views on a number of subjects. See Eisenhower, *Mandate for Change,* pp. 490–491. The president even stated that McDonald was someone he "liked personally," see Dwight D. Eisenhower, *Waging Peace, 1956–1961* (Garden City, NY: Doubleday, 1965), p. 306.

70. J. B. Delaney, "What's Behind the Steel Agreement?," *Iron Age* 174 (July 22, 1954), p. 67.

71. Hogan, *op. cit.,* pp. 4:1616–1641.

72. There were numerous hints of rank-and-file suspicions in the industry's trade press. See J. B. Delaney, "What's Behind the Steel Agreement?" *op. cit.,* p. 67; J. B. Delaney, "Steel: Expect Package Deal, Price Hike," *Iron Age* 173 (June 17, 1954), pp. 97–98; Delaney, "Price Hike Averages $3.24 a Ton," pp. 47–48. In addition, see the references to this subject in: Frank Cormier and W. J. Eaton, *Reuther* (Englewood Cliffs, NJ: Prentice-Hall, 1970), pp. 300, 317–319, 323–324; Jean Gould and Lorena Hickok, *Walter Reuther* (New York: Dodd, Mead, 1972), pp. 314–315; John Herling, *Right to Challenge, People and Power in the Steelworkers Union* (New York: Harper & Row, 1972), pp. 310–311; and Victor G. Reuther, *The Brothers Reuther and the Story of UAW* (Boston: Houghton Mifflin, 1976), pp. 362–363.

73. Delaney, "Price Hike Averages $3.24 a Ton," p. 48.

74. *Ibid.,* p. 47.

75. "Steel: Ben Fairless Turns a Corner," *Iron Age* 175 (May 12, 1955), p. 60. Also see obituary, "Roger M. Blough, Former Chief of U.S. Steel," *NYT* (October 10, 1985), p. D–21.

76. U.S. Department of Labor, *Collective Bargaining,* pp. 291–294.

77. Eisenhower, *Mandate for Change,* pp. 489–490.

78. See Minutes of the Cabinet Meeting, April 3, 1953, p. 3, Cabinet Series, Box 2,

April 3, 1953 Folder, DDE Library. This provides an early indication of the administration's position on labor–management problems. On the Burns incident, see Minutes of the Cabinet Meeting, June 24, 1954, pp. 1–2, Cabinet Series, Box 3, June 24, 1954 Folder, DDE Library. Burns, of course, was wrong in his prediction of an imminent strike; indeed, negotiations that year were relatively harmonious.

79. See T. Campbell, "Labor: Will Steel Be Struck in '56?" *Iron Age* 176 (September 22, 1955), p. 67. As early as January 1956 the president was expressing concern with the upcoming steel negotiations; see his diary entry for January 16, 1956, p. 2, DDE Diary Series, Box 12, January 1956 Diary Folder, DDE Library.

80. Vatter, *op. cit.*, pp. 98–120.

81. United States Steel Corporation, *Fifty-fourth Annual Report* [for 1955] (New York: author, 1956), p. 27.

82. "Union Says U.S. Steel Gained in Wage Rise," *NYT* (March 23, 1956), p. 43. Also see "McDonald: David and Goliath," *Iron Age* 177 (March 29, 1956), p. 25; T. Cambell, "Behind U.S. Steel's Labor Policy," *Iron Age* 177 (May 10, 1956), pp. 51–54. The White House meeting between the president, McDonald, and secretaries Humphrey and Mitchell (Treasury and Labor, respectively) was suggested by Humphrey as a mild jawboning session; see DDE Diary Series, Box 13, Phone Calls–March 1956 Folder, March 14, 1956 entry, DDE Library. For the president's summary of the meeting, see entry for March 23, 1956 in DDE Diary Series, Box 9, Diary 1955–1956 Folder, DDE Library.

83. See U.S. Department of Labor, *Collective Bargaining,* pp. 294–300. Also see J. B. Delaney, "Steel: Union Price Tag is High," *Iron Age* 177 (May 24,1956), p. 101; T. Campbell, "Negotiations: Will Joint Talks Work Out?" *Iron Age* 177 (June 7, 1956), pp. 71–72; T. Campbell, "Steel Labor: Storm Signals Are Up," *Iron Age* 177 (June 21, 1956), pp. 59–60; and Hill & Knowlton, "Background Information, the Steel–Labor Negotiations of 1956" (New York: author, July 1956 [mimeo]), a copy is in Arthur Burns Papers, Box 19, Labor–Steel Strike of 1956 Folder, DDE Library. For the industry's counter proposals to the union, see Arthur Goldberg to James Mitchell, June 18, 1956, James P. Mitchell Papers, Box 91, 1956 Steel Strike (1) Folder, DDE Library.

84. See Humphrey to the president, July 3, 1956, Administrative Series, Box 23, Humphrey–1956 (1) Folder, DDE Library; and "Pre-Strike [Log]," n.d., Mitchell Papers, Box 91, 1956 Steel Strike (2) Folder, DDE Library. These sources provide clear evidence of the administration and the president's close involvement in the steel labor negotiations. Also see U.S. Department of Labor, *Collective Bargaining,* for summary information regarding the White House position on this matter.

85. See Arthur Burns to Eisenhower, July 10, 1956, Administrative Series, Box 10, Burns–1956/1957 (3) Folder, DDE Library, for information regarding the consideration of a Taft-Hartley injunction.

86. See U.S. Department of Labor, *Collective Bargaining,* pp. 298–299; T. Campbell, "Steel: Strike Heads for Washington," *Iron Age* 178 (July 26, 1956) p. 21; "How Steel Settlement Came—And Where It Leads," *Business Week* (July 28, 1956), pp. 26–27; and T. Campbell, "Steel: Behind Strike Settlement," *Iron Age* 178 (August 2, 1956), p. 46.

87. U.S. Department of Labor, *Collective Bargaining,* pp. 298–299; J.B. Delaney, "Post-Mortem: Steel Got What It Wanted," *Iron Age* 178 (August 2, 1956), p. 47.

88. "Draft Statement by Secretary of Labor James P. Mitchell," n.d., Mitchell Papers, Box 91, 1956 Steel Strike (2) Folder, DDE Library.

89. G. H. Baker, "Are More Steel Price Hikes Coming?," *Iron Age* 178 (September 13, 1956), p. 73. It was rumored that a "second installment" of the price hike would be delayed until after the election, which was—it was alleged—part of an industry–White House "deal" to end the strike; for data concerning subsequent price hikes, see

G. J. McManus, "Costs Prod Mills Toward Steel Price Hike," *Iron Age* 178 (December 6, 1956), pp. 83–84; and "Steel Prices: On the 'Up' Escalator," *Iron Age* 179 (February 21, 1957), pp. 57–58. Another unsubstantiated rumor pointed to a deal between the union and management to sustain the strike in order to work off excess inventory then on the producers' shelves; see T. Campbell, "Was the Steel Strike a Phony?," *Iron Age* 178 (August 9, 1956), p. 7.

90. Foreign steel imports, of course, increased in 1956 owing to the strike, and they impressed some users with their generally high quality (which heretofore had been suspect). See "Aftermath: Worst Is Yet to Come," *Iron Age* 178 (August 9, 1956) pp. 35–37.

Chapter 8

1. See G. H. Baker, "Are More Steel Price Hikes Coming?" *Iron Age* 178 (September 13, 1956), p. 73; U.S. Department of Labor [E. Robert Livernash], *Collective Bargaining in the Basic Steel Industry—A Study of the Public Interest and the Role of Government* (Washington DC: author, 1961), p. 300. The administration did not appear to be unduly concerned with the increase as it had more important matters to contend with, such as the Suez crisis then in progress. In a press conference held one day after the steel-price announcement, the single question relating to steel received only a short and perfunctory response. See *Public Papers of the Presidents of the United States, Dwight D. Eisenhower, 1956* (Washington, DC: GPO, 1958), pp. 662–663.

2. See "When a Businessman-in-Politics Gets Back at the Company Helm," *Business Week* (May 31, 1958), pp. 36–39.

3. See "Steel: What Will Price Probe Show?," *Iron Age* 177 (May 17, 1956), pp. 56–57; G. H. Baker, "Fast Tax Aid for Steel in the Works," *Iron Age* 178 (July 19, 1956), p. 89; and G. H. Baker, "84th Congress: How Business Fared," *Iron Age* 178 (August 9, 1956), p. 41.

4. See N. R. Regeimbal, "Antitrust: A Bust for Politicians," *Iron Age* 175 (June 23, 1955), p. 56; and U.S. Congress, House Select Committee on Small Business, 84th Cong., 1st sess., Hearings, *Distribution Problems* (Washington, DC: GPO, 1955).

5. For the extent of the subcommittee's work in this area, see U.S. Congress, Senate Subcommittee on Antitrust and Monopoly, Committee on the Judiciary, 85th Cong., 1st sess., Hearings, *Administered Prices* (Washington, DC: GPO, 1957–61). The hearings would continue through the 87th Congress. For comment on the work of the subcommittee, see "Congress' Own Brain Trust," *Business Week* (July 20, 1957), pp. 97–106, and "GM: What's Behind Congress Probe?," *Iron Age* 176 (November 10, 1955), p. 70. Concentration in the auto industry was another subject investigated by the congressmen.

6. These as well as several other industries were probed by the Kefauver Committee during the 1950s. For the senator's recollections of these events, see Estes Kefauver, *In a Few Hands, Monopoly Power in America* (Baltimore: Penguin Books, 1965).

7. See Charles L. Fontenay, *Estes Kefauver—A Biography* (Knoxville: University of Tennessee Press, 1980), pp. 355–393. For an interesting critique of Kefauver's use of his public powers to further his personal political ambitions, see William H. Moore, *The Kefauver Committee and the Politics of Crime, 1950–1952* (Columbia: University of Missouri Press, 1974).

8. See G. Burck, "The Private Strategy of Bethlehem Steel," *Fortune* 65 (April 1962), pp. 105–112, 242ff.

9. See the testimony of Arthur B. Homer, president of Bethlehem Steel, and George McCuskey, vice president of Youngstown Sheet & Tube, in U.S. Congress,

Senate Subcommittee on Antitrust and Monopoly, Committee on the Judiciary, 84th Cong., 1st sess., Hearings, *The Proposed Bethlehem–Youngstown Merger* (Washington, DC: GPO, 1955).

10. *Ibid.* Also see William T. Hogan, *Economic History of the Iron and Steel Industry in the United States,* 5 vols. (Lexington, MA: Lexington Books, 1971), 4:1689–1692.

11. See Stanley Barnes to Bernard Shanley, August 16, 1954, and Bernard Shanley to Governor Adams, August 3, 1954, in Official File, Box 679, Folder OF 134–E–3, DDE Library. Also see Theodore P. Kovaleff, *Business and Government During the Eisenhower Administration, a Study of the Antitrust Policy of the Antitrust Division of the Justice Department* (Athens: Ohio University Press, 1980), pp. 79–80.

12. See the sources cited in n. 10. Also see Regeimbal, "Antitrust: A Bust . . . , " p. 12; and "Bethlehem–Youngstown: Controversial Engagement," *Fortune* 55 (June 1957), p. 145.

13. *United States v. Bethlehem Steel Corp.* 168 F. Supp. 576 (1958). Also See Kovaleff, *op cit.,* p. 81; and "B–Y Merger Banned," *Iron Age* 182 (November 27, 1958), p. 25.

14. Kenneth Warren, *The American Steel Industry, 1850–1970, a Geographical Interpretation* (London: Oxford University Press, 1973), pp. 314–316.

15. G. J. McManus, "Steel Takes Its Case to Congress," *Iron Age* 180 (August 1, 1957), p. 57. Also see "Who'll Probe Steelmakers First?" *Iron Age* 179 (April 25, 1957), p. 64; N. R. Regeimbal, "Big Business Probes Begin," *Iron Age* 179 (May 30, 1957), pp. 55, 152; and G. H. Baker, "Monopoly Report May Fizzle Out," *Iron Age* 180 (July 25, 1957), p. 89.

16. J. B. Delaney, "Why Steel Prices Had to Go Up," *Iron Age* 180 (July 4, 1957), pp. 38–39.

17. In a news conference held shortly after the price hike, the president expressed his belief that the action "may not have as much effect as we fear," see *Public Papers of the Presidents of the United States, Dwight D. Eisenhower, 1957* (Washington, DC: GPO, 1959), pp. 558. For more comment on the administration's position, see H. Scott Gordon, "The Eisenhower Administration: The Doctrine of Shared Responsibility," in Craufurd D. Goodwin (ed.), *Exhortation and Controls—The Search for a Wage-Price Policy 1945–1971* (Washington, DC: Brookings, 1975), pp. 129–130. On the demands for even higher prices, see Delaney, "Why Steel Prices Had to Go Up," pp. 38–39; T. Campbell, "Weir: The Lone Wolf Does It Again," *Iron Age* 177 (March 29, 1956), pp. 24–25; and "The Steel Price Rise: It Depends How You Look at It," *Business Week* (July 6, 1957), pp. 30–31

18. "Industry Assays Steel Price Rise," *NYT* (July 1, 1957), p. 31.

19. U.S. Congress, Senate, Hearings, *Administered Prices,* pt. 2, "Steel," pp. 195–416. It is perhaps worth noting that nowhere in this long record of questioning and testimony is there any mention of the potential effects of the international marketplace—and growing foreign competition—on the domestic steel industry's future.

20. *Ibid.,* p. 310.

21. *Ibid.,* p. 428.

22. U.S. Congress, Senate, Committee on the Judiciary, Subcommittee on Antitrust and Monopoly, 85th Cong., 1st sess., Report, *Administered Prices, Steel* (Washington, DC: GPO, 1958). Also see J. B. Delaney,"Steel Industry Wins First Round," *Iron Age* 180 (August 25, 1957), pp. 54–56; "How U.S. Steel Fights with Fact," *Iron Age* 180 (August 22, 1957), pp. 64–65; G. H. Baker, "Kefauver's Price Probe Fades," *Iron Age* 180 (September 19, 1957), p. 121; "Bethlehem Next in Price Probe," *Iron Age* 180 (October 10, 1957), p. 75; and "Price Probe on Again," *Iron Age* 180 (October 31, 1957), p. 25. For industry commentary on the subcommittee's steel report, see "Price Probers Hit Steel Rise," *Iron Age* 181 (March 6, 1958), p. 77.

23. AISI, *Annual Statistical Report, 1960* (New York: author, 1961), p. 9. See

McDonald to the president, September 9, 1958, for evidence of layoffs and unemployment in the industry; Name Series, Box 23, David McDonald Folder, DDE Library.

24. G. J. McManus, "Steel's '57 Earnings Misleading," *Iron Age* 181 (March 27, 1958), p. 72; and G. J. McManus, "Steel: 1958 Had Its Good Points," *Iron Age* 183 (March 26, 1959), p. 102.

25. See "Labor: McDonald Seems Victorious; His Margin Less Than 2 to 1," *Iron Age* 179 (February 21, 1957), p. 59; Richard W. Nagle, "Collective Bargaining in Basic Steel and the Federal Government, 1945–1960" (Ph.D diss., Pennsylvania State University, 1978), pp. 221–222; "The Decline of Dave McDonald," *Fortune* 58 (August 1958), pp. 169–170; John Herling, *Right to Challenge, People and Power in the Steelworkers Union* (New York: Harper & Row, 1972); and "Will Steel Prices Rise in '58?" *Iron Age* 181 (April 17, 1958), p. 65.

26. See Gordon, "Doctrine of Shared Responsibility", pp. 119–121; and Gardiner C. Means, *Pricing Power and the Public Interest, a Study Based on Steel* (New York: Harper & Row, 1962), pp. 3–8. The momentum of the subcommittee hearings gave Kefauver a showcase from which to berate the industry's behavior; steelmakers, by contrast, had no such dramatic forums through which they might respond to the senator. They did, however, attempt to better explain their position through various public relations channels, see, for example, United States Steel Corporation, *Steel and Inflation, Fact vs. Fiction* (New York: author, 1958), which was a slickly produced compendium of text, charts, graphs, and tables regarding that firm's financial situation and price decisions. Thousands of copies were distributed throughout the country by U.S. Steel. The AISI was also active during this period in promoting the industry cause to the public; see, for example, J. B. Delaney, "Steel Wins Public Support," *Iron Age* 178 (July 26, 1957), pp. 19–20; "Industry Learns to Speak Up," *Iron Age* 179 (May 23, 1957), pp. 101–102; C. M. White, "Breaking Through the Communications Barrier," in AISI, *Yearbook, 1957* (New York: AISI, 1957), pp. 79–86; and "Public Relations Session," in AISI, *Yearbook, 1958* (New York: AISI, 1958), pp. 325–328, for examples of the institute's efforts in this area.

27. Senator Kefauver to the president, June 20, 1958, Official File, Box 560, Folder OF 114 (1958), DDE Library. This message was given serious consideration by the administration; also see the other letters and telegrams from Kefauver to the president regarding the steel-price issue in the same DDE Library file.

28. The president to Senator Kefauver, June 24, 1958, Official File, Box 560, Folder OF 114 (1958), DDE Library. The president had been asked at a press conference if he intended to intervene any more forcefully in the steel-price issue, given its potential effect on national inflation, by perhaps arranging a White House conference between top steel and labor leaders; Eisenhower replied that only an "emergency" would bring this type of conference about and that one did not presently exist in steel; see *Public Papers of the Presidents of the United States, Dwight D. Eisenhower, 1958* (Washington, DC: GPO, 1959), pp. 438–439. At a subsequent news conference on August 20, 1958, he again responded to a question concerning steel prices, noting "Some slight rise in steel costs is not of itself a very great factor in living costs," and therefore (he implied) the price hike was not worth that much worry, *ibid.*, p. 630. Also see Gordon, *op. cit.*, p. 120; and Robert Griffith, "Dwight D. Eisenhower and the Corporate Commonwealth," *American Historical Review* 87 (February 1982), pp. 87–122.

29. See T. Campbell, "U.S. Steel Delays Price Boost," *Iron Age* 181 (June 26, 1958), p. 58.

30. J. B. Delaney, "Armco Sets Off Steel Price Rise," *Iron Age* 182 (August 7, 1958), pp. 44–45.

31. U.S. Congress, Senate, Hearings, *Administered Prices*, pt. 8, "1958 Steel Price Increase." These hearings were held on August 5–6. Kefauver then prepared to hold

even further public inquiries into steel's pricing behavior; see G. H. Baker, "Capital Sleuths Still Trail Steel," *Iron Age* 182 (September 11, 1958), p. 147.

32. U.S. Congress, Senate, Hearings, *Administered Prices,* pt. 11, pp. 5344–5410 (Blough's testimony), and p. 5194 (McDonald's approval of the proposal). Also see T. Cambell, "Senate Hearings Deepen Gloom in Steel Labor Crisis," *Iron Age* 183 (April 30, 1959), pp. 71–72.

33. See Chapter 5, n. 25.

34. G. H. Baker, "What Congress Plans for Steel," *Iron Age* 183 (April 16, 1959), p. 97.

35. The idea of treating the industry as a public utility had been broached in the past. During congressional investigations of U.S. Steel in the first decade of its existence, U.S. Steel Chairman E. H. Gary suggested that the government set the prices for the industry, see Ida M. Tarbell, *The Life of Elbert H. Gary, the Story of Steel* (New York: D. Appleton, 1925), pp. 231–232; and U.S. Congress, House Special Committee on Investigation of the United States Steel Corporation, 62nd Cong., 1st sess., Hearings, *United States Steel Corporation* (Washington, DC: GPO, 1911), p. 79. By the 1950s, however, there was obviously no company support whatsoever for any such public involvement in the firm's pricing policies and decision.

36. Gordon, *op. cit.,* pp. 123–124. Also see Eisenhower, *Waging Peace; 1956–1961* (Garden City, NY: Doubleday, 1965), pp. 320–321, fn. 7.

37. F. Belair, J., "President Warns on Steel Pay Rise That Lifts Prices," *NYT* (March 26, 1959), pp. 1, 12. Also see Nagle, *op. cit.,* pp. 223–225, and T. Campbell, "Behind Steel Labor Maneuvers," *Iron Age* 183 (April 16, 1959), pp. 82–83. The president was, however, generally concerned with labor legislation that would help prevent strikes in critical industries. As part of this concern, he held several meetings in late 1958 with Charles R. Hook, chairman of Armco Steel, to discuss what changes might be made in current law and how they could be successfully implemented. In 1959, of course, the Landrum-Griffin Act was passed, but this was unrelated to Hook's concerns or suggestions. See Charles R. Hook to Gerald D. Morgan, January 26, 1959, and related correspondence, Offical File, Box 639, 124 D-3 Folder, DDE Library.

38. For a useful review of these tensions, including a detailed bibliography of further sources that might be consulted, see Charles C. Alexander, *Holding the Line: The Eisenhower Era, 1952–1961* (Bloomington: Indiana University Press, 1975).

39. For Eisenhower's remarks about the labor dispute at his news conference of June 3, 1959, see *Public Papers of the Presidents of the United States, Dwight D. Eisenhower, 1959* (Washington DC: GPO, 1960) p. 429; and his defense of nonintervention, in Eisehower, *Waging Peace,* pp. 453–455. Also see T. Campbell, "Ike a Big Factor in Steel Talks," *Iron Age* 183 (May 14, 1959), pp. 108–109; T. Cambell, "Did Ike Help McDonald's Cause?" *Iron Age* 183 (May 21, 1959), pp. 104–105; and T. Campbell, "Why Steel Wage Talks Collapsed," *Iron Age* 184 (July 16, 1959), p. 57.

40. See Anthony Libertella, "The Steel Strike of 1959: Labor, Management, and Government Relations" (Ph.D diss., Ohio State University, 1972); Nagle, *op. cit.,* pp. 219–292; U.S. Department of Labor, *Collective Bargaining,* pp. 300–397; and G. J. McManus, "The Inside Story of Steel Wages and Prices 1959–1967," *Iron Age* 200 (October 19, 1967), pp. 83–87.

41. See Nagle, *op. cit.,* p. 226.

42. See G. J. McManus, "Call for Sanity in Steel Buying," *Iron Age* 183 (April 30, 1959), p. 73; and "Steel Buyers Get Strike Fever; Hedge Buying Begins," *Iron Age* 183 (January 29, 1959), pp. 47–49. The strike had long been predicted in the trade press, see "Steel Strike Likely Next Year," *Iron Age* 181 (January 16, 1958), p. 48.

43. "Record Steel Profits Intensify Strike Deadlock," *Iron Age* 184 (August 6, 1959), pp. 41–42.

44. G. J. McManus, "Steel: 1959 Was Puzzling Year," *Iron Age* 185 (April 7, 1960), p. 56.

45. Nagle, *op. cit.*, pp. 227–228.

46. *Ibid*, pp. 220–221. Stephens, it might be noted, joined Clarence B. Randall in the White House as a consultant to the CFEP.

47. See J. B. Delaney,"Steel Labor and Management Court the Public," *Iron Age* 183 (February 19, 1959), pp. 81–83; "Steel's Advertising Attacks Inflation," *Editor & Publisher* 92 (June 6, 1959), p. 24; "Adversaries in Steel Fight It Out in Advertisements," *Business Week* (June 6, 1959), pp. 25–26; Libertella, *op cit.*, pp. 78–103; National Council of Churches, Department of the Church and Economic Life, *In Search of Maturity in Industrial Relations: Some Long-Range Ethical Implications of the 1959–1960 Dispute in the Steel Industry* (New York: author, 1960), pp. 18–20; George F. Hamel, "John W. Hill, Public Relations Pioneer" (Master's thesis, University of Wisconsin, Madison, 1966), pp. 35–45; and John W. Hill, *The Making of a Public Relations Man* (New York: David McKay, 1963), *passim.* Hill was chairman of Hill & Knowlton, the public relations consultant to the AISI.

48. See McDonald to the president, June 25, 1959, and the president to McDonald, June 27, 1959, in James P. Mitchell Papers, Box 92, 1959 Steel Strike (June–July) (2) Folder, DDE Library.

49. It might be noted that this postponement was not, perhaps, because of altruistic motivations on the part of McDonald. According to Mr. Benjamin Fischer, at that time an assistant to the USWA president, the union agreed to a delay because of the pending July Fourth holiday that year. If the workers walked out on July 1 as originally planned, then the membership would miss out on this company-paid benefit. (Author's interview with Benjamin Fisher, Carnegie-Mellon University, August 17, 1984).

50. Nagle, *op. cit.*, p. 237.

51. *Ibid.*, pp. 231–233.

52. "Steel Hints Shift on Wage Freeze," *NYT* (June 11, 1959), p. 17. Also see David McDonald, *Union Man* (New York: E. P. Dutton, 1969), p. 267. McDonald did, however, offer to make some compromises on his position, for details, see Nagle, *op. cit.*, p. 237.

53. Nagle, *ibid.*, p. 268.

54. For evidence of Eisenhower's close monitoring of the situation, see his Telephone Log in DDE Diary Series, Box 43, Telephone Calls—July 1959 Folder, DDE Library, under the July 28, 1959 entry. In this source, he asks for a daily reporting on the progress of the negotiations. In addition, see the extensive reports he subsequently received, in Mitchell Papers, Box 92–93, DDE Library. For evidence that steel industry leaders did not wish for presidential intervention, see "Mitchell Confers with Nixon on Steel Strike," August 8, 1959, Mitchell Papers, Box 92, 1959 Steel Strike (August 1–15) (2) Folder, DDE Library.

55. Eisenhower to Humphrey, August 11, 1959, DDE Diary Series, Box 44, DDE Dictation (August 1959) Folder, DDE Library. Humphrey was one of the president's closest personal friends. In early July 1959, just prior to the strike's commencement, Eisenhower and his wife were house guests of the Humphreys at the latter's Georgia farm for several days. Eisenhower and Humphrey were worried about how this social visit might appear given the strike's imminence and the fact that Humphrey (the former secretary of the treasury) was now the chairman of National Steel Corporation. See Humphrey to the president, June 19, 1959, and July 6, 1959, Administrative Series, Box 23, Humphrey 1959 Folder, DDE Library.

56. Nagle, *op. cit.*, pp. 248–249.

57. For details of the White House meetings, see "Memorandum of Conversa-

tions," September 30, 1959, Official File, Box 636, 124 D Steel Strike (4) Folder, DDE Library; and "Industry Meeting," September 30, 1959, DDE Diary Series, Box 44, Staff Notes–September 1959 (1) Folder, DDE Library. For the president's subsequent official statement, see *Public Papers, Eisenhower, 1959,* p. 705. On consideration of a Taft-Hartley injunction, see Raymond Saulnier to General Persons, September 28, 1959, Official File, Box 636, 124 D Steel Strike (2) Folder, DDE Library. Also see T. Campbell, "Firm Steel Stand May Force T-H," *Iron Age* 184 (September 17, 1959), p. 65; Nagle, *op. cit.,* pp. 253–255; and Eisenhower, *Waging Peace,* p. 456.

58. Board of Inquiry, Report to the President, "The 1959 Labor Dispute in the Steel Industry" (Washington DC: author, October 19, 1959 [mimeo]), p. 37, a copy is in Administrative Series, Box 39, Steel Strike 1959 Folder, DDE Library.

59. See *Public Papers, Eisenhower, 1959,* pp. 730–731.

60. See DDE Diary Series, Box 46, Telephone Calls—November 1959 Folder, November 5, 1959 entry (for call to Mitchell), DDE Library. See Ann Whitman Diary, November 5, 1959 entry, Box 11, Diary November 1959 Folder, DDE Library, for comment on McDonald as an "actor." See "Memorandum for the Record," November 9, 1959, DDE Diary Series, Box 45, Staff Notes—November 1959 (3) Folder, DDE Library; and Minutes of the Cabinet Meeting, November 11, 1959, pp. 3–4, DDE Diary Series, Box 45, Staff Notes—November 1959 (3) Folder, DDE Library.

61. Nagle, *op. cit.,* pp. 262–268. Also see "Steel Users Were Well Prepared but Strike Starts to Hurt," *Iron Age* 184 (August 20, 1959), pp. 91–92.

62. Libertella, *op. cit.,* pp. 228–231, and Nagle, *op. cit.,* pp. 273–275.

63. Nagle, *ibid.,* p. 233. On the resistance of the USWA rank-and-file members to the initial terms of McDonald's demands, see p. 226.

64. On events surrounding the final settlement, see Eisenhower, *Waging Peace,* pp. 457–459; DDE Diary Series, Box 47, January 1960 Telephone Calls Folder, entry for January 2, 1960, DDE Library; and Nagle, *op. cit.,* pp. 277–278. Also see Board of Inquiry, Final Report to the President, "The 1959 Labor Dispute in the Steel Industry" (Washington, DC: author, January 6, 1960 [typescript]), a copy is in Official File, Box 639, 124 D-1 Steel (3) Folder, DDE Library. On the terms of the contract, see J. A. Loftus, "Steel Settlement Is Reached; Union Victor; Price Rise Seen; Nixon Mediator, Gains Stature," *NYT* (January 5, 1960), pp. 1, 18; T. E. Mullaney, "Steel Prices to Hold Now but May Rise in 3 Months," *NYT* (January 5, 1960), pp. 1, 18; A. H. Raskin, "The Rout of Big Steel," *NYT* (January 6, 1960), p. 43; J. A. Loftus, "Steel Contract Is Signed; Both Sides Pledge Amity," *NYT* (January 6, 1960), pp. 1, 43; and T. Campbell, "Steel Never Had a Real Chance," *Iron Age* 185 (January 14, 1960), pp. 26–27. For an analysis of the joint steel labor-management group assigned to study work rules, see John A. Orr, "The Rise and Fall of Steel's Human Relations Committee," *Labor History* 14 (Winter 1973), pp. 69–82.

65. *Public Papers of Presidents of the United States, Eisenhower, 1960–1961* (Washington, DC: GPO, 1961), pp. 24–25 covers the president's news conference of January 13, 1961, the first held since the strike's settlement. Also see Eisenhower's comments at the Cabinet Meeting of January 6, 1960, in Cabinet Series, Box 15, Cabinet Meeting of January 6, 1960, folder, pp. 2–4, DDE Library; and the president's comments before Republican legislative leaders on January 12, 1960, p. 4, Legislative Meeting Series, Box 3, Legislative Leaders–1960 (1) Folder, DDE Library.

66. Roger M. Blough, "Aftermath in Steel," address delivered to the National Canners Association, Miami Beach, Florida, January 18, 1960, p. 4, a copy is in Official File, Box 636, 124 D Steel Strike (4) Folder, DDE Library. Many seemed to agree that the federal government played a decisive role in the strike's settlement and thus deserved some of the blame for the highly inflationary nature of the contract terms. Over 95% of the thousands of letters received at the White House condemned the work of Nixon and Mitchell. See Don Paarlberg to General Persons, January 18, 1960, Offi-

cial File, Box 636, 124 D Steel Strike (4) Folder, DDE Library; and "Steel Dispute: Public Opinion Survey," January 1960, General File, Box 960, 126 D Folder, DDE Library.

67. U.S. Department of Labor, *Collective Bargaining,* p. 307, is at pains to make this point (perhaps because of Secretary of Labor Mitchell's central role in the resolution of the strike). Also see Nagle, *op cit.,* pp. 279–280.

68. See Eisenhower's news conference of January 13, 1960, in *Public Papers, Eisenhower, 1960–1961,* pp. 24–25. Also see Blough, "Aftermath in Steel," pp. 9–13; and "Memorandum, the SCCC to James P. Mitchell," January 6, 1960, Mitchell Papers, Box 193, 1960 Steel (January 1–13) (2) Folder, DDE Library. Regarding the rumored "deal" that brought the strike to an end, see Gordon, *op. cit.,* p. 132. Another rumor claimed that no price hikes would be made until after the November elections in order to aid the Nixon candidacy, see McManus, "The Inside Story . . . , " p. 87.

69. See "Price Boosts Not Yet in Sight," *Iron Age* 185 (March 24, 1960), p. 127; and McManus, "The Inside Story . . . , " p. 87.

70. G. H. Baker, "Congress May Query Steel Men," *Iron Age* 185 (January 21, 1960), p. 63.

Chapter 9

1. M. A. Adelman, "Steel, Administered Prices, and Inflation," *Quarterly Journal of Economics* 75 (February 1961), pp. 34–35. Also see Walter Adams and Joel B. Dirlam, "Steel Imports and Vertical Oligopoly Power," *American Economic Review* 54 (September 1964), pp. 626–655.

2. See, for example, Roger Blough, "A Talk of Two Towns." Address delivered to the Annual Meeting of the Cleveland Chamber of Commerce, April 17, 1958, a copy is in CEA Records, Box 22, Steel Folder, DDE Library. Also see R. L. Gray, "Bid for Survival," in AISI, *Yearbook, 1958* (New York: AISI, 1958), pp. 69–79; B. F. Fairless, "Facing Up to Competition," in AISI, *Yearbook, 1959* (New York: AISI, 1959), pp. 143–150; T. Campbell, "Big Steel Lays It on the Line," *Iron Age* 184 (August 6, 1959), p. 43; and "Steel Companies are United Against Union Demands," *Iron Age* 183 (June 4, 1959), p. 90.

3. USWA, "The Foreign Competition Hoax" (Washington, DC: author, 1960), a pamphlet issued by the union shortly after the end of the strike. Also see "McDonald Sounds Off on Issues," *Iron Age* 183 (March 26, 1959), p. 96.

4. As one observer commented, "World War II finally ended" for the American steelmakers. See D. A. Wells and S. Sisik, "Steel Imports: The Importance of Availability and Costs," *MSU Business Topics* 18 (Spring 1970), pp. 57–64. Also see "How U.S. Fares in World Markets," *Iron Age* 183 (February 19, 1959). p. 95.

5. See Chapter 6 for details.

6. See Burton I. Kaufman, *Trade & Aid, Eisenhower's Foreign Economic Policy, 1953–1961* (Baltimore: Johns Hopkins University Press, 1982), pp. 95–112.

7. Fairless was chosen to head the commission after international banker John McCloy turned down the assignment. A primary justification for the commission as far as the White House was concerned was to obtain a document it could use to influence Congress when consideration of the administration's 1957 foreign aid appropriation bill came up for debate. For details on the formation and operation of the commission, see CFEP File, Papers Series—Randall Series, Box 5, Fairless Commission Folder, DDE Library.

8. President's Citizen Advisers on the Mutual Security Program, *Report to the President* (Washington, DC: GPO, March 1, 1957), p. 7. For a critique of the report, see "Changes Recommended in the Mutual Security Program by the Fairless Report,"

March 5, 1957, authored by staff of the CFEP, CFEP File, Papers Series, Box 2, Fairless Report (1) and (2) Folders, DDE Library. Also see Kaufman, *op. cit.*, pp. 101–102.

9. The agencies were the Export-Import Bank, the World Bank, the International Development Bank, the International Finance Corporation, and AID. For annual compilations of assistance levels, see U.S. Congress, Senate Committee on Finance, 90th Cong., 1st sess., (Washington, DC: GPO, 1967), p. 32. Committee Print, *Steel Imports.*

10. *Ibid.* Data computed from, and provided in Tables C–1 and C–2, pp. 299–304.

11. Kiyoshi Kawahito, *The Japanese Steel Industry* (New York: Praeger, 1972), pp. 41–42.

12. For evidence of these missions, see the citations shown in various editions of the AISI *Yearbook* (New York: AISI), under the heading "Activities of the Institute."

13. See Kenneth Warren, *World Steel, an Economic Geography* (New York: Crane, Russak & Co., 1975), pp. 55–57. There were also other factors in this decision, such as the availability of certain types of coal more suitable to Bessemer steelmaking.

14. In 1957 open-hearth capacity accounted for approximately 80% of steel production in Japan. See John E. Jordan, "Memorandum to the Files: RE: Report of Meeting at Eximbank with Representatives of the Japanese Steel Industry," February 20, 1957, CFEP File, Randall Series—Subject Subseries, Box 7, Japanese Steel (6) Folder, DDE Library. It was not until the second modernization program, completed in 1960, that Japanese producers turned their emphasis from rolling operations to the manufacture of raw steel. See Kawahito, *op. cit.*, pp. 35–36.

15. See William T. Hogan, *Economic History of the Iron and Steel Industry in the United States,* 5 vols. (Lexington, MA: Lexington Books, 1971), pp. 4:1519–1521. Open-hearth accounted for approximately 90% of domestic steel production throughout the 1950s. Also see Edwin C. Barringer, *The Story of Scrap* (Washington, DC: Institute of Scrap Iron and Steel, 1954).

16. For an index of scrap prices from 1953 through 1958, see Gardiner C. Means, *Pricing Power and the Public Interest, a Study Based on Steel* (New York: Harper and Row, 1962), pp. 131–133. Also see "Scrap: National Export Barriers Rise," *Iron Age* 175 (March 17, 1955), p. 66; "Scrap: The Inside Story on Exports," *Iron Age* 176 (December 29, 1955), p. 25; and "Scrap: Prices are Acting Up Again," *Iron Age* 177 (April 16, 1956), p. 57.

17. See "A Possible Public Relations Program for the Scrap Conservation Committee of the Steel Industry—Suggestions for Consideration," September 16, 1955, Box 29, New York Office, General—1955 Folder, Hill Papers. For details on the industry's perceived problems with scrap, see "The Steel Scrap Situation—Statement of the Scrap Conservation Committee of the Steel Industry," September 12, 1955, Hill Papers (same folder as above). Regarding Clarence B. Randall's efforts to scuttle this proposed public relations program, see Randall to Dr. C. Edward Galbreath, September 21, 1955; and Randall to Gabriel Hauge, October 13, 1955, in Official Files, Box 679, Folder OF 134-E-3, DDE Library.

18. Humphrey to Weeks, March 27, 1956, Official File, Box 679, Folder OF 134-E-3, p. 1, DDE Library. In this same letter, Humphrey outlined the growth in Western Europe's commitment to open-hearth as opposed to the prevailing Bessemer process of steelmaking.

19. Joseph M. Dodge to Prescott Bush, April 24, 1956, CFEP File, Dodge Series—Subject Subseries, Box 5, Steel Scrap Folder, DDE Library. Dodge was chairman of the CFEP until the end of 1956.

20. "Excerpts from 'The World Ferrous Metallics Situation—1951–'60'" n.d., p. 2, attached to "Memorandum for Dr. Hauge" from Clarence B. Randall, January 16, 1957, Official File, Box 679, Folder OF 134-E-3, DDE Library. The study was prepared by the steel industry's scrap subcommittee (an AISI unit).

21. The Export Act of 1949 gave the Commerce Department authority to issue

licenses to firms exporting goods that were deemed critical to the national interest; thus an issue of a license effectively ends the control function, whereas the suspension of licenses effectively imposes controls. The department suspended licensing for heavy melting scrap, the grade most common to the making of steel. See "Export Curb Hits Scrap Prices," *Iron Age* 179 (March 28, 1957), p. 72. See Hogan, *op. cit.*, p. 1528, for a description of the various scrap grades and their uses. For a description of the congressionally mandated studies on the scrap situation, see "The Threat of Shortage of Ferrous Scrap Reappraised," September 9, 1957, Official File, Box 679, Folder OF 134-E-3, DDE Library.

22. The quota was not limited to Japan, however. It also would apply to the ECSC and the United Kingdom, the other major customers for U.S. scrap. For the record of correspondence and memoranda to and from the CFEP on this subject, see CFEP File, Paper Series, Box 3, CFEP 532 [export of ferrous scrap of steel] Folder; and Randall Series—Subject Subseries, Box 7, Japanese Steel (3) Folder, both in DDE Library.

23. Bud Lyon to H C. McClellan, March 20 [1957], CFEP File, Randall Series—Subject Subseries, Box 7, Japanese Steel (6) Folder, DDE Library.

24. "Report to Mr. H. B. McCoy, Administrator, BDSA, Department of Commerce," from H. W. Neblett, July 12, 1957, pp. 1, 2, 3, CFEP File, Randall Series—Subject Subseries, Box 7, Japanese Steel (3) Folder, DDE Library.

25. Ben H. Thobodeaux, Minister for Economic Affairs, to Clarence Randall, October 27, 1959, CFEP File, Randall Series—Subject Subseries, Box 7, Japanese Steel (1) Folder, DDE Library.

26. See Clarence Randall to Gardner Palmer, January 5, 1959, CFEP File, Randall Series—Subject Subseries, Box 7, Japanese Steel (1) Folder, DDE Library; Ben Thobodeaux to John Stephens, May 8, 1959, CFEP File, Staff Series, Box 8, John A. Stephens (2) Folder, DDE Library; and Gardner Palmer to Clarence Randall, November 28, 1959, CFEP File, Staff Series, Box 8, John A. Stephens (2) Folder, DDE Library.

27. See Walter Adams and Joel B. Dirlam, "Big Steel, Invention, and Innovation," *Quarterly Journal of Economics* 80 (May 1966), pp. 167–189. But also see Leonard H. Lynn, *How Japan Innovates, a Comparison with the U.S. in the Case of Oxygen Steelmaking* (Boulder, CO: Westview Press, 1982), pp. 43–118, who shows how MITI pushed Japanese firms into the BOF adoption decision based on its analysis of scrap needs as well as other factors; especially see Lynn's fn. 31 on p. 54.

28. "Memorandum of Statements by Mr. Frank Waring," July 18, 1957, CFEP File, Randall Series—Subject Subseries, Box 7, Japanese Steel (3) Folder, DDE Library. Waring was an economic consultant to the Tokyo Embassay of the United States.

29. H. W. Neblett to W. M. Braderman, July 5, 1957, "Comments on ECAFE Iron and Steel Committee at Bangkok, Thailand," CFEP File, Randall Series—Subject Subseries, Box 7, Japanese Steel (3) Folder, DDE Library.

30. "Report to Mr. H. B. McCoy, Administrator, BDSA, Department of Commerce," from H. W. Neblet, July 12, 1957, *op. cit.*, p. 4.

31. *Public Papers of the Presidents of the United States, Dwight D. Eisenhower, 1957* (Washington, DC: GPO, 1959), pp. 277–278.

32. Robert A. Pastor, *Congress and the Politics of U.S. Foreign Economic Policy, 1929–1976* (Berkeley: University of California Press, 1980), pp. 102–104. Also see "Tariffs: Metal Import Duties Cut Under GATT," *Iron Age* 177 (June 21, 1956), p. 83; and "Tariff Hearings—Metal Duty List Will be Simplified," *Iron Age* 184 (July 9, 1959), p. 61.

33. See "Imports Threaten More Products," *Iron Age* 182 (December 4, 1958), p. 93. Also see "Imports Threaten Wire Market," *Iron Age* 179 (April 25, 1957), pp. 65–66; "Wire Makers Fight Import Challenge," *Iron Age* 180 (October 3, 1957), p. 55; K. W. Bennett, "Imports Hit Oil Country Market," *Iron Age* 180 (November 21, 1957), p. 90; "Wire Products Make Sales Gains," *Iron Age* 181 (March 27, 1958), p. 164;

"What Buyer Patterns Tell About Market," *Iron Age* 181 (April 24, 1958), p. 192; and "Is Wire Import Threat Easing?" *Iron Age* 182 (August 21, 1958), p. 63. Also see U.S. Congress, Senate, *Steel Imports,* pp. 123–124.

34. R. R. Kay, "Steel Import Hurt Coast Mills," *Iron Age* 181 (February 13, 1958), p. 101. Also see R. R. Kay, "Why Imports Increase in Farwest," *Iron Age* 182 (December 18, 1959), p. 73; R. R. Kay, "Nineteen Pct More Steel in '59," *Iron Age* 183 (April 16, 1959), p. 99; and R. R. Kay, "Why Steel Imports Worry West," *Iron Age* 184 (August 13, 1959), p. 67.

35. See AISI, *Steel Imports—A National Concern* (Washington, DC: author, July, 1970), p. 65, Table 6. Also see K. W. Bennett, "Japanese Push Sales in Midwest," *Iron Age* 185 (March 17, 1960), p. 64.

36. R. W. Hardy and J. S. Revis, "New Trends in World Steel Brighten U.S. Outlook," *Iron Age* 186 (September 15, 1960), p. 163.

37. "Why U.S. Navy Buys Japanese Plate," *Iron Age* 183 (January 22, 1959), p. 27.

38. See, for example, R. H. Eshelman, "Nail Automation Challenges Foreign Competition," *Iron Age* 185 (March 31, 1960), pp. 149–151.

39. "Grim Forecast on Steel Imports," *Iron Age* 181 (March 20, 1958), p. 64. Also see U.S. Congress, House Committee on Ways and Means, 85th Cong., 2nd sess., Hearings, *Budget Appropriations* (Washington, DC: GPO, 1958).

40. See the concern expressed in AISI, Committee on Foreign Relations, *Foreign Trade Trends—American Iron and Steel Industry, 1953–1957* (New York: author, 1958), p. 5.

41. Gray, "Bid for Survival," p. 76.

42. "World Steel Market Is Tougher," *Iron Age* 182 (December 25, 1958), p. 22. The same sentiments were expressed by AISI Chairman Benjamin Fairless at the group's 1959 annual meeting, see Fairless, "Facing Up to Competition," p. 146.

43. "Memorandum—Foreign Competition in Steel," November 4, 1960, attached to correspondence from Roger Blough to Rev. Cameron P. Hall, November 4, 1960, in turn attached to Roy Blough to Cameron P. Hall, n.d., Papers of Roy Blough, Box 18, National Council of Churches—Steel Committee (1) Folder, HST Library. Roy Blough—no relation to Roger Blough—was a member of President Truman's CEA from 1950 to 1952, after which he served on the faculty of Columbia University; in 1959 he served as a consultant to the National Council of Churches' study of the steel strike.

44. The industry's capital expenditure data is in U.S. Congress, Senate, *Steel Imports,* p. 420, Table H-3. Also see: "Steel Capacity," *Iron Age* 181 (February 13, 1958), p. 90; G. J. McManus, "Steel's Capital spending Plans," *Iron Age* 181 (June 5, 1958), pp. 82–83; "Steel Capacity," *Iron Age* 183 (February 5, 1959), p. 90; G. J. McManus, "Are Steel Mills Set for Another Capital Spending Spree?" *Iron Age* 183 (March 5, 1959), pp. 37–39; G. J. McManus, "Steel Spending Due to Climb," *Iron Age* 185 (January 7, 1960), pp. 164–165; and "Steel Capacity," *Iron Age* 185 (January 28, 1960), p. 60.

45. See K. W. Bennett, "Exports: U.S. Faces Tougher Selling," *Iron Age* 173 (March 18, 1954), p. 75; K. W. Bennett, "Exports: The Competition is Rough," *Iron Age* 177 (March 15, 1956), pp. 50–51; "Can U.S. Hold World Markets?" *Iron Age* 183 (February 19, 1959), pp. 93–97; "How Serious Is Trade Problem?" *Iron Age* 184 (October 8, 1959), p. 81; "Exports: Some Good News," *Iron Age* 184 (December 31, 1959), p. 25; and K. W. Bennett, "Why U.S. Has Export Trouble," *Iron Age* 186 (July 7, 1960), p. 43.

46. See Kaufman, *Trade & Aid,* pp. 172–196.

47. G. H. Baker, "U.S. Moves to Increase Exports," *Iron Age* 185 (February 18, 1960), pp. 71–73; and "Steel Firms Study Export Problems," *Iron Age* 185 (April 28, 1960), p. 74.

48. This point was emphasized in late 1960 when the German government temporarily encouraged imports of American steel into its markets in order to assist inflation-control measures then in effect. Clarence B. Randall of the CFEP organized the domestic campaign to line up participation by American producers. He notes, "The big companies like U.S. Steel and Bethlehem do selling [abroad] intelligently, but that is not true of smaller companies." See "Excerpts from Jounral of 9/28/60," CFEP File, Randall Series—Subject Subseries, Box 11, Steel (3) Folder, DDE Library.

49. See Jack G. Kaikati, "The Anti-Export Policy of the U.S." *California Management Review* 23 (Spring 1981), pp. 5–19.

50. See "Deliveries Shorten for Heavy Steel," *Iron Age* 181 (January 30, 1958), p. 132; "Heavy Steel Market Continues Slow," *Iron Age* 181 (February 27, 1958), p. 142; and "Sheet Sales Display Signs of Life," *Iron Age* 181 (May 22, 1958), p. 182. These citations chronicle a decline in the export prices posted by U.S. Steel Export Association (U.S. Steel's export subsidiary), probably in response to cuts in export prices posted by European producers. Also see "New Trends in World Steel Brighten U.S. Outlook," *Iron Age* 186 (September 15, 1960), pp. 162–165 for information on export price trends during the 1950s. U.S. Congress, Senate, *Steel Imports,* pp. 119–127 also provides interesting information on this subject. Finally, note the comment:

> Although most steel producers agree that price changes have relatively little effect on total steel consumption in a given market, producers in the ECSC and Austria and those in Japan nevertheless seem to be prepared to try to expand their share of the export market by making price sacrifices in order to keep their plant in operation. This policy is in marked contrast to that followed in the United States, and, it would seem, in the United Kingdom, where the steel industries seem less disposed to offer heavy cuts in prices to overseas consumers. Producers in these countries rely to a lesser extent on exports and are, therefore, less readily influenced by conditions on the export market than producers in the ECSC, Austria, and Japan, who export a consistently higher proportion of their output. This difference in policy is particularly apparent in a period of recession.

The years 1958–62, of course, were recessionary for world steel markets, the quotation is from OEEC, *The Iron and Steel Industry in Europe, 1958–1959* (Brussels: author, 1960), pp. 95–97.

51. G. J. McManus, "World Steelmaking Heads into an Era of Major Change," *Iron Age* 184 (December 31, 1959), p. 17.

52. Jacques Singer, "Trade Liberalization in the Canadian Steel Industry," in H. Edward English (ed.) *The Impact of Trade Liberalization: Volume 2* (Toronto: University of Toronto Press, 1969), p. 62.

53. "How a U.S. Warehouseman Looks at Steel Import Problem," *Iron Age* 184 (July 16, 1959), p. 69. This same view was often expressed by others, see, for example, U.S. Department of Labor [E. Robert Livernash], *Collective Bargaining in the Basic Steel Industry—A Study of the Public Interest and the Role of Government* (Washington, DC: author, 1961), pp, 173–174.

54. See "May Steel Exports Top Imports," *Iron Age* 186 (July 14, 1960), p. 64; and "Steel Exports Gain," *Iron Age* 186 (August 18, 1960), p. 13.

55. AISI, *Steel Imports—A National Concern,* p. 62, Table 3.

56. "Crucible Steel Joins Common Market," *Iron Age* 186 (September 29, 1960), p. 13, and "Allegheny Ludlum Plans New Belgian Mill," *Iron Age* 186 (October 27, 1960), p. 13.

57. See Mira Wilkins, *The Maturing of Multinational Enterprise: American Business Abroad from 1914 to 1970* (Cambridge: Harvard University Press, 1974), p. 305.

58. "Is Investing Abroad a Trend in Steel?," *Iron Age* 186 (November 3, 1960), p. 45.

59. Wilkins, *op. cit.*, p. 378.

60. See "Interest Keeps Growing in Foreign Steel Sites," *Iron Age* 186 (November 10, 1960), p. 13; G. J. McManus, "Revitalized World Steel Industry Challenges U.S. Steelmakers," *Iron Age* 186 (December 29, 1960), p. 27; and Logan T. Johnston, "Opportunity International," in AISI, *Yearbook, 1961* (New York: AISI, 1961), pp. 260–263.

61. On the extent of multinational activity by American firms at this time, see Wilkins, *op. cit.*, pp. 245–408.

62. See Michael E. Porter, *Competitive Strategy* (New York: Free Press, 1980), pp. 237–274.

63. Benjamin F. Fairless, "Problems of Steel in New Decade," in AISI, *Yearbook, 1960* (New York: AISI, 1960), p. 166. The need for liberalized depreciation policies to allow steel firms to increase the cash flows necessary to finance foreign expansion was also emphasized by Logan Johnston, president of Armco Steel, in his AISI address on multinationalism at the 1961 annual meeting; see Johnston, *op. cit.*, pp. 256–259.

64. This was often the case because the smaller firms usually produced the less costly, commodity-type items, whereas the larger steel firms produced those products having a higher value-added content. In the late 1950s most imports of steel products into the United States were commodity-type goods, thus putting them into direct competition with the smaller producers.

65. See "Imports Threaten More Products," *Iron Age* 182 (December 4, 1958), p. 93; and "Wire Import Probe," *Iron Age* 183 (March 12, 1959), p. 85. This involved the U.S. Trade Commission Investigations No. 74 and No. 75 into nails, spikes, brads, and staples, hearings on which were held March 3, 1959 in Washington, DC. This particular case was terminated by the commission without formal findings of injury on March 12, 1959; see John M. Leddy and Janet L. Norwood, "The Escape Clause and Peril Points Under the Trade Agreements Program," in William B. Kelly, Jr. (ed.), *Studies in United States Commercial Policy* (Chapel Hill: University of North Carolina Press, 1963), p. 169. It might be noted that the USWA opposed the industry position before the Trade Commission, testifying that the steel industry should not be allowed to win a dumping case such as this. See "Statement of Meyer Bernstein," representative of the USWA, before the commission, a copy of this testimony is in AISI Vertical Files, Box S/T 1, Tariffs Folder, Eleutherian Mills-Hagley Foundation Historical Library, Wilmington, DE.

66. Thus in the Dillon round of tariff negotiations that began in the fall of 1960 between the United States and the Common Market, the AISI—reflecting the interests of the larger producers—urged only that "no new concessions should be granted on American tariffs applicable to steel products." Higher tariffs were not advocated, nor was there a call for reductions in foreign trade barriers. See Max Howell, Executive Vice President of the AISI, to chairman, Committee for Reciprocity Information, August 5, 1960, p. 3, and attached AISI "Statement: Tariffs and the Foreign Trade of the United States in Steel"—both, in turn, attached to Richard Storch to Clarence Randall, August 19, 1960, CFEP file, Randall Series—Subject Subseries, Box 11, Steel (1) Folder, DDE Library. Also see J. D. Baxter, "Time to Speak Up on Tariffs," *Iron Age* 185 (June 30, 1960), pp. 114–115; R. W. Crosby, "New Laws Don't Impress Industry," *Iron Age* 186 (July 14, 1960), p. 79; "A Voice on Tariffs," *Iron Age* 186 (July 21, 1960), p. 111; R. W. Crosby, "Call Comes for Higher Tariffs," *Iron Age* 186 (August 4, 1960), p. 75; and E. C. Beaudet, "U.S. Trade Policy for the '60s," *Iron Age* 186 (December 22, 1960), pp. 31–32. Also see Hogan, *op cit.*, 5:2038–2039.

67. Richard E. Mooney, "Showdown Nears on Dumping Cases," *NYT* (April 15, 1963), p. 45.

68. "Treasury Rejects Steelmakers' Contention that Japan is 'Dumping' Wire Rods in US," *Wall Street Journal* (May 7, 1963), p. 3. For further discussion of these

events, see James A. Kohn, "The Antidumping Act: Its Administration and Place in American Trade Policy," *Michigan Law Review* 60 (February 1963), pp. 407–438; Alexis C. Coudert, "The Application of the United States Antidumping Law in the Light of a Liberal Trade Policy," *Columbia Law Review* 65 (February 1965), pp. 189–231; Note, "The Antidumping Act—Tariff or Antitrust Law?," *Yale Law Journal* 74 (March 1965), pp. 707–724; and Ronald L. Styn, "The Antidumping Act: Problems of Administration and Proposals for Change," *Stanford Law Review* 17 (April 1965), pp. 730–749. Also see Robert R. Miller, "United States Antidumping Policy and the Steel Industry Experience" (Ph.D. diss., Stanford University, 1967).

69. For one point of view regarding this development, see William E. Scheuerman, *The Steel Crisis: The Economics and Politics of a Declining Industry* (Westport, CT: Greenwood Press, 1986).

Chapter 10

1. See Grant McConnell, *Steel and the Presidency, 1962* (New York: W. W. Norton, 1963); and Roy Hoopes, *The Steel Crisis* (New York: John Day, 1963) for analyses of the well-publicized 1962 Kennedy–steel confrontation.

2. See the remarks of T. F. Patton, chairman of the AISI following the death of Fairless, in "Steel and the Public Interest," in AISI *Yearbook, 1962* (New York: AISI, 1962), pp. 235–242.

3. See Ferdinand L. Molz, "The Political Economy of Steel Import Quotas," *Journal of Economic Issues* 4 (June–September 1970), pp. 60–76. For an overview of the economic changes confronting the industry after 1960, see Donald F. Barnett and Louis Schorsch, *Steel, Upheaval in a Basic Industry* (Cambridge, MA: Ballinger, 1983).

4. The recent literature surrounding industrial policy is voluminous, to say the least. For a convenient overview, see Chalmers Johnson (ed.), *The Industrial Policy Debate* (San Francisco: ICS Press, 1984).

5. For a persuasive explication of the power of ideology to motivate American businesspeople, see David Vogel, "Why Businessmen Distrust Their State: The Political Consciousness of American Corporate Executives," *British Journal of Political Science* 8 (1978), pp. 45–78.

6. The author is grateful to Professor Louis Galambos for most convincingly making this point clear to him. Also see Louis Galambos, *America in Middle Age* (New York: New Press, n.d.)

7. Robert Griffith, "Dwight D. Eisenhower and the Corporate Commonwealth," *American Historical Review* 87 (February 1982), p. 103.

8. Thomas K. McCraw, *Prophets of Regulation* (Cambridge: Harvard University Press, 1984), pp. 300–309.

9. Richard H. K. Vietor, *Energy Policy in America Since 1945* (New York: Cambridge University Press, 1984), pp. 345–354.

10. For commentary on an industrial policy for steel, see Hon. Charles A. Vanik, "The Development of an Industrial Policy for Steel in the United States During a Period of Trade Frictions," in OECD (ed.), *Steel in the 80s* (Paris: author, 1980), pp. 232–244; and Joel S. Hirschhorn, "Troubles and Opportunities in the United States Steel Industry," in M. E. Dewar (ed.), *Industry Vitalization—Toward a National Industrial Policy* (New York: Pergamon Press, 1982), pp. 3–41. This is a far from exhaustive listing of studies on this subject.

11. See Anthony Solomon, "Report to the President: A Comprehensive Program for the Steel Industry" (Washington, DC: U.S. Department of the Treasury, December 6, 1977 [mimeo]). The Solomon Report recommended a number of substantive provisions to assist the steel industry to recover from the effects of foreign competition

besetting the industry in late 1977. Among these was formation of a tripartite commission consisting of representatives from labor, industry, and government to coordinate the longer-term needs of the industry; this can perhaps be viewed as an attempt to formulate and implement an industrial policy for steel in the United States. Unfortunately, little if any lasting benefit derived from the committee's work (or Solomon's Report): The participating interests were unable to overcome their entrenched prior positions sufficiently to provide the basis for a working (or politically legitimate) industrial policy. Solomon himself expressed disappointment with the eventual outcome of his recommendations, noting their wide divergence from his original proposals. See Hirschhorn, *op. cit.,* for details of the program. (Solomon's views on the outcome of his recommendations were obtained in an interview with him by Paul A. Tiffany in New York City, New York Federal Reserve Bank, June 17, 1980.)

12. Warren I. Susman, *Culture as History* (New York: Pantheon Books, 1984).

Bibliography

Manuscript and Original Collection Sources

AISI Vertical Files, Eleutherian Mills-Hagley Foundation Historical Library, Wilmington, DE.

Bethlehem Steel Corporation Historical Files, Charles Schwab Memorial Library, Bethlehem Steel Corporation, Bethlehem, PA.

Dwight D. Eisenhower Presidential Library, Abilene, KS.

John W. Hill Papers, Mass Communication History Center of the State Historical Society of Wisconsin, Madison.

Thomas W. Lamont Papers, Baker Library, Harvard Business School, Boston, MA.

Harry S. Truman Presidential Library, Independence, MI.

Government Documents

OEEC, *The Iron and Steel Industry in Europe, 1958 to 1959* (Brussels: author, 1960).

Public Papers of the Presidents of the United States, Harry S. Truman, 8 Vols., 1945–1953 (Washington, DC: GPO, 1962).

Public Papers of the Presidents of the United States: Dwight D. Eisenhower, 8 Vols., 1953–1961 (Washington, DC: GPO, 1960–61).

United Nations, ECE, *European Steel Trends in the Setting of the World Market* (New York: author, November 1949).

United Nations, ILO, Iron and Steel Committee, *Regularization of Production and Employment at a High Level* (Geneva: author, 1947).

U.S. Bureau of Corporations, *Report of the Commissioner of Corporations on the Steel Industry, Part I: Organization, Investment, Profits, and Position of the United States Steel Corporation* (Washington, DC: GPO, 1911).

U.S. Bureau of Labor, *Report on Conditions of Employment in the Iron and Steel Industry,* 4 Vols. (Washington, DC: GPO, 1911–13).

U.S. Comptroller General, General Accounting Office, *New Strategy Required for Aiding Distressed Steel Industry* (Washington, DC: author, January 8, 1981).

———, *Report to the Congress of the United States, Examination of 100 Million Dollar Loan to ECSC, June 30, 1957* (Washington, DC: author, March 13, 1958).

U.S. Congress, *Congressional Record,* 1945–60. Washington, DC.

U.S. Congress, Joint Committee on the Economic Report, 80th Cong., 2nd sess., Hearings, *Increases in Steel Prices* (Washington, DC: GPO, 1948).

———, Joint Committee on the Economic Report, 81st Cong., 1st sess., Report, *Joint Economic Report* (Washington, DC: GPO, 1949).

———, Joint Committee on the Economic Report, 81st Cong. 2nd sess., Report, *Basic Data Relating to Steel Prices* (Washington, DC: GPO, 1950).

———, 81st Cong., 2nd sess., Hearings, *December 1949 Steel Price Increases* (Washington, DC: GPO, 1950).

———, 81st Cong., 2nd sess., Report, *December 1949 Steel Price Increases* (Washington, DC: GPO, 1950).

U.S. Congress, JEC, 86th Cong., 1st sess., *Materials Prepared in Connection with the Study of Employment, Growth, and Price Levels,* Study Paper No. 2, Otto Eckstein and Gary Fromm, "Steel and Postwar Inflation" (Washington, DC: GPO, 1959).

U.S. Congress, Office of Technology Assessment, *Technology and Steel Industry Competitiveness* (Washington, DC: GPO, June 1980).

U.S. Congress, TNEC, 76th Cong. 3rd sess., Hearings, *Investigation of Concentration of Economic Power,* pt. 26 (Washington, DC: GPO, 1940).

U.S. Congress, House, Committee on Investigation of the United States Steel Corporation, 62nd Cong., 1st sess., Hearings, *United States Steel Corporation* (Washington, DC: GPO, 1911).

———, Subcommittee on Study of Monopoly Power, Committee on the Judiciary, 81st Cong., 2nd sess., Hearings, *Study of Monopoly Power,* Pts. 4A, 4B, "Steel" (Washington, DC: GPO, 1950).

———, Committee on the Judiciary, Antitrust Subcommittee, 84th Cong. 1st sess., Hearings, *WOC's and Government Advisory Groups,* Pts. I–III (Washington, DC: GPO, 1955).

———, Committee on Public Works, Subcommittee to Investigate Questionable Trade Practices, 80th Cong., 2nd sess., Interim Report, *Investigating Questionable Trade Practices* (Washington, DC: GPO, 1948).

———, Select Committee on Foreign Aid, 80th Cong., 1st sess., *United States Steel Requirements and Availabilities as They Affect European Needs for Interim Aid,* Preliminary Report Six (Washington, DC: GPO, 1947).

———, Select Committee on Foreign Aid, 80th Cong., 2nd sess., *The Role of Steel in the European Recovery Program,* Supplement to Preliminary Report Six (Washington, DC: GPO, 1948).

———, Select Committee on Small Business, 81st Cong., 2nd sess., Hearings, *Steel, Acquisitions, Mergers, and Expansion of 12 Major Companies, 1900 to 1950* (Washington, DC: GPO, 1950).

———, Select Committee on Small Business, 81st Cong., 1st sess., Hearings, *Patterns of Steel Distribution in the States of Arkansas, Louisiana, Oklahoma, and Texas* (Washington, DC: GPO, 1949).

———, Special Committee on Post-War Economic Policy and Planning, Subcommittee on Foreign Trade and Shipping, 78th Cong., 2nd sess. and 79th Cong.,

1st sess., Hearings, *Post-War Economic Policy and Planning,* pt. 4 (Washington, DC: GPO, 1945).

U.S. Congress, Senate, Committee on Finance, 90th Cong., 1st sess. (Washington, DC: GPO, 1967), (Washington, DC: GPO, 1967), Committee Print, *Steel Imports.*

———, Committee on the Judiciary, Subcommittee on Antitrust and Monopoly, 85th Cong., 1st sess., Hearings, *Administered Prices,* "Steel" (Washington, DC: GPO, 1958).

———, Committee on the Judiciary, Subcommittee on Antitrust and Monopoly, 84th Cong., 1st sess., Hearings, *The Proposed Bethlehem–Youngstown Merger* (Washington, DC: GPO, 1955).

———, Committee on Military Affairs, Subcommittee on Surplus Property and Special Committee on Postwar Economic Policy and Planning, Subcommittee on Industrial Reorganization, 79th Cong., 1st sess., Joint Hearings, *War Plants Disposal—Iron and Steel Plants* (Washington, DC: GPO, 1946).

———, Special Committee to Study Problems of American Small Business, 80th Cong., 1st sess., Hearings, *Problems of American Small Business* (Washington, DC: GPO, 1947).

———, Special Committee to Study Problems of American Small Business, 80th Cong., 2nd sess., *Steel Supply and Distribution Problems,* Final Report (Washington, DC: GPO, 1949).

U.S. Department of Agriculture, Gladys L. Baker *et al., Century of Service—The First 100 Years of the United States Department of Agriculture* (Washington, DC: author, 1963).

U.S. Department of Commerce [G. Lyle Belsley], National Production Authority, Historical Reports on Defense Production, Report No. 19, *Consultation with Industry—History of the Office of Industrial Advisory Committees of the National Production Authority* (Washington, DC: author, 1953).

———, National Production Authority, Historical Reports on Defense Production, Report No. 28, *Iron and Steel—History of the Iron and Steel Division of the National Production Authority* (Washington, DC: author, 1953).

U.S. Department of Commerce, *Origin and Development of the Continental Steel Entente,* Trade Information Bulletin No. 484 (Washington, DC: GPO, 1927).

———, *International Iron and Steel,* vol. 1, no. 1 (March 1955) through vol. 3, no. 1 (March 1957). Publication ended with March 1957 issue.

U.S. Department of Interior, *National Resources and Foreign Aid. Report of J. A. Krug, Secretary of the Interior* (Washington, DC: author, October 9, 1947).

U.S. Department of Labor [E. Robert Livernash], *Collective Bargaining in the Basic Steel Industry—A Study of the Public Interest and the Role of Government* (Washington, DC: author, 1961).

U.S. Department of State, Bureau of Intelligence and Research [Corwin D. Edwards], *Cartelization in Western Europe* (Washington, DC: author, June, 1964).

———, *Foreign Relations of the United States,* (Washington, DC: GPO, various years), annual volumes for 1948, 1949, 1950, 1952–54.

———, Office of Public Affairs, *The United States Reciprocal Trade Agreements Program and the Proposed International Trade Organization, A Summary of Background Information* (Washington, DC: author, March, 1948).

U.S. Secretary of the Treasury, *Final Report on the Reconstruction Finance Corporation* (Washington, DC: GPO, 1959).

[———, Office of the U.S. Secretary of the Treasury], Anthony Solomon, Undersecretary of the Treasury, *Report to the President: A Comprehensive Program for the Steel Industry* (Washington, DC: author, December 6, 1977), mimeo.

U.S. ECA, Industrial Advisory Committee, *Report on Plants Scheduled for Removal*

as Reparations from the Three Western Zones of Germany* (Washington, DC: author, 1949).
———, *European Recovery Program, Iron and Steel Commodity Study* (Washington, DC: author, 1949).
U.S. FTC, *A List of 1,000 Large Manufacturing Companies, Their Subsidiaries and Affiliates, 1948* (Washington, DC: GPO, 1951).
———, *Monopolistic Practices and Small Business* (Washington, DC: GPO, 1952).
———, *Practices of the Steel Industry Under the Code* (Washington, DC: GPO, 1934).
———, *Report of the Federal Trade Commission on the Concentration of Productive Facilities, 1947—Total Manufacturing and 26 Selected Industries* (Washington, DC: GPO, 1950).
———, *Report on International Steel Cartels* (Washington, DC: GPO, 1948).
[U.S.] President's Citizen Advisors on the Mutual Security Program, *Report to the President* (Washington, DC: GPO, March 1, 1957).
[U.S. President's] Commission on Foreign Economic Policy, *Report to the President and the Congress* (Washington, DC: author, January 23, 1954).
———, *Minority Report* (Washington, DC: author, January 30, 1954).
[U.S. President's] CEA, *The Midyear Economic Report of the President to the Congress, July 21, 1947,* in *The Economic Reports of the President* (New York: Reynal and Hitchcock, 1948).
[U.S. President's] Materials Policy Commission, *The Outlook for Key Commodities, Resources for Freedom, a Report to the President,* 2 vols. (Washington, DC: GPO, 1952).
[U.S. President's] Public Advisory Board for Mutual Security, *A Trade and Tariff Policy in the National Interest* (Washington, DC: GPO, 1953).
[U.S. President's] Steel Industry Board, *Report to the President of the United States on the Labor Dispute in the Basic Steel Industry* (Washington, DC: GPO, September 10, 1949).
[U.S. President's] Steel Tripartite Advisory Committee, *Report to the President by the Steel Tripartite Advisory Committee on the United States Steel Industry* (Washington, DC: author, September 25, 1980), mimeo.
U.S. Tariff Commission, *Iron and Steel, War Changes in Industry Series,* Report No. 15 (Washington, DC: GPO, 1946).

Judicial Decisions

Federal Trade Commission v. Cement Institute, 333 U.S. 683 (1948).
FTC, *In the Matter of United States Steel Corporation et al.,* Docket 760, 8 FTC Decisions 1 (1924).
United States v. United States Steel Corporation et al., 251 U.S. 417 (1920).
Youngstown Sheet & Tube Co. v. Sawyer, 343 U.S. 579 (1952).

Books

Abel, I. W., *Collective Bargaining—Labor Relations in Steel: Then and Now.* New York: Columbia University Press, 1976.
Alexander, Charles C., *Holding the Line: The Eisenhower Era, 1952–1961.* Bloomington: Indiana University Press, 1975.
Asch, Peter, *Economic Theory and the Antitrust Dilemma.* New York: John Wiley & Sons, 1970.

Bach, G. L., *Inflation: A Study in Economics, Ethics, and Politics.* Providence, RI: Brown University Press, 1958.
———, *Making Monetary and Fiscal Policy.* Washington, DC: Brookings, 1971.
Backer, John H., *The Decision to Divide Germany.* Durham, NC: Duke University Press, 1978.
Bailey, Stephen K., *Congress Makes a Law—The Story Behind the Employment Act of 1946.* New York: Columbia University Press, 1950.
Baldwin, David A., *Foreign Aid and American Foreign Policy.* New York: Praeger, 1966.
Baran, P. A., and Sweezy, P. M., *Monopoly Capital.* New York: Monthly Review Press, 1966.
Barnett, Donald F., and Louis Schorsch, *Steel, Upheaval in a Basic Industry* (Cambridge, MA: Ballinger, 1983).
Bartless, Robert V., *The Reserve Mining Controversy.* Bloomington: Indiana University Press, 1980.
Batty, Peter, *The House of Krupp.* London: Secker and Warburg, 1966.
Bauer, Raymond A.; De Sola Pool, Ithiel; and Dexter, Lewis A., *American Business and Public Policy, the Politics of Foreign Trade.* New York: Atherton Press, 1963.
Beasley, W. G., *The Modern History of Japan,* 2nd ed. New York: Praeger, 1974.
Becker, William H., *The Dynamics of Business-Government Relations, Industry & Exports, 1893–1921.* Chicago: University of Chicago Press, 1982.
Berglund, Abraham, *The United States Steel Corporation.* New York: Columbia University Press, 1907.
Berglund, Abraham, and Wright, Philip G. *The Tariff on Iron and Steel.* Washington, DC: Brookings, 1929.
Berkowitz, Edward, and McQuaid, Kim, *Creating the Welfare State.* New York: Praeger, 1980.
Bernstein, Irving, *Turbulent Years: A History of the American Worker, 1933–1941.* Boston: Houghton Mifflin, 1970.
Bidwell, Percy W., *What the Tariff Means to American Industries.* New York: Harper & Brothers, 1956.
Blackman, John L., Jr., *Presidential Seizure in Labor Disputes.* Cambridge: Harvard University Press, 1967.
Borden, William S., *The Pacific Alliance: United States Foreign Economic Policy and Japanese Trade Recovery, 1947–1955.* Madison: University of Wisconsin Press, 1984.
Brady, Robert A., *The Rationalization Movement in German Industry.* Berkeley: University of California Press, 1933.
Brandes, Joseph, *Herbert Hoover and Economic Diplomacy.* Pittsburgh: University of Pittsburgh Press, 1962.
Brody, David, *Steelworkers in America, the Nonunion Era.* Cambridge: Harvard University Press, 1960.
Broude, Henry W., *Steel Decisions and the National Economy.* New Haven: Yale University Press, 1963.
Brown, William A., Jr. *The United States and the Restoration of World Trade.* Washington, DC: Brookings, 1950.
Bellush, Bernard, *The Failure of the NRA.* New York: W. W. Norton, 1975.
Burn, Duncan, *The Economic History of Steelmaking, 1867–1939.* Cambridge: Cambridge University Press, 1940.
———, *The Steel Industry, 1939–1959.* Cambridge: Cambridge University Press, 1961.
Butler, Joseph G., Jr., *Fifty Years of Iron and Steel.* Cleveland: Penton Press Co., 1923.

Campbell, John C., *The United States in World Affairs, 1945–1947.* New York: Harper & Brothers, 1947.
Carosso, Vincent P., *Investment Banking in America, a History.* Cambridge: Harvard University Press, 1970.
Carr, J. C. and Taplin W., *History of the British Steel Industry.* Cambridge: Harvard University Press, 1962.
Catton, Bruce, *The War Lords of Washington.* New York: Harcourt, Brace, 1948.
Clawson, Marion, *New Deal Planning, the National Resources Planning Board.* Baltimore: Johns Hopkins University Press, 1981.
Clay, Lucius D., *Decision in Germany.* Garden City, NY: Doubleday, 1950.
Clemens, Diane S., *Yalta.* New York: Oxford University Press, 1970.
Cochran, Bert, *Harry Truman and the Crisis Presidency.* New York: Funk & Wagnalls, 1973.
Cohen, Jerome B., *Japan's Economy in War and Reconstruction.* Minneapolis: University of Minnesota Press, 1949.
Collins, Robert M., *The Business Response to Keynes, 1929–1964.* New York: Columbia University Press, 1981.
Cooling, Benjamin F., *Gray Steel and Blue Water Navy, the Formative Years of America's Military-Industrial Complex, 1881–1917.* Hamden, CT: Shoe String Press, Archon Books, 1979.
Corey, Lewis, *The House of Morgan.* New York: G. Howard Watt, 1930.
Cormier, Frank, and Eaton, W. J., *Reuther.* Englewood Cliffs, NJ: Prentice-Hall, 1970.
Craig, Gordon A., *Germany, 1866–1945.* New York: Oxford University Press, 1978.
Crandall, Robert W., *The U.S. Steel Industry in Recurrent Crisis.* Washington, DC: Brookings, 1981.
Dam, Kenneth W., *The GATT, Law and International Economic Organization.* Chicago: University of Chicago Press, 1970.
Daugherty, C. R.; DeChazeau, M. G.; and Stratton, S. S., *The Economics of the Iron and Steel Industry,* 2 vols. New York: McGraw-Hill, 1937.
Dewar, Margaret E. (ed.), *Industrial Vitalization—Toward a National Industrial Policy.* New York: Pergamon Press, 1982.
Diebold, William, Jr., *The Schuman Plan, a Study in Economic Cooperation, 1950–1959.* New York: Praeger, 1959.
Divine, Robert A., *Eisenhower and the Cold War.* New York: Oxford University Press, 1981.
Donovan, Robert J., *Conflict and Crisis.* New York: W. W. Norton, 1977.
———, *Tumultuous Years, the Presidency of Harry S. Truman, 1949–1953.* New York: W. W. Norton, 1982.
Edwards, Corwin D. (ed.), *A Cartel Policy for the United Nations.* New York: Columbia University Press, 1945.
Eggert, Gerald G., *Steelmasters and Labor Reform, 1886–1923.* Pittsburgh: University of Pittsburgh Press, 1981.
Eiselen, M. R., *The Rise of Pennsylvania Protectionism.* Philadelphia: author, 1932.
Eisenhower, Dwight, D., *Mandate for Change, 1953–1956.* Garden City, N.Y.: Doubleday, 1963.
———, *Waging Peace, 1956–1961.* Garden City, N.Y.: Doubleday, 1965.
Engler, Robert, *The Politics of Oil.* Chicago: University of Chicago Press, 1961.
Evans, John W., *The Kennedy Round in American Trade Policy.* Cambridge: Harvard University Press, 1971.
Ewald, William B., Jr., *Eisenhower the President—Crucial Days, 1951–1960.* Englewood Cliffs, NJ: Prentice-Hall, 1981.
Eysenbach, Mary L., *American Manufactured Exports, 1879–1914.* New York: Arno Press, 1976.

Feis, Herbert, *Churchill, Roosevelt, Stalin: The War They Waged and the Peace They Sought.* Princeton: Princeton University Press, 1957.

Ferrell, Robert H. (ed.), *Off the Record—The Private Papers of Harry S. Truman.* New York: Harper & Row, 1980.

Fontenay, Charles L., *Estes Kefauver—A Biography.* Knoxville: University of Tennessee Press, 1980.

Ford, Joseph W., *The Steel Import Problem.* New York: Fordham University Press, 1961.

Friden, Lennart, *Instability in the International Steel Market.* Stockholm: Beckmans, 1972.

Friedel, Frank, *Franklin D. Roosevelt, the Apprenticeship.* Boston: Little, Brown, 1952.

Friedrich, Carl J., *American Experiences in Military Government in World War II.* New York: Rinehart, 1948.

Gaddis, John L., *The United States and the Origins of the Cold War, 1941–1947.* New York: Columbia University Press, 1972.

Galambos, Louis, *America at Middle Age.* New York: New Press, n.d.

Gardner, Richard N., *Sterling-Dollar Diplomacy,* Expanded ed. New York: McGraw-Hill, 1969.

Gimbel, John, *The American Occupation of Germany, Politics and the Military, 1945–1949.* Stanford: Stanford University Press, 1968.

Gold, Bela; Rosegger, G.; and Boylan, M. G., Jr., *Evaluating Technological Innovations.* Lexington, MA: Lexington Books, 1980.

Gottlieb, Manuel, *The German Peace Settlement and the Berlin Crisis.* New York: Paine–Whitman, 1960.

Gould, Jean, and Hickok, Lorena. *Walter Reuther.* New York: Dodd, Mead, 1972.

Graham, Otis L., *Toward a Planned Society, from Roosevelt to Nixon.* New York: Oxford University Press, 1976.

Greenstein, Fred I., *The Hidden-Hand Presidency, Eisenhower as Leader.* New York: Basic Books, 1982.

Gulick, Charles A., *Labor Policy in the United States Steel Corporation.* New York: Columbia University Press, 1924.

Hadley, Eleanor M., *Antitrust in Japan.* Princeton: Princeton University Press, 1970.

Hamby, Alonzo L., *Beyond the New Deal: Harry S. Truman and American Liberalism.* New York: Columbia University Press, 1973.

Hansen, Alvin H., *The American Economy.* New York: McGraw-Hill, 1957.

Harris, Howell, J., *The Right to Manage.* Madison: University of Wisconsin Press, 1982.

Hartley, Fred A., *Our New National Labor Policy: The Taft-Hartley Act and the Next Steps.* New York: Funk & Wagnalls, 1948.

Hartmann, Susan, *Truman and the 80th Congress.* Columbia: University of Missouri Press, 1971.

Hawley, Ellis W., *The New Deal and the Problem of Monopoly, a Study in Economic Ambivalence.* Princeton: Princeton University Press, 1966.

Henderson, W. O., *The Genesis of the Common Market.* Chicago: Quandrangle, 1962.

Hendriksen, Eldon S., *Capital Expenditures in the Steel Industry, 1900–1953.* New York: Arno Press, 1978.

Herling, John, *Right to Challenge, People and Power in the Steelworkers Union.* New York: Harper & Row, 1972.

Herring, George C., Jr., *Aid to Russia 1941–1946: Strategy, Diplomacy, and the Origins of the Cold War.* New York: Columbia University Press, 1973.

Hexner, Ervin, *The International Steel Cartel.* Chapel Hill: University of North Carolina Press, 1943.

Hill, John, *The Making of a Public Relations Man.* New York: David McKay, 1963.

Himmelberg, Robert F., *The Origins of the National Recovery Administration,* New York: Fordham University Press, 1976.
Hogan, William T., *Economic History of the Iron and Steel Industry in the United States,* 5 vols. Lexington, MA: Lexington Books, 1971.
Hoopes, Ray, *The Steel Crisis.* New York: John Day, 1963.
Iida, Ken-ichi, *History of Steel in Japan.* Tokyo: Nippon Steel Corporation, 1973.
Janeway, Eliot, *The Struggle for Survival: A Chronicle of Economic Mobilization During World War II.* New Haven: Yale University Press, 1951.
Johnson, Chalmers (ed.), *The Industrial Policy Debate.* San Francisco: ICS Press, 1984.
———, *MITI and the Japanese Miracle, The Growth of Industrial Policy, 1925–1975.* Stanford: Stanford University Press, 1982.
Johnson, William A., *The Steel Industry of India.* Cambridge: Harvard University Press, 1966.
Kaufman, Burton I., *Efficiency and Expansion: Foreign Trade Organization in the Wilson Administration, 1913–1921.* Westport, CT: Greenwood Press, 1974.
———, *The Oil Cartel Case: A Documentary Study of Antitrust Activity in the Cold War Era.* Westport, CT: Greenwood Press, 1978.
———, *Trade & Aid, Eisenhower's Foreign Economic Policy, 1953–1961.* Baltimore: Johns Hopkins University Press, 1982.
Kawahito, Kiyoshi, *The Japanese Steel Industry.* New York: Praeger, 1972.
Keeling, B. S., and Wright, A.E.G., *The Development of the Modern British Steel Industry.* London: Longmans, 1964.
Kefauver, Estes, *In a Few Hands, Monopoly Power in America.* Baltimore: Penguin Books, 1965.
Kovaleff, Theodore P., *Business and Government During the Eisenhower Administration, a Study of the Antitrust Policy of the Antitrust Division of the Justice Department.* Athens: Ohio University Press, 1980.
Lamoreaux, Naomi R., *The Great Merger Movement in American Business, 1895–1904.* New York: Cambridge University Press, 1985.
Lauderbaugh, Richard A., *American Steel Makers and the Coming of the Second World War.* Ann Arbor:University Microfilms International, UMI Research Press, 1980.
Lewis, Cleona, *America's Stake in International Investments.* Washington, DC: Brookings, 1938.
Lister, Louis, *Europe's Coal and Steel Community, an Experiment in Economic Union.* New York: Twentieth Century Fund, 1960.
Lochner, Louis P., *Herbert Hoover and Germany.* New York: Macmillan, 1960.
Lockwood, William, *The Economic Development of Japan.* Princeton: Princeton University Press, 1954.
Lynn, Leonard H., *How Japan Innovates, a Comparison with the U.S. in the Case of Oxygen Steelmaking.* Boulder, CO: Westview Press, 1982.
Maier, Charles S., *Recasting Bourgeois Europe.* Princeton: Princeton University Press, 1975.
Manchester, *The Arms of Krupp, 1587–1968.* Boston: Little, Brown, 1968.
Manners, Gerald, *The Changing World Market for Iron Ore, 1950–1980.* Baltimore: John Hopkins University Press, 1971.
Mansfield, *Industrial Research and Technological Innovation.* New York: Norton, 1968.
Marcus, Maeva, *Truman and the Steel Seizure Case.* New York: Columbia University Press, 1977.
Martin, James S., *All Honorable Men.* Boston: Little, Brown 1950.

Mayall, K . L., *International Cartels, Economic and Political Aspects.* Rutland, VT: Charles E. Tuttle, 1951.

McConnell, Grant, *Private Power and American Democracy.* New York: Alfred A. Knopf, 1967.

———, *Steel and the Presidency, 1962.* New York: W. W. Norton, 1963.

———, *The Steel Seizure of 1952.* Indianapolis, IN: Bobbs-Merrill, 1960.

McCraw, Thomas K., *Prophets of Regulation.* Cambridge: Harvard University Press, 1984.

McDonald, David, *Union Man.* New York: E. P. Dutton, 1969.

McEachern, Doug, *A Class Against Itself, Power and the Nationalisation of the British Steel Industry.* Cambridge: Cambridge University Press, 1980.

McQuaid, Kim, *Big Business and Presidential Power, from FDR to Reagan.* New York: William Morrow, 1982.

Means, Gardiner C., *Pricing Power and the Public Interest, a Study Based on Steel.* New York: Harper & Row, 1962.

Merkle, Judith, *Management and Ideology.* Berkeley: University of California Press, 1980.

Millis, Harry A., and Brown, Emily C., *From the Wagner Act to Taft-Hartley: A Study of National Labor Policy and Labor Relations.* Chicago: University of Chicago Press, 1950.

Moore, William H., *The Kefauver Committee and the Politics of Crime, 1950–1952.* Columbia: University of Missouri Press, 1974.

Morgenthau, Henry, Jr., *Germany is Our Problem* New York: Harper & Brothers, 1945.

Nakamura, Takafusa, *Economic Growth in Prewar Japan.* Translated by Robert A. Feldman. New Haven: Yale University Press, 1983.

———, *The Postwar Japanese Economy, Its Development and Structure.* Translated by J. Kaminski. Tokyo: University of Tokyo Press, 1981.

Nelson, Ralph, *Merger Movements in American Industry, 1895–1956.* Princeton: Princeton University Press, 1959.

Neustadt, Richard E., *Presidential Power.* New York: John Wiley & Sons, 1980.

Norton, Hugh S., *The Employment Act and the Council of Economic Advisers, 1946–1976.* Columbia: University of South Carolina Press, 1977.

Parmet, Herbert S., *Eisenhower and the American Crusades.* New York, Macmillan, 1972.

Pastor, Robert A., *Congress and the Politics of U.S. Foreign Economic Policy, 1929–1976.* Berkeley: University of California Press, 1980.

Pollard, Robert A., *Economic Security and the Origins of the Cold War, 1945–1950.* New York: Columbia University Press, 1985.

Porter, Michael E., *Competitive Strategy.* New York: Free Press, 1980.

Pounds, Norman J. G., and Parker, William N., *Coal and Steel in Western Europe.* Bloomington: Indiana University Press, 1957.

Randall, Clarence B., *A Creed for Free Enterprise.* Boston: Little, Brown, 1952.

Rayback, Joseph G., *A History of American Labor,* Rev. ed. New York: Free Press, 1966.

Reuther, Victor G., *The Brothers Reuther and the Story of the UAW.* Boston: Houghton Mifflin, 1976.

Root, Franklin R., University of Maryland, Bureau of Business and Economic Research, *The European Coal and Steel Community,* Part II. College Park: author, June 1956.

Ross, George W., *The Nationalisation of Steel.* London: Macgibbon & Kee, 1965.

Rowley, Charles K., *Steel and Public Policy.* London: McGraw-Hill, 1971.

Russell, Charles S., and Vaughan, William J., *Steel Production: Processes, Products, and Residuals.* Baltimore: Johns Hopkins University Press, 1976.

Sawyer, Charles, *Concerns of a Conservative Democrat.* Carbondale: Southern Illinois University Press, 1968.

Scherer, F. M., *Industrial Market Structure and Economic Performance.* 2nd ed. Chicago: Rand McNally, 1980.

Scheuerman, William, *The Steel Crisis: The Economics and Politics of a Declining Industry.* Westport, CT: Greenwood Press, 1986.

Schlesinger, Arthur M., Jr., *The Coming of the New Deal.* Boston: Houghton Mifflin, 1958.

Schriftgiesser, Karl, *Business and Public Policy.* Englewood Cliffs, NJ: Prentice-Hall, 1967.

Schroeder, Gertrude G., *The Growth of Major Steel Companies, 1900–1950.* Baltmore: Johns Hopkins University Press, 1953.

Schumpeter, Joseph A., *History of Economic Analysis.* New York: Oxford University Press, 1954.

———, *The Theory of Economic Development.* Cambridge: Harvard University Press, 1934.

Seager, H. R., and Gulick, C. A., Jr., *Trust and Corporation Problems.* New York: Harper & Brothers, 1929.

Seidman, Joel, *American Labor from Defense to Reconversion.* Chicago: University of Chicago Press, 1953.

Southard, Frank A., Jr., *American Industry in Europe.* Boston: Houghton Mifflin, 1931.

Stanwood, Edward, *American Tariff Controversies in the Nineteenth Century,* 2 vols. Boston: Houghton Mifflin, 1903.

Stein, Herbert, *The Fiscal Revolution in America.* Chicago: University of Chicago Press, 1969.

Stigler, George J., and Kindahl, James K., *The Behavior of Industrial Prices.* New York: Columbia University Press, 1970.

Stocking, George W., *Basing Point Pricing and Regional Development—A Case Study of the Iron and Steel Industry.* Chapel Hill: University of North Carolina Press, 1954.

———, *Workable Competition and Antitrust Policy.* Nashville, TN: Vanderbilt University Press, 1961.

Stocking, George W. and Watkins, M. W., *Monopoly and Free Enterprise.* New York: Twentieth Century Fund, 1951.

Tarbell, Ida M., *The Life of Elbert H. Gary.* Boston: D. Appleton, 1925.

Taussig, F. W., *Some Aspects of the Tariff Question.* Cambridge: Harvard University Press, 1915.

———, *The Tariff History of the United States,* 8th ed. New York: G. P. Putnam's Sons, 1931.

Temin, Peter, *Iron and Steel in Nineteenth Century America.* Cambridge: MIT Press, 1964.

Truman, Harry S., *Year of Decisions, Memoirs by Harry S. Truman,* Vol. 1. Garden City, NY: Doubleday, 1955.

———, *Years of Trial and Hope, Memoirs by Harry S. Truman, 1946–1952,* vol. 2. Garden City, NY: Doubleday, 1956.

Ulman, Lloyd, *The Government of the Steel Workers' Union.* New York: John Wiley & Sons, 1962.

Urofsky, Melvin I., *Big Steel and the Wilson Administration.* Columbus: Ohio State Universtiy Press, 1969.

Vatter, Harold G., *The U.S. Economy in the 1950s, an Economic History*. New York: W. W. Norton, 1963.

Vietor, Richard H. K., *Energy Policy in America Since 1945*. New York: Cambridge University Press, 1984.

Wallich, Henry C., *Mainsprings of the German Revival*. New Haven: Yale University Press, 1955.

Warren, Kenneth, *The American Steel Industry 1850–1970, a Geographical Interpretation*. London: Oxford University Press, 1973.

———, *World Steel, an Economic Geography*. New York: Crane, Russak & Co., 1975.

Weiss, Leonard W., *Case Studies in American Industry*, 3rd ed. New York: John Wiley & Sons, 1980.

Westin, Alan F., *The Anatomy of a Constitutional Law Case*. New York: Macmillan, 1958.

———, *Economics and American Industry*. New York: John Wiley & Sons, 1961.

White, Gerald T., *Billions for Defense: Government Financing by the Defense Plant Corporation During World War II*. University: University of Alabama Press, 1980.

Wilkins, Mira, *The Emergence of Multinational Enterprise: American Business Abroad from the Colonial Era to 1914*. Cambridge: Harvard University Press, 1970.

———, *The Maturing of Multinational Enterprise: American Business Abroad from 1914 to 1970*. Cambridge: Harvard University Press, 1974.

Williams, William Appleman, *The Tragedy of American Diplomacy*, revised ed. New York: Delta, 1962.

Williamson, Oliver E., *Markets and Hierarchies*. New York: Free Press, 1975.

Wilson, Joan Hoff, *Herbert Hoover, Forgotten Progressive*. Boston: Little, Brown, 1975.

Wolff, Leon, *Lockout: The Story of the Homestead Strike, 1892*. New York: Harper & Row, 1965.

Yergin, Daniel, *Shattered Peace, the Origins of the Cold War and the National Security State*. Boston: Houghton Mifflin, 1977.

Ziegler, Harmon, *The Politics of Small Business*. Washington, DC: Public Affairs Press, 1961

Scholarly Articles

Adams, Walter, "The Steel Industry," in W. Adams (ed.) *The Structure of American Industry*, 5th ed. New York: Macmillan, 1977, pp. 86–129.

Adams, Walter and Dirlam, Joel B., "Big Steel, Invention, and Innovation," *Quarterly Journal of Economics* 80 (May 1966), pp. 167–189.

———, "Steel Imports and Vertical Oligopoly Power," *American Economic Review* 54 (September 1964), pp. 626–655.

Adelman, M. A., "Steel, Administered Prices, and Inflation," *Quarterly Journal of Economics* 75 (February 1961), pp. 16–40.

"The Antidumping Act—Tariff or Antitrust Law?" *Yale Law Journal* 74 (March 1965), pp. 707–724.

Barloon, Marvin, "The Question of Steel Capacity," *Harvard Business Review* 27 (March 1949), pp. 209–236.

Berglund, Abraham, "The United States Steel Corporation and Industrial Stabilization," *Quarterly Journal of Economics* 38 (August 1924), pp. 607–630.

———, "The United States Corporation and Price Stabilization," *Quarterly Journal of Economics* 38 (November 1923), pp. 1–30.

Bernstein, Barton, "Economic Policies of the Truman Administration," in R. S. Kirkendall (ed.), *The Truman Period as a Research Field.* Columbia: University of Missouri Press, 1967, pp. 87–149.

———, "The Truman Administration and the Steel Strike of 1946," *Journal of American History* 52 (March 1966), pp. 791–803.

Blair, John M., "Administered Prices: A Phenomenon in Search of a Theory," *American Economic Review* 49 (May 1959), pp. 431–450.

Caves, Richard, and Uekusa, Masu, "Industrial Organization," in H. Patrick and H. Rosovsky (eds.), *Asia's New Giant—How the Japanese Economy Works.* Washington, DC: Brookings, 1976, pp. 459–523.

Collins, Robert M., "Positive Business Responses to the New Deal: The Roots of the Committee for Economic Development, 1933–1942," *Business History Review* 52 (Autumn 1978), pp. 369–391.

Coudert, Alexis C., "The Application of the United States Antidumping Law in the Light of a Liberal Trade Policy," *Columbia Law Review* 65 (February 1965), pp. 189–231.

Crowther, Don, Q., and Staff, "Work Stoppages Caused by Labor-Management Disputes in 1946," *Monthly Labor Review* 64 (May 1947), pp. 780–800.

Cutler, Addison T., "Price Control in Steel," in OPA, Office of Temporary Controls, *Studies in Industrial Price Control.* Washington, DC: GPO, 1947, pp. 37–85.

Dalkin, Allin W., "Foreign Securities in the American Money Market, 1914–1930," *Harvard Business Review* 10 (January 1932), pp. 227–240.

Denison, E. F., and Chung, W. K., "Economic Growth and Its Sources," in H. Patrick and H. Rosovsky (eds.), *Asia's New Giant—How the Japanese Economy Works.* Washington, DC: Brookings, 1976, pp. 63–151.

DiBacco, Thomas B., "American Business and Foreign Aid: The Eisenhower Years," *Business History Review* 41 (Spring 1967), pp. 21–35.

Diebold, William, Jr., "The End of the I.T.C., " in Princeton University, Department of Economics and Social Institutions, *Essays in International Finance,* No. 16 (October 1952), pp. 1–37.

Dilley, D. R., and McBride, D. L., "Oxygen Steelmaking—Fact vs. Folklore," *Iron and Steel Engineer* 44 (October 1967), pp. 131–152.

Dimond, T., "This New Round of Steel Expansion," *Harvard Business Review* 34 (May–June 1956), pp. 85–93.

Gerrity, John A., "The United States Steel Corporation versus Labor: The Early Years," *Labor History* 1 (Winter 1960), pp. 3–38.

Gideonse, Max, "Foreign Trade, Investments, and Commercial Policy," in H. F. Williamson (ed.), *The Growth of the American Economy,* 2nd ed. Englewood Cliffs, NJ: Prentice-Hall, 1951, pp. 785–808.

Goodwin, Craufurd D., and Herren, R. Stanley, "The Truman Administration: Problems and Policies Unfold," in Craufurd D. Goodwin (ed.), *Exhortation and Controls—The Search for a Wage-Price Policy, 1945–1971.* Washington, DC: Brookings, 1975, pp. 9–93.

Gordon, H. Scott, "The Eisenhower Administration: The Doctrine of Shared Responsibility," in Craufurd D. Goodwin (ed.), *Exhortation and Controls—The Search for a Wage-Price Policy, 1945–1971.* Washington, DC: Brookings, 1975, pp. 95–134.

Gordon, Lincoln, "The Organization for European Economic Cooperation," *International Organization* 10 (February 1956), pp. 1–11.

Greer, Edward, "The Political Economy of U.S. Steel Prices in the Postwar Period," in Paul Zarembka (ed.), *Research in Political Economy,* vol. 1. Greenwich, CT: JAI Press, 1977, pp. 59–86.

Griffith, Robert, "Dwight D. Eisenhower and the Corporate Commonwealth," *American Historical Review* 87 (February 1982), pp. 87–122.

Hah, Chong-do, and Lindquist, R. M., "The 1952 Steel Seizure Revisited: A Systematic Study in Presidential Decision Making," *Administrative Science Quarterly* 20 (December 1975), pp. 587–605.

Hawkins, Harry C. and Norwood, Janet L., "The Legislative Basis of United States Commercial Policy," in William B. Kelly, Jr. (ed.), *Studies in United States Commercial Policy.* Chapel Hill: University of North Carolina Press, 1963, pp. 69–103.

Hexner, Ervin, "American Participation in the International Steel Cartel," *Southern Economic Journal* 8 (July 1941), pp. 54–79.

Hirschhorn, Joel S., "Troubles and Opportunities in the United States Steel Industry," in Margaret E. Dewar (ed.), *Industry Vitalization—Toward a National Industrial Policy.* New York: Pergamon Press, 1982.

Hogan, Michael J., "The Search for a 'Creative Peace': The United States, European Unity, and the Origins of the Marshall Plan," *Diplomatic History* 6 (Summer 1982), pp. 267–285.

Hughes, Thomas P., "Technological Momentum in History: Hydrogenation in Germany, 1898–1933," *Past & Present* 44 (August 1969), pp. 106–132.

Jones, Byrd L., "The Role of Keynesians in Wartime Policy and Postwar Planning, 1940–1946," *American Economic Review* 62 (May 1972), pp. 125–133.

Kaikati, Jack G., "The Anti-Export Policy of the U.S.," *California Management Review* 23 (Spring 1981), pp. 5–19.

Keehn, N. H., "A World of Becoming: From Pluralism to Corporatism," *Polity* 9 (Fall 1976), pp. 19–39.

Kohn, James A., "The Antidumping Act: Its Administration and Place in American Trade Policy," *Michigan Law Review* 60 (February 1962), pp. 407–438.

Koistinen, Paul A. C., "Mobilizing the World War II Economy: Labor and the Industrial-Military Alliance," *Pacific Historical Review* 42 (November 1973), pp. 443–478.

Latham, Earl, "The Politics of Basing Point Legislation," *Law and Contemporary Problems* 15 (Spring 1950), pp. 272–310.

Lauderbaugh, Richard, "Business, Labor, and Foreign Policy: U.S. Steel, the International Steel Cartel, and Recognition of the Steel Workers Organizing Committee," *Politics and Society* 6 (1976), pp. 433–457.

Leddy, John M., and Norwood, Janet L., "The Escape Clause and Peril Points Under the Trade Agreements Program," in William B. Kelly, Jr. (ed.), *Studies in United States Commercial Policy.* Chapel Hill: University of North Carolina Press, 1963, pp. 124–173.

Lightner, M. W., and McBride, D. L., "Basic Open-Hearth Steelmaking in the U.S.A.," *Journal of the Iron and Steel Institute* 189 (July 1958), pp. 205–216.

Maier, Charles S., "Between Taylorism and Technocracy: European Ideologies and the Vision of Industrial Productivity in the 1920s," *Journal of Contemporary History* 5 (1970), pp. 27–61.

Mancke, Richard B., "The Determinants of Steel Prices in the United States: 1947–1965," *Journal of Industrial Economics* 16 (April 1968), pp. 147–160.

Marx, Fritz M., "The Bureau of the Budget: Its Evolution and Present Role, I," *American Political Science Review* 39 (August 1945), pp. 653–684.

McAdams, A. K., "Big Steel, Invention, and Innovation: Reconsidered," *Quarterly Journal of Economics* 81 (August 1967), pp. 457–474.

McCraw, Thomas K., "Rethinking the Trust Question," in T. K. McCraw (ed.), *Regulation in Perspective.* Cambridge: Harvard University Press, 1981, pp. 1–55.

Meade, Edward S., "The Price Policy of the United States Steel Corporation," *Quarterly Journal of Economics* 22 (1908), pp. 452–466.

Molz, Ferdinand L., "The Political Economy of Steel Import Quotas," *Journal of Economic Issues* 4 (June–September 1970), pp. 60–76.

Orr, John A., "The Rise and Fall of Steel's Human Relations Committee," *Labor History* 14 (Winter 1973), pp. 69–82.

Parker, William N., "The Schuman Plan—A Preliminary Prediction," *International Organization* 6 (August 1952), pp. 381–395.

Parrish, John B., "Iron and Steel in the Balance of World Power," *Journal of Political Economy* 64 (October 1956), pp. 369–388.

Pittman, Steuart L., "The Foreign Aid Programs and the United States Government's Antitrust Policy," in Kingman Brewster, Jr., *Antitrust and American Business Abroad.* New York: McGraw-Hill, 1958, app. A, pp. 459–473.

Reagan, Michael D., "The Business and Defense Services Administration, 1953–57," *Western Political Quarterly* 14 (June 1961), pp. 569–586.

Robinson, Maurice H., "The Gary Dinner System: An Experiment in Cooperative Price Stabilization," *Southwestern Political and Social Science Quarterly* 7 (September 1926), pp. 128–159.

Rosegger, G., "Steel Imports and Vertical Oligopoly Power: Comment," *American Economic Review* 57 (September 1967), pp. 913–917.

Rosen, Martin J., "The Brussels Entente: Export Combination in the World Steel Market," *University of Pennsylvania Law Review* 106 (June 1958), pp. 1079–1116.

Singer, Jacques, "Trade Liberalization in the Canadian Steel Industry," in H. Edward English (ed.), *The Impact of Trade Liberalization: Volume 2.* Toronto: University of Toronto Press, 1969, pp. 3–126.

Slesinger, Reuben E., "Steel Imports and Vertical Oligopoly Power: A Comment," *American Economic Review* 56 (March 1966), pp. 152–155.

Stebbins, Phillip E., "Truman and the Seizure of Steel: A Failure in Communication," *Historian* 34 (November 1971), pp. 1–21.

Styn, Ronald L., "The Antidumping Act: Problems of Administration and Proposals for Change," *Stanford Law Review* 17 (April 1965), pp. 730–749.

Susman, Warren I., "History and the American Intellectual: The Uses of a Usable Past," in Warren I. Susman, *Culture as History.* New York: Pantheon Books, 1984, pp. 7–26.

Sweezy, Alan, "The Keynesians and Government Policy, 1933–1939," *American Economic Review* 62 (May 1972), pp. 116–124.

Taylor, Graham D., "The Rise and Fall of Antitrust in Occupied Germany, 1945–1948," *Prologue* 11 (Spring 1979), pp. 23–39.

Tiffany, Paul A., "Corporate Management of the External Environment: Bethlehem Steel, Ivy Lee, and the Origins of Public Relations in the American Steel Industry," *Essays in Economic and Business History* 5 (1987), pp. 1–18.

———, "Opportunity Denied: The Abortive Attempt to Internationalize the American Steel Industry, 1903–1929," *Business and Economic History,* 2nd series, 16 (1987).

Tower, Walter S., "The New Steel Cartel," *Foreign Affairs* 5 (January 1927), pp. 249–266.

Vanik, Charles A., "The Development of an Industrial Policy for Steel in the United States During a Period of Trade Frictions," in OECD (ed.), *Steel in the 80s.* Paris: author, 1980, pp. 232–244.

Vernon, Raymond, "Trade Policy in Crisis," Princeton University, International

Finance Section, *Essays in International Finance,* No. 29 (March 1958), pp. 1–13.

Vogel, David, "Why Businessmen Distrust Their State: The Political Consciousness of American Corporate Executives," *British Journal of Political Science* 8 (1978), pp. 45–78.

Warnecke, Steven J., "The American Steel Industry and International Competition," in S. Strange and R. Tooze (eds.), *The International Politics of Surplus Capacity.* London: Allen & Unwin, 1981, pp. 137–149.

Wells, D. A., and Sisik, S., "Steel Imports: The Importance of Availability and Costs," *MSU Business Topics* 18 (Spring 1970), pp. 57–64.

Woolcock, Stephen, "The International Politics of Trade and Production in the Steel Industry," in John Pinder (ed.), *National Industrial Strategies and the World Economy.* London: Allanheld, Osmun, 1982, pp. 53–84.

Popular Articles

Bell, J., "The Comeback of Krupp," *Fortune* 53 (February 1956), pp. 101–108, 200ff.

"Bethlehem–Youngstown: Controversial Engagement," *Fortune* 55 (June 1957), p. 145.

Blough, Roger M., "My Side of the Steel Price Story," *Look* 27 (January 29, 1963), pp. 19–23.

Brown, A. C., "Prosperity but no Profits," *Nation's Businesss* 15 (August 1927), pp. 15–17.

Burck, Gilbert, "The Private Strategy of Bethlehem Steel," *Fortune* 65 (April 1962), pp. 105–112, 242ff.

———, "The Transformation of U.S. Steel," *Fortune* 53 (January 1956), pp. 88–95, 198ff.

Cheney, O. H., "America in the World Steel Markets—A Warning," *Annalist* 30 (November 25, 1927), pp. 819–820.

"The Decline of Dave McDonald," *Fortune* 58 (August 1958), pp. 169–170.

Friendly, Alfred, "American Industry's Grand Strategy," *The Nation* 162 (January 19, 1946), pp. 62–63.

Grace, Eugene G., "Industry and the Recovery Act," *Scribner's Magazine* 95 (February 1934), pp. 96–100.

Hill, John W., "Is United States Financing German Competition?," *Iron Trade Review* 76 (January 29, 1925), p. 339.

"It Happened in Steel," *Fortune* 15 (May 1937), pp. 91–94, 176ff.

Lee, J. R., "U.S. to Revamp Industry Consultant Methods," *The Journal of Commerce* (December 16, 1953), pp. 1, 8.

"The Manifesto," *Fortune* 15 (May 1937), p. 91.

McDonald, John, "Steel Is Rebuilding for a New Era," *Fortune* 74 (October 1966), pp. 130–137, 216ff.

"Myron Charles Taylor," *Fortune* 13 (June 1936), pp. 117–120, 172ff.

Randall, Clarence B., "European Steel: Monopoly in the Making," *Atlantic Monthly* 188 (October 1951), pp. 34–38.

———, "A Steel Man Looks at the Schuman Plan," *Atlantic Monthly* 186 (October 1950), pp. 35–37.

———, "Steel: The World's Guinea Pig," *Atlantic Monthly* 190 (December 1952), pp. 31–34.

Ruttenberg, Harold J., "End the Steel Famine," *Harper's Magazine* 96 (February 1948), pp. 111–117.

Smith, Richard A., "Bethlehem Steel and the Intruder," *Fortune* 47 (March 1953), pp. 100–105ff.
"U.S. Steel: I," *Fortune* 13 (March 1936), pp. 59–67, 152ff.
"U.S. Steel: IV," *Fortune* 13 (June 1936), pp. 113–116, 164ff.
Welles, Sumner, "Pressure Groups and Foreign Policy," *Atlantic Monthly* 180 (November 1947), pp. 63–67.

Company, Trade Association, and Miscellaneous Publications

AISI, *Annual Statistical Report* (various issues).
———, *Charting Steel's Progress,* 1955 ed. New York: author, 1956.
———, Committee on Foreign Relations, *Foreign Trade Trends—American Iron and Steel Industry, 1953–1957.* New York: author, 1958.
———, *Monthly Bulletin* (various issues).
———, *Steel Facts* (various issues).
———, *Steel Imports—A National Concern.* Washington, DC: author, July, 1970.
———, *Yearbook of the AISI.* New York: author, 1910–65.
Barringer, Edwin C., *The Story of Scrap.* Washington, DC: Institute of Scrap Iron and Steel, 1954.
E. W. Axe and Co., Research Department, *The Postwar Outlook for the Steel Industry.* Tarrytown, NY: author, 1944.
Fisher, Douglas A., *Steel in the War.* New York: USS, 1946.
———, *Steel Serves the Nation, 1901–1951, The Fifty Year Story of United States Steel,* New York: USS, 1951.
National Council of Churches, Department of the Church and Economic Life, *In Search of Maturity in Industrial Relations: Some Long-Range Ethical Implications of the 1959–1960 Dispute in the Steel Industry.* New York: author, 1960.
Smith, Bradford B., *America's Steel Capacity—What It Is, What It Does.* New York: AISI, 1948.
USS, *Annual Report* (various years).
———, *Steel and Inflation, Fact vs. Fiction.* New York: author, 1958.
USWA, *The Foreign Competition Hoax.* Washington, DC: author, 1960.
Who's Who in America (various volumes).

Newspapers and Popular Periodicals

American Metal Market
Barron's Weekly
Business Week
Commercial and Financial Chronicle
Economist (London)
Editor & Publisher
Forbes
Iron Age (1930–60)
Metal Bulletin (London)
NYT (1940–60)
Steel
Time
Wall Street Journal

Unpublished Manuscripts

Branyan, Robert L., "Antimonopoly Activities During the Truman Administration." Ph.D. diss., University of Oklahoma, 1961.

Cahn, Linda A., "National Power and International Regimes: The United States Commodity Policies 1930–1980." Ph.D. diss., Stanford University, 1980.

Ciscel, David H., "The Decline of the United States Steel Industry: A Study of Market Entropy." Ph.D. diss., University of Houston, 1971.

Hamel, George F., "John W. Hill, Public Relations Pioneer." Master's thesis, University of Wisconsin, Madison, 1966.

Hill, William S., Jr., "The Business Community and National Defense: Corporate Leaders and the MIlitary, 1943–1950." Ph.D. diss., Stanford University, 1979.

Hodin, Michael W., "A National Policy for Organized Free Trade, or, How to Cope with Protectionism: The Case of United States Foreign Trade Policy for Steel, 1976–1978." Ph.D. diss., Columbia University, 1979.

Kuklick, Bruce R., "American Foreign Policy, Economic Policy and Germany, 1939–1946." Ph.D. diss., University of Pennsylvania, 1968.

Libertella, Anthony, "The Steel Strike of 1959: Labor, Management, and Government Relations." Ph.D. diss., Ohio State University, 1972.

Lynn, Leonard H. and McKeown, Timothy, "The Development of Trade Associations in the U.S. and Japan." Working Paper, Carnegie-Mellon University, Department of Social and Decision Sciences, College of Humanities and Social Sciences, 1987.

Miller, Robert R., "United States Antidumping Policy and the Steel Industry Experience." Ph.D. diss., Stanford University, 1967.

Moody, Jesse C., Jr., "The Steel Industry and the National Recovery Administration: An Experiment in Industrial Self-Government." Ph.D. diss., Univeristy of Oklahoma, 1965.

Motter, David C., "Government Controls over the Iron and Steel Industry During World War II: Their Development, Implementation, and Economic Effect." Ph.D. diss., Vanderbilt University, 1958.

Nagle, Richard W., "Collective Bargaining in Basic Steel and the Federal Government, 1945–1960." Ph.D. diss., Pennsylvania State University, 1978.

Paterson, Thomas G., "The Economic Cold War: American Business and Economic Foreign Policy, 1945–1960." Ph.D. diss., University of California, Berkeley, 1968.

Reagan, Patrick D., "The Architects of Modern American National Planning." Ph.D. diss., Ohio State University, 1982.

Tedesco, Paul H., "Patriotism, Protection, and Prosperity: James Moore Swank, The American Iron and Steel Association, and the Tariff, 1873–1913." Ph.D. diss., Boston University, 1970.

Tiffany, Paul A., "The Steel Industry Responds: The Rise of Public Relations." Working Paper, University of California, Berkeley, 1980.

List of Acronyms

AID	Agency for International Development
AISI	American Iron and Steel Institute
BDSA	Bureau of Defense Services Administration
BOF	Basic Oxygen Furnace
CEA	Council of Economic Advisors
CED	Committee for Economic Development
CEEC	Committee for European Economic Cooperation
CFEP	Council on Foreign Economic Policy
CIA	Central Intelligence Agency
CIO	Congress of Industrial Organizations
CPI	Consumer Price Index
DPA	Defense Production Administration
ECA	Economic Cooperation Administration
ECAFE	Economic Commission for Asia
ECE	Economic Commission for Europe
ECSC	European Coal and Steel Community
EDC	European Defense Community
EEC	European Economic Community
ESA	Economic Stabilization Agency
FBI	Federal Bureau of Investigation
FRB	Federal Reserve Board
FTC	Federal Trade Commission
GAO	General Accounting Office
GATT	General Agreement on Tariffs and Trade

GNP	gross national product
GOP	Grand Old Party
GPO	Government Printing Office
HA	High Authority (ECSC)
IAC	Industry Advisory Committee
ILO	International Labor Organization
ISC	International Steel Cartel (Entente internationale de l'acier)
ITO	International Trade Organization
JEC	Joint Economic Committee
MITI	Ministry of International Trade and Industry (Japan)
NIRA	National Industrial Recovery Act
NLRA	National Labor Relations Act
NRA	National Recovery Administration
NSRB	National Security Resources Board
NYT	New York Times
ODM	Office of Defense Mobilization
OECD	Organization for Economic Cooperation and Development
OEEC	Organization for European Economic Cooperation
OPA	Office of Price Administration
OPS	Office of Price Stabilization
RFC	Reconstruction Finance Corporation
ROW	Rest of the World
RTAA	Reciprocal Trade Agreements Act
SCAP	Supreme Commander Allied Powers (Japan)
SCCC	Steel Companies Coordinating Committee
SEA	Steel Export Association
SUB	supplemental unemployment benefits
SWOC	Steel Workers Organizing Committee
TNEC	Temporary National Economic Committee
UAWA	United Automobile Workers of America
UMWA	United Mine Workers of America
USS	United States Steel Corporation
USWA	United Steelworkers of America
WPI	Wholesale Price Index
WSB	Wage Stabilization Board

Index

Administered pricing, steel, 16, 48, 138, 145–146, 156. *See also* Gardiner C. Means
Agency for International Development (AID), 180
Agriculture, U.S. Department of, 24, 29, 43, 91
Alsop, Joseph, 201 n.50
Alsop, Stewart, 207 n.55
American Institute of Imported Steel, 223 n.67
American Iron and Steel Association, 193 n.9
American Iron and Steel Institute (AISI), 9, 10, 12, 14, 18, 21, 28, 30, 31, 43, 51–52, 62, 94, 100, 104, 107, 121, 128, 169, 170, 172, 181, 185
  formation of, 7, 193 n.9
  and Randall Commission, 110–115, 118
American Steel Warehouse Association, 180
Antidumping statutes, U.S., 183
Antitrust. *See also* Emanuel Celler; Estes Kefauver; Temporary National Economic Committee (TNEC)
  House, U.S., studies of, 88–90
  Industrial Advisory Committees (IAC), 113
  International Steel Cartel (ISC), 16
  Senate, U.S., studies of, 154–155, 155–160
  steel industry, 1940s, 54
  steel industry, 1950s, 155–157
  United States Steel Corp., 8, 48, 78
Armco Steel Company, 15, 71, 159, 177, 246 n.37
Atlantic Steel Company, 177

Baker, Howard, Jr., 92
Baruch, Bernard, 54
Basic oxygen furnace (BOF). *See* Technology, steel
Basing-point pricing, steel, 7, 11, 37, 51–52, 58, 93, 209 n.87. *See also* Pittsburgh Plus pricing, steel
Bean, Louis H., 24, 29–30, 31, 33, 91, 131, 198 n.14. *See also* Capacity expansion, steel
Bethlehem Steel Corporation 12, 14, 49, 56, 70, 71, 77, 103, 118, 141, 143, 157, 178, 183, 195 n.26
  capacity expansion, 93, 133–135
  capital goods manufacture, 11
  executive compensation, 239 n.45
  foreign trade (1930s), 17

Bethlehem Steel Corporation (*continued*)
International Steel Cartel (ISC), 15–16
merger with Youngstown Sheet & Tube Company, 155–156
public relations, 57
strike (1949), 85
Blair, John, 155
Block, Joseph L., 27
Blough, Roger M., 144, 149, 157, 158, 160, 165, 177, 178, 180
Blough, Roy, 252 n.43
Broude, Henry W., 131
Brownell, Herbert, 156
Brubaker, Otis, 158, 200 n.49
Brussels Entente, 121–122, 233–234 n.91
Bureau of the Budget (U.S.), Fiscal Division, 24
Burns, Arthur, 145, 149, 242 n.78
Bush, Prescott, 171
Business Council, 51
Business and Defense Services Administration, 230 n.45
Business-government relations in steel,
1901, 5
1920, 8
1940, 17
World War II, 1940s, 19–20, 51
1947–48, 58
1952, 96–102
1953, 103–104
American Iron and Steel Institute (AISI), 94
government policy toward steel, 126–127, 204 n.90
pricing controversies, 40–41, 55–56, 86–88, 138–152
small business sector, 26–31
summary, 185–190
Truman, Harry, 60–63, 205 n.24
Byrnes, James F., 71

Capacity expansion, steel
conflict, 1940, 17, 18
conflict, World War II, 19
conflict, 1940s, 21–41, 61–63, 65–66, 82
conflict, 1950, 91–94
growth, 129, 132–137
"rounding out" and "greenfield" expansion, 131, 142–144
Capacity utilization rate, steel, 27, 136
Celler, Emanuel, 88–90
Celler-Kefauver Act, 156, 221 n.39
Cement Institute, 52, 57, 58, 209 n.85. *See also* Basing-point pricing
Certificate of necessity program, steel, 92–93, 132–133, 141–142, 146, 240 n.50
Clay, Lucius D., 69–70, 78
Commerce, (U.S.) Department of, 32, 33, 35, 43, 48, 54, 55, 62, 72, 91, 99, 112–113, 170, 173–175, 178
Commission on Foreign Economic Policy, 109. *See also* Randall Commission
Committee for Economic Development (CED), 51
Committee for Reciprocity Information, 107, 111–112
Congress of Industrial Organization (CIO), 45, 148
Cooper, R. Conrad, 162–163, 165
Council of Economic Advisors (U.S.), 32, 39, 43, 49, 51, 55, 62, 97–98, 145, 207 n.55
Council on Foreign Economic Policy (CFEP), 115, 116, 173–175

Daniels, Josephus, 195 n.26
Davidson, C. Girard, 32
Dawes Plan, 10
Defense, (U.S.) Department of, 43, 98, 99, 101, 136
Defense Production Administration (DPA), 95
Dewey, Thomas E., 59–60
Diebold, William, Jr., 120
Dillon, Read and Company, 69
Dixon, Paul Rand, 155
Dodge, Joseph, 115
Draper, William H., Jr., 69–70
Dulles, John Foster, 218 n.88

Economic Commission for Europe (ECE), 216 n.63
Economic Cooperation Administration (ECA), 43, 74, 76, 77
Economic Stabilization Agency (ESA), 95, 100
Economist, 72
Eisenhower, Dwight D., 103, 105, 107, 117, 118, 127, 133, 136, 155, 178, 183, 188
antitrust, steel industry, 156
business-government relations, 188–189
domestic economic policy, 126, 128–129, 160, 170
European Coal and Steel Community (ECSC), 120, 171–172
foreign economic policy, 108–110, 114–116, 168–169, 176, 229 n.27

inflation, domestic, 145–147
Japan, 175–176
labor disputes, intervention into, 159, 161–165, 245 n.28
prices, steel industry, 147–151
steel industry, attitude toward, 137, 142, 146–147, 154, 171–172
Entente internationale de l'acier. *See* International Steel Cartel (ISC)
European Coal and Steel Community (ECSC), 78–79, 82, 105, 119, 120–122, 124, 177, 217 n.78, 233 n.88
European Defense Community, 120
European Economic Community (EEC), 79
European steel industry
World War I, 10
1920s, 12
1930s, 14, 17–18
post–World War II reconstruction, 66–67, 86
cartel arrangements, 121–122
modernization plans, 110–111
price policy regarding U.S. markets, 144
Schuman Plan and ECSC, 76–79, 124–125
scrap steel needs, 170–172
U.S. government assistance to, 118–120, 169
Exports, American steel, 178–180
Export-Import Bank, 111

Fairless, Benjamin, 45–46, 50, 56, 72, 84–85, 88–89, 96–97, 100–101, 129, 135, 141, 144, 147–148, 169, 181, 185, 205 n.24
Fairless Commission, 169
Fairless Works, 93–94, 131, 133, 155. *See also* United States Steel Corporation
Federal Bureau of Investigation (FBI), 55
Federal Mediation and Conciliation Service, 150
Federal Trade Commission (FTC), 43, 51–52, 57, 59, 87, 90, 158
Feinsinger, Nathan, 98
Financial performance, American steel industry
1945–52, 34
1947, 39
1947–52, 40–41, 50
1953–60, 130
Fischer, Benjamin, 247 n.49
Flanders, Ralph, 55
Ford Motor Company, 149
Gary, Elbert H., 7–8 9, 10, 12, 88, 189
General Agreement on Trade and Tariffs (GATT), 106, 114, 116, 122
General Electric Corporation, 98
General Motors Corporation, 46
Geneva steelworks, 206–207 n.46
German steel industry, 33, 67–70, 72–76, 86
Girdler, Thomas, 24, 50
Golden, Clint, 201 n.49
Gore, Albert, 160
Grace, Eugene G., 12, 49, 70, 77

M. A. Hanna Company, 74
Hansen, Alvin, 23
Harding, Warren G., 13
Harriman, W. Averill, 54, 215 n.51
Hartley, Fred A., 99. *See also* Taft-Hartley Act
Herter Committee, 211 n.3
High Authority (HA), *See* European Coal and Steel Community (ECSC)
Hill & Knowlton, 31
Hoffman, Paul G., 74, 77
Homer, Arthur B., 56, 103, 135
Homestead strike, 13
Hook, Charles R., 246 n.37
Hoover, Herbert, 215 n.51
Hull, Cordell, 13
Humphrey, George M., 74–75, 142, 150, 154, 163, 170–171, 215 n.54, 247 n.55

Industrial Advisory Committee (IAC), 54–55, 74–75, 112–113, 189–190
Industrial policy, steel, 19, 102, 108, 171–172, 190, 211–212 n.3, 255–256 n.11
Inland Steel Company, 27, 28, 71, 109, 110, 118
Interior, (U.S.) Department of, 32, 43, 112
International Labor Organization (ILO), 200–201 n.49
International Steel Cartel (ISC), 12, 15–16, 16, 17–18, 23, 25, 69–70, 74, 78, 88–89, 196 n.47
International Trade Organization (ITO), 106, 114
International trade, steel, 18, 67–79, 104, 105–111, 117–118, 139, 141, 151–152, 167–168, 169–172, 177–182, 233 n.88. *See also* European steel industry; German steel industry;

International trade, steel (*continued*)
Japanese steel industry; United Kingdom steel industry
Investment bankers and steel, 10–11
Investment, steel, 139–144, 178–179, 180–182, 239 n.41
Iron Age, 31, 55, 56, 59, 61, 66, 70, 81, 85, 86, 146, 180

Japanese steel industry
- 1930s, 18, 217 n. 84
- post-World War II, 79–81, 126
- cartels, 234 n.91
- export growth, 177
- GATT, 114
- Korean War boom, 223 n.67
- prices, 144
- U.S. government assistance to, 169–176, 218 n.88

Joint Economic Committee (U.S. Congressional), 55–57, 62, 87–88, 147, 154–155, 160
Jones & Laughlin Steel Company, 30, 133
Justice, (U.S.) Department of, 8, 43, 48, 54, 89, 90, 113, 156, 158

Kaiser, Henry J., 207 n.46
Kefauver, Estes, 146, 155–159, 166, 245 n.26
Kennedy, John F., 185
Keynes, John Maynard, 23, 24, 31, 43
Khrushchev, Nikita, 164
Korean War, 90–92, 94–96, 98, 101, 104, 113, 121, 126, 132, 134, 137, 146, 170, 223–224 n.67
Krug, Julius, 32
Krupp Steel Company, 75, 234 n.95

Labor, (U.S.) Department of, 43, 150, 151, 163
Labor, steel
- strikes, 44–47, 84–86, 96–102, 147–152, 160–168
- union recognition, 11, 13–14
- wages, 58, 83–87, 158–159

Lamont, Thomas W., 49, 50, 51, 72
Levi, Edward H., 91
Lewis, John L., 13, 14
Lone Star Steel Company, 92–93, 202 n.60, 222 n.53
Lovett, Robert A., 224 n.74

MacArthur, General Douglas, 80–81, 224 n.71
McCarthy, Joseph, 96
McCloy, John J., 249 n.7
McCraw, Thomas K., 188
McDonald, David J., 147–151, 158, 160, 162–165. *See also* Labor, steel; United Steel Workers of America (USWA)
McGrath, J. Howard, 224 n.72
McKinley, William, 109
McLouth Steel Company, 133
Marcus, Maeva, 100
Marshall Plan, 71, 74, 107, 109, 214 n.47
Martin, Edward, 26, 27
Martin, James S., 78, 221 n.33
Means, Gardiner C., 138, 198 n.14. *See also* Administered pricing, steel
Midvale Steel Company, 195 n.26
Mitchell, James, 150, 151, 163–165, 248 n.66
Moreel, Ben, 30
Morgan, J. P., 5, 7
Morgan, J. P. & Company, 10, 49, 52–53
Morgenthau, Henry J., Jr., 18, 68, 69
Murray, James, 31, 62, 200 n.36
Murray, Philip, 45, 85, 100–101, 147–148. *See also* United Steel Workers of America (USWA)

Nasser, Gamal Abdul, 151
National Farmers Union, 24
National Industrial Recovery Act (NIRA), 12
National Labor Relations Act (NLRA), 13
National Petroleum Council, 112
National Planning Associates, 24
National Recovery Administration (NRA), 12, 13
National Resources Planning Board, 24
National Security Resources Board (NSRB), 91–93
National Steel Company, 15, 62, 77, 94, 142, 154, 157, 163
Nixon, Richard, 101, 164–166, 248 n.66
North American Steel Company, 92
Nourse, Edwin G., 51

Office of Defense Mobilization (ODM), 95, 98, 99, 132
Office of Price Administration (OPA), 45, 46, 47

Office of Price Stabilization (OPS), 95, 97, 98, 99, 100
Olds, Irving S., 18, 49, 50, 61
O'Mahoney, Joseph, 56, 87, 155, 160, 209 n.87
Organization for European Economic Cooperation (OEEC), 76–77
Organization for Trade Cooperation (OTC), 114

Patman, Wright, 62, 222 n.53
Pauley, Edwin J., 80
Pittsburgh Plus pricing, steel, 7, 52. *See also* Basing-point pricing, steel
Potsdam Conference, 69, 74, 80
Prices, steel
  controversies, 96–101, 104, 137–152, 153–154, 156–159, 165–166
  index, 35, 36, 38
  inflation, 239 n.39
  U.S. and foreign practices, 253 n.50
Production, steel, 5–6, 33
Public Advisory Board for Mutual Security, 107
Public relations, steel, 31, 53, 57, 86, 100, 104, 226 n.106
Putnam, Roger, 100

Randall, Clarence B., 109, 110, 115, 118–119, 121, 170, 172, 173–174, 183, 226 n.108, 253 n.48
Randall Commission, 109–114
Raw materials, steel, 35–36, 92–93, 118, 169–170, 203 n.74
Reciprocal Trade Agreements Act (RTAA), 13, 105–110, 114, 176
Reconstruction Finance Corporation (RFC), 92
Republic Steel Company, 24, 50, 69, 71, 159
Reuther, Walter, 30, 148
Rogers, William P., 164
Roosevelt, Franklin D., 11, 12, 18, 19, 42, 47, 49, 68, 69, 105, 106, 195 n.26

St. Lawrence Seaway, 37, 110, 177
Salt Producers Association, 52
Saulnier, Raymond J., 145–146
Sawyer, Charles, 32, 62, 91, 210–211 n.112
Schroeder, Gertrude G., 7
Schuman, Robert, 78–79
Schuman Plan, 111, 116, 118–121, 217 n.78
Schumpeter, Joseph, viii, 7
Scrap, steel, 36–37, 169–174, 203 n.77
Seizure case, steel, 96–102
Small business, 25–26, 28, 52, 64, 136, 199 n.24
Smith, Bradford B., 31, 32
Smoot-Hawley Tariff, 109
Snyder, John W., 45, 198 n.16
Soviet Union, 68–69, 71, 72, 79, 117, 136
State, (U.S.) Department of, 68, 72, 106
Steel Export Association, 12. *See also* Webb-Pomerene Act
Steel Industry Board, 84–86. *See also* Labor, steel
Steel Workers Organizing Committee (SWOC), 13, 14, 16, 45. *See also* United Steel Workers of America (USWA)
Steelman, John R., 96–100, 203 n.77, 213 n.32, 224 n.70
Stephens, John A., 162–163
Stettinius, Edward R., Jr., 49, 68, 206 n.43
Stigler, George, 89
Stimson, Henry, 68
Suez Canal, 151
Supplemental Unemployment Benefits (SUB), 150, 151
Susman, Warren I., 190
Sykes, Wilfred, 28–29
Symington, W. Stuart, 91–92

Taft-Hartley Act, 60, 84, 99–100, 150, 162–165, 226 n.116
Taft, Robert, 55
Taft, William H., 8
Tariffs, steel, 11, 13, 107–108, 111–112, 176–177
Tariff Commission, U.S., 67, 79, 111, 183
Tata Iron and Steel Company, 231 n.65
Taylor, Myron, 14, 15–17, 42, 88
Technology, steel
  basic oxygen furnace (BOF), 39, 133–134, 136, 174–175
  Japanese steel industry, 172–176
  scrap supply, 169–172
  technology change, steel, 37
  "technological momentum," 134
Temporary National Economic Committee (TNEC), 16–17, 56, 89, 155
Tennessee Steel Company, 92

Tower, Walter, 18, 22, 30, 43, 57, 107, 128–129
Treasury, (U.S.) Department of, 68, 150, 154, 170
Truman, Harry, 33, 52, 58, 59, 60, 68, 71, 72, 84, 92, 112, 145, 155, 188, 205 n.24
  basing-point pricing, 57–58, 93
  business-government relations, 60, 63
  domestic economic policy, 44, 47–49, 51, 54, 95
  European reconstruction, 69, 74
  international trade, 105–109
  labor, 45–46, 84, 96–102, 105
  leadership, 46–47, 48 53, 96
  small business, 25–26, 55
  steel capacity expansion, 30, 32, 60–62, 129
  steel industry, 42–43, 47, 48–50, 55–57, 59, 72, 83, 88, 91, 139, 141
  steel seizure case, 96–102, 105

United Auto Workers of America (UAWA), 30, 148–149
United Kingdom (UK) steel industry, 61, 76–78, 125–126
United Mine Workers of America (UMWA), 13, 151
United States Steel Corporation (USS), 42, 45, 51, 58, 61, 84, 96, 144, 156, 158, 169, 185, 189. *See also* Roger Blough; Benjamin Fairless; Fairless Works; Elbert H. Gary; Geneva steel works
  business-government relations, 49–50, 88, 159, 187, 195 n.26
  capacity expansion, 17–19, 31–32, 35, 63, 93–94, 129, 131, 133, 135, 143, 155
  foreign trade, 9, 15–16, 71, 72, 75, 78, 81, 107, 118, 134, 151, 178, 180, 183
  formation, 5–8, 10
  investigations by government, 48, 52, 55–56, 78, 88–90
  labor relations, 13, 14, 45, 59, 84–86, 100–101, 147–151, 160, 162–165, 178
  pricing activities, 12, 15–17, 49, 51–52, 55, 57, 59, 87–88, 89, 137, 139, 153, 157, 159
  Roosevelt, Franklin D., 11
United States Steel Export Corporation, 215 n.54
United Steel Workers of America (USWA), 31, 45, 58, 83, 84, 85, 96–101, 147, 157, 158–160, 163–165, 168, 182, 185. *See also* Labor, steel
United Steelworks Union, 75. *See also* Vereinigte Stahlwerke

Vereinigte Stahlwerke, 75. *See also* United Steelworks Union
Vertical integration, steel, 33
Vietor, Richard H. K., 189
Vinson, Fred, 99

Wage Stabilization Board (WSB), 95, 97, 98, 100
Wallace, Henry, 48, 53, 54
Webb-Pomerene Act, 12, 15–16, 179
Weeks, Sinclair 170–171
Weir, Ernest T., 62, 77
Wherry, Kenneth S., 26, 28–29, 31
Wilson, Charles E., 98–99
Wilson, Woodrow, 9
de Witt, Epinard, 202 n.60
Wysor, Rufus, 69

Yalta Conference, 68–69, 74
Youngstown Sheet & Tube Company, 155–156